Return to Energy

Return to Energy

Lived Experience as God Designed

How Competing Metabolisms
Shape Health and Disease

MICHAEL MACDONALD

COUNTERCURRENTS
PRESS

Return to Energy: Lived Experience as God Designed
returntoenergy.org
© 2026 Michael MacDonald

COUNTERCURRENTS
P R E S S

Published by Countercurrents Press
Mt. Pleasant, MI

To request permissions, contact the author.

Printed in the United States of America.
First edition 2026.

Cover and layout design by G Sharp Design, LLC.
www.gsharpmajor.com

ISBN 979-8-9959012-1-1 (paperback)
ISBN 979-8-9959012-2-8 (hardcover)
ISBN 979-8-9959012-0-4 (ebook)

Library of Congress Control Number: 2026911722

To my wife, Carol

We walked this path together

To my daughter, Megan

You inspire me every day

To my son, Stuart

For stepping into the unknown,
and for your unwavering optimism

CONTENTS

PART I

The Turning Point

PART II

Rediscovering the Energy System

PART III

Movement, Performance, and Practice

READER NOTE

This book is shared for educational purposes and personal reflection, not as medical advice. The ideas, experiences, and scientific discussions presented here reflect my own journey and understanding of human metabolism, health, and energy. Everybody is different, and responses to dietary change, fasting, or exercise can vary widely. Readers—especially those with medical conditions, metabolic disorders, or who take medications—should consult a qualified healthcare professional before making significant changes to diet, fasting practices, or physical activity. Nothing in this book is intended to replace personalized medical care.

AUTHOR'S NOTE

We live in a world that moves fast and burns people out even faster. Most of us are doing our best to keep up. We follow the guidance we're given. We push harder, try more, and assume that effort alone will eventually restore what feels lost.

For me, something didn't seem right. I began asking questions I had never asked before; questions Carol and I kept returning to as we watched our health, energy, and focus drift further from what we believed was possible: Is the body truly healing? Is our energy supporting our lives, or quietly draining them? Is this really the best we can expect from modern living?

From childhood, I was wired to question what didn't make sense, to look for patterns and test whether explanations actually matched reality. That instinct shaped my professional life. I became a certified public accountant, trained to audit, reconcile, and verify claims. You don't assume something is true; you learn to examine it, test it, and reconcile it with evidence.

That same lens eventually turned inward. Carol and I followed the recommendations. We did what conventional wisdom said should work. Yet energy declined. Migraines worsened. Weight crept upward. I developed unrecognized prediabetes. And my wife carried her own burdens: persistent fatigue and the quiet strain of moving through daily life without reliable energy. None of this matched the promises of the modern health paradigm. The numbers and our lived experience didn't align.

We began to recognize the same pattern in the people around us: family, friends, clients, strangers. People exhausted. You could see it in posture, mood, and pace. Fatigue had become part of our culture. When it became clear that the framework wasn't producing the results it promised, we began asking better questions, together. If outcomes were failing, perhaps the assumptions were incomplete. That search led us to a deeper understanding of energy—where it comes from, how it is created, preserved, and lost, and why so many people feel as though they are running on empty despite sincere effort and good intentions.

This became a return to first principles: how we ate, how we lived, and how closely our lives aligned with the biological design that shaped human physiology long before the modern world. Healing, we discovered, did not begin with pushing harder. It began with restoring access to energy. Our recovery unfolded in places modern medicine rarely emphasizes: metabolic health, movement in nature, nourishment, rhythm, and spiritual grounding. Over time, it became clear that these were not new discoveries at all, but ancient truths written into human biology from the beginning.

For my wife, Carol, the change was gentle but deeply meaningful. As energy returned, the constant sense of strain eased. Days felt more manageable. Mood steadied. Healing didn't arrive as a dramatic moment, but as a gradual return; clarity where there had once been fog, calm where there had been exhaustion, and the quiet reassurance that her body was once again supporting the life she was living.

As we applied these principles together, the changes became unmistakable. Migraines faded. Energy returned. Blood sugar normalized. Mental clarity sharpened. Strength rose to levels we hadn't felt in decades. What surprised us most was not only the transformation itself, but how quickly the body responded once it was placed back into the environment it was designed for. This journey was not one we traveled alone. It unfolded through years of learning, listening, and guidance from those who helped us see what we could not see on our own. As the pieces came together, the truth became clear—and once we saw it, we couldn't keep it to ourselves.

This book exists because so many people are searching for answers but are rarely offered a framework that helps them understand why healing succeeds or fails. They are given fragments—symptoms and labels—but seldom an understanding of the energy that underpins human health. My hope is simple: to bridge lived experience with how the body actually works, and help others rediscover a capacity for healing that has always existed, yet is rarely understood.

Each chapter blends story, science, and practical insight. The story shows what this looks like in real life: shared, imperfect, and unfolding over time. The science explains what is happening beneath the surface—not just what to do, but why it works. The practical guidance exists so these principles can be applied with clarity and confidence. Whether you read straight through or explore the sections that resonate most, my hope is that this book restores something essential: the understanding that energy is not merely fuel. It is the foundation of vitality, clarity, healing, and intentional living.

Thank you for beginning this journey. What follows is both a shared story and an exploration of human physiology. A return to the energy that has always been available to us, as we were designed from the beginning.

HOW TO USE THIS BOOK

This book is designed to be read from beginning to end, as the ideas build gradually from lived experience toward deeper scientific understanding and practical application. At the same time, readers are welcome to move through chapters in whatever way serves them best, returning to sections that speak most directly to their interests or questions.

The structure blends story, science, and practical insights so you can engage with it at your own pace and in your own way. While each chapter can stand on its own, the book is organized to move from foundational ideas toward deeper understanding and application.

Throughout the book, research, historical work, and clinical perspectives are woven into the narrative to support the ideas presented and to invite deeper exploration. Reading lists and references are gathered in the back of the book for those who wish to explore the source material more deeply. Additional references, updates, and supporting materials can also be found at **returntoenergy.org**.

INTRODUCTION

Modern healthcare did not arrive at its current place by accident. It evolved through a series of pivotal ideas—some brilliant, some incomplete, all influential. Over time, these ideas shaped not only how we treat disease, but how we understand the body itself.

One of the most defining moments came with the rise of germ theory. When Louis Pasteur argued that pathogens—bacteria, viruses, and other microbes—were the primary cause of disease, medicine adopted a worldview centered on external threat. Illness became something that arrived from the outside, something to be feared, fought, and eradicated.

At roughly the same moment in history, a competing framework existed. Antoine Béchamp proposed that disease was not determined solely by exposure, but by the internal terrain—the metabolic state, resilience, and biological environment of the host. In this view, the body was not merely a passive victim of invasion, but an active participant in its own health.

Over time, Pasteur's germ theory gained global dominance, directing medicine toward contamination, infection, and external causation. Yet what truly transformed public health in the 20th century was not germ theory alone, but the sweeping advances in sanitation, clean water systems, refrigeration, food safety, and waste management—followed by the lifesaving introduction of aseptic and antibiotic therapies. Together, these breakthroughs sharply reduced infectious disease rates and reshaped what was possible in healthcare. These victories deserve recognition and should not be diminished.

Even so, the germ-based model remained the central organizing framework of medicine long after the challenges it originally addressed had largely been resolved. As infectious diseases waned, another pattern rose in prominence: chronic, slow-moving disorders rooted not in invasion but in metabolic imbalance. Yet the dominant perspective persisted, applying a model built for epidemics to conditions with entirely different origins.

To understand those origins, we have to step outside the clinic and look at how humans once lived. For most of our existence, survival required metabolic agility. Food was unprocessed, nutrient-dense, and seasonally variable. Movement was constant. Fasting and refeeding created natural cycles of repair. Mitochondria thrived under these oscillating rhythms, insulin remained sensitive, and autophagy functioned as part of normal life.

Then, in a blink of evolutionary time, everything changed. The modern dietary landscape—abundant food, continuous eating, refined carbohydrates, engineered hyper-palatability—placed humans in a biochemical environment the body was never designed for. Instead of metabolic flexibility, we developed metabolic rigidity. Instead of mitochondrial renewal, we generated mitochondrial overload. Instead of intermittent energy scarcity, we lived in perpetual excess.

The shift was subtle enough to go unnoticed but powerful enough to reshape the biology—and the emotional lives of entire populations. Many began feeling tired, foggy, stressed, and unwell without understanding why. Beneath it all, the metabolism quietly strained under the weight of modern living.

That strain does not announce itself immediately. At the cellular level, its consequences begin long before symptoms appear. Hyperinsulinemia and insulin resistance start as adaptive responses—protective mechanisms guarding cells from glucose-driven oxidative stress.

But over time, these compensations escalate. Insulin levels rise. Mitochondria become overloaded. Autophagy declines. Oxidative stress builds. Cellular signaling distorts.

From these upstream imbalances flow the conditions that now dominate modern life: metabolic syndrome, cardiovascular disease, fatty liver, cancer, neurodegeneration, and cognitive decline. What appear to be separate diseases are, in many cases, different expressions of the same underlying metabolic disruption. **This realization raises a single, powerful emotional and scientific question:**

What if cholesterol is not the origin of atherosclerosis?

What if amyloid is not the origin of Alzheimer's disease?

What if mutations are not the primary origin of cancer? What if these findings are not root causes at all, but downstream responses, signals of cells struggling under chronic metabolic strain? And what if much of our modern suffering is not inevitable, but instead reflects biological imbalance that can be corrected?

Even as evidence accumulates, healthcare remains anchored to an older framework. We target cholesterol, glucose, amyloid, blood pressure, and tumor markers as if they were invaders. We chase metrics instead of mechanisms. We treat the symptoms of metabolic dysfunction while overlooking the forces that created them.

Medications can alter pathways, but they rarely restore the underlying breakdown in cellular energy. They often manage numbers without addressing the mitochondrial dysfunction at the heart of chronic illness. Patients sense this mismatch even when they can't articulate it—the feeling that treatments help, yet healing never fully arrives.

This gap between the problems we face and the tools we rely on defines the chronic disease crisis of the modern world. But we are not bound to the original road. The evidence for metabolic restoration through nutrition, fasting, mitochondrial renewal, circadian alignment, and energy flexibility—is no longer emerging; it is established. A different path exists, one that reconnects health to metabolism and metabolism to the conditions under which human physiology evolved.

This approach does not reject modern medicine. It expands it. It honors genuine triumphs—such as antiseptics and antibiotics—while acknowledging their limits. It reintroduces the concept of terrain. It recognizes that internal balance is not a philosophical abstraction, but a measurable biological reality. And emotionally, it does something else. It gives people their agency back. It restores the idea that health is not random. That our outcomes are not predetermined. That our bodies are not mysteries beyond influence.

We stand again at a fork in the road. One direction continues the disease-centric model, endlessly labeling and managing downstream effects. The other looks upstream—to mitochondrial health, metabolic stability, and the energy systems that have sustained humanity for millions of years. This book chooses the upstream road. It begins with the historical arc that shaped modern healthcare, examines the metabolic roots of chronic illness, and returns to a simple but powerful realization: The terrain matters again. Energy matters again. And the path toward healing may be far closer than we ever expected—yet strikingly different from the version we were raised to accept.

PART I

The Turning Point

CHAPTER ONE

A MOMENT OF RECKONING

*"When we are no longer able to change a situation,
we are challenged to change ourselves."*
VIKTOR E. FRANKL, *MAN'S SEARCH FOR MEANING*

Sitting there undressed in a gown, struggling to stay awake—despite the fact it was only mid-morning—I felt strangely exposed. My mind kept circling the same thought: my plan had failed again. The fluorescent lights buzzed overhead, and the sterile smell of the exam room clung to my skin. For nearly ten years, I had tried different ways to regain my fitness and return to that race—the one that once defined my youth and passion. But now, fast approaching fifty, the starting line felt farther away each year, like a mirage retreating the closer I moved toward it. Long tax seasons, the steady demands of family life, and weekends that never truly allowed for recovery had become a constant juggling act—one that slowly eroded the life I imagined for my health. The race hadn't just faded into the distance; it had been eclipsed by an exhaustion that now filled every waking moment. With no answers, my focus wasn't on goals or finish lines—it was on staying awake long enough to finish the exam without passing out.

For me, that race was never really about running. For a decade, it represented something deeper: a refusal to let age quietly redefine who I was. Signing up meant I was healthy enough, prepared enough, and still in control. I'd always believed the hardest part wasn't race day—it was getting to the starting line ready for the event you committed to. Preparation was success. Health was success. But year after year, when registration opened, I knew the conditions weren't moving in the right direction. The desire was there. The motivation was there. My body, increasingly, was not.

This wasn't my annual wellness exam. I had been here just six months earlier. I scheduled this appointment because something felt wrong—urgent enough that I couldn't ignore it. Summer was already halfway over, and despite lighter work demands and more time to exercise, my health hadn't improved. It had worsened. What I finally recognized was an emergency of a different kind: an energy crisis.

Alone in the exam room, waiting for Dr. Parks, I felt the weight of that realization. For years, my exams had followed a familiar pattern— more questions, cautious optimism, and the quiet hope that this time would be different. I believed if I asked the right questions, made the right adjustments, I could keep a long tax season from resetting my progress to zero. Winter always meant overtime, irregular meals, and disrupted sleep. Still, I was determined to regain control of my hunger and health. Everything I thought mattered came down to three basics: growing in my profession, providing for my family, and running that damn race.

As the minutes ticked by in the silent room, my anxiety grew. When Dr. Parks finally opened the door, my heart skipped, snapping me awake. My hands were sweaty, my speech slurred—a far cry from the person I wanted to be. His eyes, usually calm and professional, lingered on mine longer than normal. Concern settled in as he took in my appearance. I saw myself reflected in his expression—puffy, pale, unfamiliar. The longer he looked, the more I felt like a shadow of my former self, a man trapped in a body that no longer felt like his own. Dr. Parks' usual greeting "How are you doing?" was different this time. His voice remained calm, but it carried an urgency I hadn't noticed before. The silence that followed felt heavy with unspoken concern. I didn't have to say much for him to see that something was wrong.

When I began to speak, I stumbled, almost slurring my words. I tried to explain the constant low energy that followed me like a dense fog—morning, noon, and night. I described my struggles at work, the growing difficulty staying focused, and how my days had started to revolve around food and timing. I was eating snacks at certain times to avoid the next migraine, planning lunch around driving home just to take a nap. Meals weren't optional anymore; they had become essential to simply getting through the day. It wasn't uncommon to miss lunch by an hour and immediately feel the familiar pressure of another headache building.

Everything I believed should have been making my days more productive seemed to be doing the opposite. I ate what most would consider a reasonable breakfast: apples, whole-grain granola bars, nothing extreme. Lunch was usually a peanut butter and jelly sandwich on whole-wheat bread with a tall glass of skim milk. An afternoon snack followed, then dinner, and often dessert before bed. By conventional standards, it all looked reasonable.

I exercised three or four times a week at the gym. But the running—the thing that mattered most—wasn't happening. Each attempt to run quickly dissolved into a walk. Weight gain, knee pain, and the relentless ache in my joints made it impossible to build momentum. Even my gym workouts felt heavy and unrewarding, difficult at best. My appearance and weight felt stuck, unchanged no matter how much effort I applied.

This wasn't a bad stretch or a temporary setback. This was my life now. Each day felt like a negotiation with a body that no longer responded to my commands, hunger always creeping in at the edges. The work I was putting in—diet, exercise, discipline—no longer felt like a path forward, only a way to slow the downward slide. I thought I was doing everything "right," yet everything felt unmistakably wrong.

Dr. Parks listened quietly, his expression growing more serious as I spoke. We had been tracking my health together since I turned forty—something my father strongly encouraged, given our family history, insisting I begin prostate exams early. Now, nearly a decade later, it was July 2019. I was two months shy of forty-nine, staring down another year of unmet health goals.

He began reviewing my charts, walking through the labs the way he always did. He explained my CBC, the complete blood count, then noted that my PSA remained in the normal range. No concerns there. He paused briefly before pointing out anemia, a low red blood cell count. Given my fatigue, it made sense. We had only begun to discuss possible reasons for my low energy when his attention shifted to the next numbers on the page—my weight and fasting blood sugar.

His tone changed slightly, more cautious now. "Your weight has climbed to 232," he said. "That puts you in the obese range. And your fasting blood sugar is 107—just above normal." He tried to strike a balance, emphasizing the numbers while placing them in context: age, stress, busy lives. "It's not something to be overly concerned about at this point." But I couldn't shake the feeling that something was wrong. As Dr. Parks spoke, my mind went back through the numbers from earlier years. Normal fasting blood sugar is under 100 mg/dL. Mine had started in the 80s, climbed into the mid-90s, then the upper 90s. Now it was over 100—107. The pattern was unmistakable, and it didn't sit right with me. Why wasn't this raising more concern? The visit was already moving beyond conversation and toward wrapping up, but my mind stayed fixed on that number. I felt like I knew my chart better than anyone in the room. This wasn't a random fluctuation. Something in my body was changing.

Finally, I said it out loud. I pointed to the trend, the steady rise in my blood sugar over the years. Dr. Parks looked again, his attention sharpening. "Would you be open to additional testing?" he asked after a moment. "It's definitely something we should keep an eye on." He suggested adding an A1C test to the labs already ordered for anemia. Unlike a single fasting number, he said, the A1C would reflect my average blood sugar over the past three months.

We agreed that my fasting blood sugar had been rising each year, and it was time to look deeper. The A1C would become the new measure, one that reflected more than a single morning's snapshot. This wasn't just about today's number; it was about trajectory, the slow, quiet rise that could spell trouble if left unchecked.

As I left Dr. Parks' office that day, the weight of those numbers followed me. This was no longer just about a race I couldn't run—it was about time itself, and the gradual erosion of my health that I hadn't yet learned how to stop. Still, for the first time, I felt a sense of direction. I needed to understand what was happening, and find a way to turn those numbers around.

What was happening to me, and what was happening to all of us? Everywhere I looked, health seemed to be slipping—especially as people got older. Friends, colleagues, and family members were carrying more medical labels, more prescriptions, more limitations, year after year. Yet so much of it was accepted as normal, almost expected, as if declining energy and health were simply the cost of time passing.

I told myself I wouldn't accept that path. The gradual accumulation of weight, poorer lab results, chronic conditions, and diminishing energy had become routine—something managed rather than questioned. In exam rooms, these changes were often met with reassurance and a prescription, not curiosity. The message was quiet but clear: this is how aging works. Learn to live with it.

But I wasn't ready to accept that fate. There had to be another way. The questions kept circling: *How do I fix this? How do I take back control?* What struck me most was the exhaustion I kept seeing—not just in myself, but in so many people around me. This wasn't only about weight or lab results; it was about life itself. People were working harder simply to get through the day, juggling family, children, bills, and constant stress, only to fall back on the comforts of modern habits, foods engineered to taste good in the moment, yet leave you hungry again, more depleted, and wanting more.

How many of us have set out to lose weight, only to be told that if we fail, it's our fault? *Calories in, calories out*—that's what we're told. Simple math. Personal responsibility. But what about the other messages we've absorbed over a lifetime, the ones that quietly shaped how we eat, move, and live?

Breakfast is the most important meal of the day, it jump-starts your metabolism. Your brain won't function without sugar. Low fat is heart-healthy. Carbohydrates are necessary for energy. Cardio and strength training will burn off excess calories. Whole wheat, grains, fiber, fruit—bananas, apples—all essential for digestion and hunger control. *Don't restrict food groups. Take a cheat day.*

Then there's the world we live in: fast food on every corner, bagels and donuts in the break room, barbecue and beer with friends. Add the constant reassurance of *everything in moderation*, and the result is a landscape where discipline and restraint are quietly undermined. The challenge becomes overwhelming. Many of us end up trapped in a 24-hour cycle of effort and relapse, blaming ourselves for a system never designed for metabolic health.

I remember the race—not just any race, but the one that defined my early teenage years. I trained on my own, but we also trained together, my father and my two older brothers. It wasn't structured or overly planned, just basic three and five-mile jogs, enough to prepare for local community road races. These were the kinds of races meant to support charities, raise funds, and encourage fitness. We didn't run to win. We ran for the thrill of competition, to push ourselves, to chase personal records, and to feel that unmistakable sense of being alive.

My earliest memory of running goes back to when I was five or six years old, living in a suburban town outside Pontiac, Michigan. My father would head out for a jog at night, a couple of hours after dinner. He'd slip out through the garage and start down the road, looping through the neighborhood on a familiar two or three-mile route.

One evening, I decided to follow him, but I left a few moments after he did, trailing him just far enough to remain unseen. I wanted to test myself, maybe even prove that I could keep up. The sky was darkening, streetlights flickering one by one, glowing only at the corners of each block. The cool night air felt exhilarating against my face as I turned corners and ran down the next street. It felt easy, almost effortless, yet fast. The darkness made the moment surreal, as if I had stepped outside myself. That night was my first real jog, and the feeling stayed with me. I didn't have words for it then, but I was fully absorbed—moving freely, attention narrowed, thoughts drifting, carried forward by the simple joy of motion.

A few years passed, and the race—the pinnacle of our summer training, became the Bobby Crim 10 Mile, held each August in downtown Flint, Michigan. Our family returned every year for what felt like the next decade. It may have started with just my father, but soon my two brothers and I joined in. It became our thing, end-of-summer family tradition.

My father ran that race at the same age I was now, well into his forties, proof that it wasn't a young man's event. People ran it in their forties, fifties, and beyond. Yet here I was, struggling just to get back to the starting line. As I grew older, college, work, and family gradually took over, and the running stopped. Nearly twenty years passed before I finally asked myself, *Could I get back in shape to register for this race again?*

But now, in my late forties, my body kept answering no. Each summer I tried to train enough to sign up, but I was never quite ready. Life, or at least my understanding of it, always seemed to get in the way.

Replaying the visit in my head, I drove back to work. Could I really take back control? Could I defy the odds and avoid becoming another quiet statistic—someone who never quite found his way back to health? The road ahead felt uncertain, even daunting. Still, something inside me refused to go dark. A flicker of hope remained. I sensed that something important was standing in my way, and that progress wouldn't come from trying harder, but from understanding what my body was asking for. If I could uncover what had shifted beneath the surface, it might finally open a path forward. Whatever it took, I was determined to find it.

CHAPTER TWO

CROSSING THE THRESHOLD

"In the middle of difficulty lies opportunity."
ALBERT EINSTEIN

My life fell into a familiar cycle. Winter meant long tax seasons—extended hours, compressed days, irregular meals, and sleep that never quite caught up. I worked through it the way I always had, telling myself that summer would restore what winter had taken.

And in some ways, it did.

Summer brought more daylight, more movement, and more intention. I had more time for the gym and more time to walk the trail. I paid closer attention to what I ate. One summer I gave up hotdogs and hamburgers. Another summer I did what I call the starvation diet—forcing small meals despite my body wanting more fuel. The migraines were brutal. I treated those months as a reset, an opportunity to recover ground lost during the winter stretch.

For a while, it worked. But each year, something subtle changed. Progress slowed. Recovery lagged. What once came back easily required more effort. Weight loss stalled sooner. Energy returned partially, then stalled. No matter how disciplined the summer became, it never fully erased the winter.

When I finally stopped looking at individual seasons and started looking across years, the pattern was unmistakable. I gained more than I lost, not dramatically, not all at once, just enough to matter. Ten pounds gained. Seven lost. A small remainder carried forward. Baseline plus three.

Over time, those small remainders accumulated. The weight was visible, but the deeper change wasn't. What I felt most was less energy and more fatigue—less energy for the gym, slower recovery, less resilience when life demanded more. I could no longer fuel my way out of fatigue or train my way back to equilibrium. Effort still mattered, but it no longer translated the way it once had. The system itself seemed less responsive, as if something fundamental had shifted out of view.

By the time the numbers on my labs began to move in the wrong direction, the internal change had already been underway for years. Prediabetic. Hypoglycemic. Those words did not compute at first. This was years in the making. It had been building quietly, season by season, year by year, while I was focused on managing around it. All I knew at the time was that I wanted resolution, not maintenance.

My wife, Carol, has always carried herself quietly—putting family, work, and the needs of others ahead of her own health, often without drawing attention to herself. It wasn't something she announced or even consciously chose. It was simply how she moved through the world. Her path looked different from mine, but it was moving in the same direction.

In her thirties, she was diagnosed with hypertension and placed on medication. It was presented as manageable, common, and controlled—something to monitor, something to live with. In her forties, breast cancer arrived. It was diagnosed early. Surgery was still required. She endured it with resolve and without drama. Tamoxifen was prescribed for five years, a hormone-blocking drug. The immediate crisis passed, and life moved forward again. From the outside, it looked like a story of survival, and in many ways, it was. But survival came with a cost that didn't disappear once treatment ended.

Then came grief.

Walking had always been her release. Even on the hardest days, she'd lace up her shoes and head out the door, not for distance or fitness, but for a few minutes of quiet. She rarely talked about what she was carrying. She just kept moving.

In 2017, Carol lost her mother to pancreatic cancer. The loss was profound in a way that resisted explanation. She later described it as *love that had nowhere to go*—an emotional weight carried without release. There was no timeline for it, no clear endpoint. It simply became part of the background of her life.

Through all of this, Carol continued to function. She worked. She showed up. She cared for others. She followed medical advice. She did what was asked of her. When concerns arose, they were addressed individually—one appointment, one recommendation, one step at a time. But nothing ever addressed the whole.

Carol tried to break through on her own. She committed to doing what she was told, counting calories, joining Weight Watchers, tracking steps and daily walks—only to encounter the same resistance time and again. Nothing fundamental changed. The only real outcome was a growing denial of hunger. By 2019, the concern had escalated. Carol was advised to undergo a colonoscopy after something abnormal appeared. It felt relentless. *What was going on? How could so many warning signs be treated as isolated events?* By that summer, it was clear this wasn't something either of us could solve alone. Years of effort had brought us to the same place. Like me, Carol reached the same conclusion.

We both lost our mothers within a few years of each other. The losses came differently and landed differently, but they shared a quiet finality that altered how we saw time. What once felt distant suddenly felt close. What once felt postponable no longer did.

I processed my mother's death internally. It sharpened my awareness and narrowed my tolerance for excuses. Her absence made the fragility of health and momentum impossible to ignore. I became more alert to decline, less willing to dismiss warning signs or explain them away. My mother passed in April 2019. She was seventy-five. There were no goodbyes, no closure, no final chapter. She died from complications of pneumonia. My grief didn't arrive the way I expected it to. There were no tears, no overwhelming pain—just reflection on how abruptly life can end. Too soon.

I did cry later, while holding Rita, my mother's chihuahua. After my mom passed, the family decided Rita would be best cared for with me. At the time, it felt like a practical decision. Only later did I understand how much it mattered. Rita became my connection to my mother—a small, living thread that carried more weight than I anticipated. She brought laughter into the house, and with it, memories. Affectionate, funny, and quirky, demanding attention one moment, then barking and pretending to bite if you leaned in too close. It was strange and hilarious all at once, and we loved her for it.

Having Rita with me didn't erase the grief, but it gave it somewhere to go. Through the laughter she brought, the memories she stirred, and the sadness that surfaced alongside them, I was finally able to say goodbye.

At the same time, I was changing in a different way. I was no longer numb to my decline, I could feel it clearly now. The slow erosion of energy, resilience, and capacity had become impossible to ignore. I knew I had to make intentional moves, even if I didn't yet understand what they needed to be. This time, I refused to put it off and wait for another summer. It was time to commit.

We processed the loss of our mothers differently, but we moved through it together. Change became something we shared. Neither of us collapsed—and that, too, was part of the problem. We continued working, parenting, showing up, doing what was required. From the outside, nothing appeared urgent. From the inside, things were breaking down.

For years, we had interpreted fatigue, hunger, weight gain, and mental fog as separate problems, each something to manage, negotiate, or override. Eat to fix the headache. Snack to prevent the crash. Train harder to counter the weight. Rest more to compensate for fatigue. Every response assumed the same thing: that energy had to be supplied from the outside, again and again, or everything would fall apart.

But when we stepped back far enough, the pattern became impossible to miss. The problem wasn't lack of effort. It wasn't lack of discipline. It wasn't supply.

It was access.

Each issue was treated individually, but none addressed the signal coordinating energy across the entire system. We began to understand that the body doesn't simply store energy and release it on demand. Access to stored energy is regulated, and the signal at the center of that regulation turned out to be a hormone most of us had never really understood: insulin.

Until then, insulin had existed in our minds as something given to diabetics to manage blood sugar, important, but distant, and rarely explained. What we hadn't understood was that insulin is not just about blood sugar. It is the primary signal that tells the body whether to store energy or release it, and different foods influence that signal in very different ways.

When insulin is high, the body is in storage mode. Energy is trapped. When insulin falls, the body can open the door. Energy is released.

It is a way home.

Not a shortcut or a hack, but a return to something familiar. A return to a body that knows how to sustain itself, without constant input. Hunger softens. Energy steadies. The frantic search for fuel quiets. What once felt inaccessible becomes available again, not through force, but through permission.

The word for releasing stored fat is lipolysis. It sounds technical, but the meaning is simple: fat being broken down so it can be used for energy. It's what the body is designed to do between meals, overnight, and during long stretches of activity. But lipolysis only happens when insulin is low enough to allow access. This was the missing piece.

We weren't tired because we lacked energy. We were tired because the energy we carried was locked away. We weren't hungry because we needed calories—we were hungry because insulin prevented us from accessing what we already had. The body was asking for fuel not because the tank was empty, but because the fuel line was blocked. The energy was there, but it couldn't reach the engine. So the body kept asking for more fuel, even while carrying plenty. That analogy explained everything we had been feeling. The headaches made sense. They arrived with insulin surges after rising glucose, then lingered as glucose fell but insulin remained high. The fatigue followed the same arc, brief energy after eating, followed by a crash that demanded another input. Even the weight gain finally fit the picture.

Calories in, calories out had taught us to think of the body like a bank account. But insulin changes the rules. Two people can eat the same number of calories and have very different outcomes depending on insulin signaling. When insulin stays elevated, fat is directed into storage and held there. When insulin falls, fat becomes available again.

In that context, calories don't disappear, but they don't determine direction either. The body is not a simple ledger. It is a regulated system.

What we had been doing for years, eating frequently, choosing "healthy" carbohydrates, managing portions—kept insulin elevated enough that fat release never fully occurred. We were living in a constant in-between state: never fully fed, never fully fasted, never accessing the energy reserve meant to carry us. That explained why effort stopped working. We could train harder, but recovery lagged.

We could eat less, but the system resisted. From the inside, it felt like personal failure. From the outside, it looked like aging or stress. What it really was: a body stuck.

The brain felt this too. The brain depends on steady energy. When glucose rises and falls under constant insulin pressure, mental clarity suffers. Focus narrows. Mood becomes reactive. Concentration fades. Emotional steadiness erodes. What we had attributed to stress increasingly looked metabolic.

The more we learned, the clearer the story became. This wasn't about finding the perfect diet or chasing weight loss. It was about changing the signal that determined whether energy could be accessed or remain trapped. If insulin stayed high, nothing else mattered— not calories, not willpower, not intention. The body would remain in storage mode. If insulin was allowed to fall, access would return and energy would stabilize without force. That realization removed the last illusion we had been clinging to: that moderation was neutral. In this context, moderation wasn't balance. It was interference.

Continuing to eat foods that reliably raised insulin, while hoping for a different outcome—no longer felt responsible. It felt like pressing the same button again and again, expecting the system to change.

So we chose a direction. We decided to remove the foods most responsible for keeping insulin elevated and allow it to fall long enough to see what the body would do when access was restored. We didn't frame this as a diet. We framed it as an experiment in physiology, a way to test whether insulin really was the signal tying everything together. If it was, we would feel it. And if it wasn't, we would keep looking. For the first time, we had a plan.

CHAPTER THREE
CHANGING THE CONDITIONS

*"If you want to understand function,
change the environment."*
SYSTEMS BIOLOGY

Once we understood the problem, waiting was no longer an option.** This wasn't a moment for gradual adjustment or symbolic change. If insulin was the signal controlling access to energy—if it determined whether we were storing fuel or allowed to use it, then the environment had to change, fully, and right away.

We didn't start with motivation. We started with removal. The goal was simple: lower insulin long enough to see what happened when the body was allowed to access stored energy again. That meant changing the foods that drove insulin highest and most often.

At the time, we didn't yet have the language for what was happening. Our research led us to believe that something fundamental was wrong, not because energy was absent, but because access to it was being restricted by the environment we had created. That idea is captured in a simple principle—one that would later guide everything we did: *"If you want to understand function, change the environment."*

It was our theory, unproven at the time. That we hadn't failed to generate energy, but were failing to access it. And that as long as the environment remained unchanged: constant carbohydrate intake, frequent eating, persistently elevated insulin—we were never actually observing normal human function. We were observing a system behaving exactly as it was being instructed to behave.

If this was true, then many of the symptoms people live with every day, fatigue, constant hunger, brain fog, mood instability, and metabolic disease—were not signs of broken bodies, but predictable outcomes of an environment that never allows access to stored energy. Not because the system is defective, but because the signals never change. The environment stays the same. The instructions never stop.

So the goal wasn't to push harder or try to override biology with discipline. It was to change the conditions that kept energy locked away. Lower insulin. Remove the inputs that distorted signaling. Then watch what happened when the environment finally allowed stored energy to return.

Only then could we begin to understand how the body was meant to function.

But there was another weight we carried into that decision— one unrelated to hunger or energy. It was fear, deeply learned and rarely examined.

We grew up in the 1980s with very clear messages about how responsible people were supposed to eat. Those ideas were taught early, reinforced often, and carried forward without much reason to question them. They shaped our instincts long before we ever began thinking seriously about metabolism or energy.

For many of us, that fear had a name: Dr. Robert Atkins. Atkins was a physician who spoke directly to everyday people, arguing that the way many of us were eating, especially our heavy reliance on refined carbohydrates, was driving weight gain and poor health. His message was simple and practical: change what you eat, change how your body behaves. He didn't speak primarily to academics or policymakers. He spoke to families, patients, and people who were struggling.

But growing up, his approach was treated as dangerous, extreme, even reckless. It wasn't framed as an alternative to be evaluated, but as a warning. The message was clear: this was not how responsible people ate. Fat, especially saturated fat, was something to avoid. Carbohydrates, by contrast, were familiar and reassuring. They were presented as safe, necessary, even protective. What we planned to do wasn't the same, but it was close enough to trigger the same unease. Like the Atkins Diet, we knew that eating this way already carried a cultural warning label. People had been taught to fear it long before they ever understood it.

Even before we began, we understood how it would sound. Family and friends would dismiss the idea as radical. Some would be polite but unconvinced. Others would be openly concerned for our well-being—not because they had examined the evidence, but because this way of eating violated what responsible people were supposed to do. And yet, that was exactly what we were doing. We were increasing fat. We were eating foods we had been taught to limit or avoid. We were stepping directly into an environment that contradicted decades of messaging, not out of rebellion, but because our research suggested it was the only way to test the theory honestly.

In practical terms, this meant changing the proportions of what we ate. The dietary pattern we'd grown up with, reinforced by the food pyramid and decades of public guidance, placed carbohydrates at the center of the plate, often making up half or more of daily calories. Grains, starches, and sugars were presented as the foundation of a healthy diet. Protein was included, but moderated, something to be consumed in controlled portions rather than relied upon for energy. Fat, by contrast, was treated with suspicion. It was limited, cautioned against, and often framed as something to avoid altogether. Energy was expected to come primarily from carbohydrates, not from fat.

We decided to move in the opposite direction. We couldn't even remember where we'd heard it, but the idea was clear: *flip the damn pyramid.* That was essentially what we were doing.

Carbohydrates were reduced substantially. Protein became adequate and consistent. Fat—once treated as something to restrict became the primary source of energy. Not because it was indulgent, but because it was stable. Vegetables and fruit remained, but in smaller proportions. Meals were built around real food. Single-ingredient food.

Still, the question lingered. Were the warnings right? Had decades of public health guidance been protecting us—or steering us away from something essential? We had read the literature. We had listened to clinicians and researchers who challenged the prevailing narrative. But belief and certainty are not the same thing.

This was the moment where theory met reality. This wasn't ideology. It was an experiment. And this time, it was ours. With that context in place, we could finally begin to understand what was actually driving the system.

This is how we understood it. The glycemic index became useful as a simple way to describe how quickly a food raises blood sugar, which is one of the main drivers of insulin release. Foods with a high glycemic index produce rapid spikes, while foods with a low glycemic index produce slower, smaller responses. Frequent spikes mean frequent insulin release, and frequent insulin release keeps the body locked in storage mode.

Once we understood that, the plan became clearer and more measurable. We had a roadmap to what would become our own clinical trials.

The first and most important step happened in our kitchen. The truth was straightforward: in the context of metabolic dysregulation, sugar and starch converge on the same outcome—they become glucose in the body. Sugar went first, not because it was unique, but because it was obvious. Bread followed, then pasta, rice, cereal, potatoes, and anything made primarily from flour or starch. Different foods, same result. We didn't try to moderate these foods. We didn't substitute "healthier" versions. We didn't keep them around just in case. They came out of the house entirely.

At the same time, we stopped grazing. Constant eating: snacks, bars, small meals meant to "keep energy up" was quietly reinforcing the same problem. Each eating event required insulin. Frequent insulin meant no window for fat release. So we simplified. Meals became fewer, clearer, and more deliberate.

Once carbohydrates were removed, something else had to become central: protein. Eggs. Beef. Poultry. Fish. Foods that satisfied hunger instead of provoking it. Foods that didn't require insulin surges to manage. Foods that allowed long stretches between meals without anxiety or collapse. Alongside protein, we stopped fearing fat. Butter. Olive oil. Avocado oil. Animal fats. These weren't extras. They were necessary. Fat provided energy without demanding insulin. It allowed meals to be complete rather than stimulating.

We didn't track calories or measure portions. We relied instead on a principle we'd come to understand through functional medicine: when meals are built around protein and fat, the body responds with fullness, making overeating increasingly difficult. We ate until satisfied and stopped when hunger ended. That was the point.

The first change wasn't weight. It was hunger. The constant pull to eat softened. The urgency faded. Meals no longer triggered the desire to eat again shortly afterward. Waiting became possible, not forced, just natural.

We hadn't stopped believing in calories; they just stopped being useful. Direction mattered more than arithmetic. We weren't eating less because we tried to. We were eating less because we needed less. The system had changed.

This is often misunderstood. We were eating foods that were more calorie-dense, yet total intake fell—not because of restraint, but because the body was finally responding to the food itself rather than chasing energy through volume. When energy is inaccessible, hunger persists regardless of how much food is consumed. Meals satisfy briefly, then fail, prompting the body to ask for more—not because it needs calories, but because it can't reach them. Eating becomes a repeated attempt to close an energy gap that never quite resolves.

The old paradigm was carbohydrate-centric, built on managing calories after hunger appeared. The new paradigm was metabolic-centric: choosing foods that restored satiety, stabilized energy, and allowed appetite to regulate itself.

You might expect immediate reward in the first week. What we experienced instead were breaking waves—a normal response to switching fuel systems. We were struggling to stay afloat. The first week did not feel like progress. It felt like discomfort and doubt. Once carbohydrates were removed and insulin began to fall, the body had to confront something it hadn't done in years: run without constant incoming glucose. For decades, maybe longer, energy had arrived on schedule. Breakfast. Snacks. Lunch. Afternoon rescue. Dinner. Late-night insurance. Now the supply line was gone. The body was being asked to do something it had forgotten how to do, release stored energy and rely on it.

Those breaking waves were the surface expression of something deeper. This period has a name, though we didn't know it at the time. It's often referred to as keto flu, but that language obscures what's actually happening. This wasn't an illness. It wasn't an infection. It was a metabolic awakening, the reactivation of systems that had gone quiet after years of reliance on incoming glucose. At its core, this awakening isn't about deprivation. It's a withdrawal from the fuel the body had grown dependent on. For years, often decades, energy arrived on demand from the outside. When that supply stopped, the body didn't fail; it paused—not because energy was gone, but because it hadn't been asked to access it from within. As liver and muscle glycogen depleted, energy dropped quickly—not as a sign of danger, but as a sign of change. The body was being reintroduced to something it hadn't practiced in years: drawing from stored fuel again. Those pathways weren't lost. They were simply quiet. And in the early days, the adjustment can feel like weakness—when it's really the body reactivating its own stored-energy system. The instinct in that first week is to quit, to assume discomfort means the approach is flawed. But we didn't quit. We adjusted.

The kidneys responded almost immediately. For years they had been told to hold the line—to sodium and to water—by the same signal, delivered day after day. Insulin wasn't just hovering in the background. It was the quiet authority behind the curtain, sending the same message day after day: retain. And sodium—faithful, literal, easily recruited—became the enforcer. Wherever sodium was ordered to stay, water followed. That's what water does. It never travels alone. So the kidneys held the line. They didn't do it because they were broken. They did it because they were being pressured into obeying instructions. And the longer that message ran, the more the bloodstream began to feel like a system under constant load—more fluid kept in circulation than the moment truly required. Pressure rose as fluid shifted. The vessels tightened and adapted, and the heart was asked to work harder, pushing blood forward as if the normal current of life had become a headwind.

The vascular system became the front line—tasked with carrying more volume, more force, more demand—minute after minute, meal after meal, year after year. It wasn't dramatic at first. It was subtle: a slow tightening, a constant push, a body living as if it had to brace itself. The kidneys weren't under strain in the same way. They were the gatekeepers—tasked with maintaining balance—forced to enforce rules that no longer served the system. They held the line not because they were failing, but because they were obeying orders that never stopped coming. And then the orders changed.

Insulin, the quiet authority behind it all, eased its grip. Not suddenly. Not dramatically. But enough for the kidneys to notice. The call to hold the line was no longer constant. The gatekeepers recognized the shift and were finally able to stand down. Sodium was the first to respond. No longer conscripted to enforce retention, it was freed to move again. And as it did, water followed—just as faithfully as it always had. What had been trapped in circulation began to leave. Volume fell.

The vascular system felt it immediately. Vessels that had spent years bracing against constant force were finally free to relax. The headwind softened. The heart no longer had to lean into every beat. Blood moved forward again, not by force, but by flow. Nothing needed to be overridden. The system didn't require willpower or correction, only the right orders. Once those changed, every character returned to its original role. The kidneys balanced. Sodium regulated. Water moved where it was needed and left when it wasn't. The vessels stopped fighting, and the whole system began to settle.

What came next required us to do the opposite of what we'd been told our entire lives. Conventional wisdom said to cut salt to ease pressure. But by then the pressure was already easing. The real burden had never been salt—it was the constant flood of high-glycemic foods that kept the system locked in defense. Once those foods were removed, the orders changed, and the system relaxed.

Now the body needed what it had been deprived of. We gave it salt—not recklessly, but intentionally. Salt to restore balance. Salt to support circulation. Salt to allow nerves to fire, muscles to contract, and cells to communicate the way they were designed to. We drank water when we were thirsty, guided by the body's natural thirst rather than fixed recommendations. We rested when needed. We supported the system instead of fighting it.

Over time, equilibrium returned. The kidneys settled into a new pattern. Fuel became dependable rather than urgent. Fat and ketones moved from backup systems to primary sources of energy. What once felt like weakness revealed itself for what it was: a period of recalibration. The body wasn't failing. It was relearning how to run on what it had stored all along.

By the end of the first week, the disruption had passed. By the second, energy returned—quietly, steadily, without the swings we'd grown used to. It wasn't the sharp urgency of stimulation, but something more reliable. Energy that stayed. By the third week, life began to feel normal again, only better. Carol moved through her days with clarity and resilience. She could walk, work, and engage without drawing down a limited reserve. Fatigue no longer dictated the shape of her day. I felt it too. Exercise no longer required negotiation. Recovery followed effort naturally. Movement stopped feeling costly. What surprised us most wasn't the improvement, it was how little effort it took to sustain it. Nothing felt fragile. Nothing demanded constant management. The system seemed to support itself.

Time, in particular, returned. Our days were no longer organized around fueling, around breakfast schedules, snack planning, or the low-grade anxiety of wondering when energy would run out. We weren't living from meal to meal. We were living our lives. It didn't take long to notice that breakfast was no longer the most important meal of the day. Mornings unfolded without urgency. A small lunch was enough when we wanted one, and sometimes it wasn't needed at all. Dinner became something we looked forward to again—not as rescue, but as enjoyment. We cooked. We ate together. There was no emergency takeout, no last-minute delivery ordered out of starvation. Food returned to its proper place. It supported life, but it no longer dictated it.

That freedom was subtle at first, but unmistakable. When a system stops demanding constant input, space opens—space to think, to move, to plan, to be present. What we gained wasn't just steady energy. We gained time that had been quietly consumed for years by the need to refuel.

And the change wasn't only physical. As energy stabilized, so did our mental capacity. Focus deepened. Curiosity widened. The constant background noise that had dulled attention for years began to fade. We found ourselves reading again, and then reading more. Not casually, but with focus and intent. My mind wanted information the way a healthy body wants movement.

Carol described it in the simplest terms. "My brain is lit," she'd say. "On fire." Not anxious or overstimulated—just awake. Fully online. The kind of clarity where work demands don't feel like weight anymore, and you realize you've been moving through the day without thinking about hunger at all.

That mental clarity didn't feel like stimulation; it felt like access. Access to thought, to attention, to ideas that had once felt just out of reach. For years, we'd said the same things without really questioning them. I want to read, but I'm too tired. I want to learn more, but I don't have the energy. If only I could stay awake. If only I could wake up earlier. We assumed those limits were fixed—part of getting older, part of being busy, part of life.

As that clarity returned, those quiet constraints fell away. Reading no longer felt like work reserved for rare moments of alertness. It became something we could do because the energy was there. Mornings opened up. Evenings stayed present. The mind no longer shut down early, asking to be left alone.

If energy could be this stable without constant refueling—if the body and brain could function this well under such simple conditions—then the model we'd been taught was incomplete. There had to be another system at work. Not a new one, but an older one. A system designed to carry humans through uncertainty, scarcity, and sustained effort without collapse.

Understanding the system became more than curiosity; it became a responsibility. We had experienced one of the most profound transformations of our lives in an unexpectedly short period of time. Yes, we were still healing from years of dysregulated energy. But something undeniable had awakened. The body was responding. The mind was clearing. A system we hadn't been taught to trust was revealing itself as part of a far older design.

That recognition brought confidence, not arrogance, not certainty, but assurance. We could feel the system coming back online, working in harmony with the way the body was created to function. Day by day, it felt less like discovery and more like remembrance.

We wanted to understand it fully. Not to dominate the system, but to respect it, to learn how it worked, where it came from, and why it had been so easy to overlook for so long. In doing so, we took back control, not by force or restriction, but by restoring order. Hunger no longer dictated our days. Energy no longer felt borrowed or fragile. We felt a quiet gratitude in that process, a sense of touching something sacred in ordinary life, of moving through our days with energy returned and awareness restored. We hadn't created healing. We had simply removed the interference, and watched what God built into the human body begin to work again. And we wanted others to feel that same joy, that same steadiness, that same freedom. Understanding that system became our next step.

PART II

Rediscovering the Energy System

CHAPTER FOUR

THE FORGOTTEN ENERGY SYSTEM

*"What we observe is not nature itself, but nature
exposed to our method of questioning."*
WERNER HEISENBERG

We are taught a simple story about energy. We eat, food is broken down into a readily available fuel, and that fuel powers the body. When energy runs low, we eat again. When health declines, we assume the solution is more fuel, or greater discipline in how we manage it.

This story is familiar, intuitive, and deeply ingrained. It shapes how we eat, how we treat illness, how we interpret hunger, and how we judge our own bodies. It is reinforced by cultural norms, medical education, and daily routine, and few people ever stop to question it.

But beneath that familiar story lies another: older, quieter, and far more resilient. Long before nutrition labels, meal schedules, or dietary guidelines, the human body evolved in an environment where food was intermittent and energy demand was unpredictable. Survival did not depend on constant intake; it depended on adaptability.

The body learned not only how to store energy, but how to access it when external fuel was unavailable. That capacity did not disappear when agriculture arrived, when cities formed, or when modern abundance took hold. It remained embedded in our physiology, waiting.

Ketones belong to that deeper story. They are not a modern invention, nor were they discovered to sell diets, optimize performance, or provoke controversy. They emerged gradually, often reluctantly, through scientific attempts to answer a more fundamental question: *how does the body function without abundance?*

Again and again, across decades of research, ketones appeared at moments of stress, scarcity, and survival. They surfaced in studies of fasting and starvation, in investigations of brain metabolism, and in disease states where energy delivery to cells became unreliable. Each time, they revealed the same truth: the body is not dependent on a single fuel source. It has a built-in capacity to shift, adapt, and sustain life when conditions change.

Yet despite their repeated appearance, ketones never entered the broader cultural understanding of health. They were framed as secondary, pathological, or extreme; associated with illness, starvation, or emergency states. Meanwhile, modern life moved steadily in the opposite direction, toward constant feeding and the assumption that energy must always come from outside the body, on demand.

At multiple points in history, researchers recognized that ketones were not merely a backup, but a regulated, efficient, and protective energy system. Those insights, however, remained largely confined to academic literature, rarely taught in medical schools or carried into routine clinical practice, and almost never integrated into everyday thinking about health.

The result is a defining tension of modern life. We live in bodies capable of remarkable metabolic flexibility, yet most of us never experience it. Hunger is interpreted as danger. Fatigue during metabolic transition is interpreted as failure. Without constant intake, we assume energy must collapse. And so an entire fuel system—one designed to stabilize energy when glucose cannot—remains unfamiliar, unused, and misunderstood.

It's like swimming out into the ocean, where the hardest strokes come closest to shore. The breaking waves are loud and disorienting, white water surging and collapsing in every direction. Discomfort feels unsafe, progress uncertain. Many turn back here, convinced the water ahead must be worse. But beyond the break, the surface settles. Movement smooths. Breathing steadies. The ocean reveals a calmer, more settled state, one that can only be reached by passing through the turbulence first.

It is about rediscovering a forgotten capacity—one that science has encountered repeatedly, that culture has consistently overlooked, and that modern health crises are now forcing us to confront. Ketones are not new. What is new is the urgency to remember them.

Otto Warburg and the First Warning
About Fragile Energy

In the early twentieth century, a German scientist named Otto Warburg began noticing something that did not fit the prevailing understanding of biology. While studying cancer cells, he observed that many of them relied heavily on the fermentation of glucose, even when oxygen was readily available.

This was puzzling. Fermentation was a crude, inefficient way to generate energy, typically used only when oxygen was scarce. Healthy cells, when given the choice, relied instead on a far more efficient process, one carried out inside mitochondria, the specialized structures designed to extract maximal energy from fuel. Yet here were cells choosing the short, wasteful path even when better options were available.

Warburg proposed a controversial idea for his time: perhaps disease was not driven solely by faulty instructions or runaway growth, but by a breakdown in how cells produce energy. In his view, something fundamental had gone wrong with the cell's ability to breathe and generate power efficiently.

Many scientists disagreed with how far Warburg took this conclusion, and debates over its implications continue to this day. But even critics acknowledged the significance of his central observation. Warburg had revealed a vulnerability that extended beyond cancer: when cells lose access to efficient energy pathways, they become dependent on inferior ones.

This dependence comes at a cost. Fermentation can keep a cell alive in the short term, but it is energetically expensive and unstable. Cells locked into this mode must consume fuel continuously just to keep going. They survive, but they do not thrive.

In hindsight, Warburg's work offered an early warning. A biological system that relies on a single, fragile energy source is vulnerable to breakdown. Flexibility matters. Efficiency matters. And when those qualities are lost, dysfunction follows.

What makes this moment especially striking is what was happening outside the laboratory at the same time. As Warburg was uncovering the dangers of glucose dependence at the cellular level, modern society was moving rapidly in the opposite direction: building food systems, medical assumptions, and cultural norms around constant glucose availability. Regular meals became expected. Snacks became routine. Hunger became something to avoid rather than understand.

The parallel went largely unnoticed. Warburg's insight remained confined to scientific circles, while everyday life reinforced the idea that steady glucose supply was both normal and necessary. The possibility that reliance on a single fuel could create fragility rather than strength was rarely considered. The warning was there early. But it was quiet, uncomfortable, and easy to ignore.

Hans Krebs and the Blueprint for Metabolic Choice

While Warburg focused on what went wrong, another scientist was quietly documenting how energy was meant to work when the system was healthy. Hans Krebs was not searching for disease; he was mapping normal function. In doing so, he uncovered something profound about human design: energy metabolism was not linear, and it was not exclusive. It was integrative.

The citric acid cycle—now commonly called the Krebs cycle—revealed that different fuels all converge on the same central engine. Energy derived from glucose flowed into it. Energy derived from fat flowed into it. And critically, so did ketones. The body was not built to depend on a single input. It was built to translate multiple forms of fuel into usable energy, depending on availability and need. Ketones were not an emergency detour or a metabolic mistake. They were built into the system.

This insight carried a quiet but powerful implication. Metabolism was designed for flexibility. The body could shift between fuels without distress, drawing on internal reserves when external supply waned. Efficiency and adaptability were not afterthoughts; they were core features of human physiology.

Yet as this biology became clearer, the surrounding culture moved in the opposite direction. Nutrition guidance increasingly emphasized a single dominant fuel. Medical education favored simplified, linear explanations that were easier to teach and standardize. Metabolic flexibility, though foundational, proved difficult to measure, difficult to explain, and difficult to fit into modern frameworks. What biology designed as optional and responsive, society began to treat as fixed and narrow.

Over time, this shift shaped expectations. The idea that the body could move smoothly between fuels faded from public awareness. Energy came to be understood as something that required constant input, careful timing, and external control rather than internal capacity. Krebs had revealed a system built for choice. Modern life quietly replaced that choice with routine, and in doing so, one of the body's most powerful advantages, its ability to adapt, slipped further out of view.

Scarcity Revealed What Abundance Concealed

By the mid–twentieth century, scientists studying fasting and starvation began noticing something that quietly challenged prevailing assumptions about human biology. People deprived of food did not steadily weaken or unravel in the way many expected. After an initial period of adjustment, many experienced something counterintuitive: energy stabilized, muscle was spared, and mental clarity often improved. Rather than collapsing, the body appeared to reorganize itself.

Physiologist George Cahill helped explain why. As glucose availability declined, insulin levels fell, opening access to stored energy.

The liver began converting fat into ketones, and the body shifted into a different metabolic state, one designed not for growth, but for efficiency and preservation. Most surprising was the brain's response. Long believed to depend exclusively on glucose, it adapted readily, deriving a substantial portion of its energy from these alternative molecules.

Ketones changed the rules. They reduced the need to break down muscle for fuel, sparing protein. They provided a steady, internally generated energy supply, stabilizing the highs and lows that accompany constant feeding. They ensured that vital organs, especially the brain, continued to function even when external intake stopped. What had long been labeled starvation began to look less like failure and more like strategic conservation. Only later did scientists begin to recognize that scarcity was doing more than preserving energy. It was also initiating repair.

Periods of reduced intake did not merely shift fuel sources; they activated internal maintenance programs that modern abundance rarely allows to run. Damaged proteins were cleared away. Dysfunctional cellular components were recycled. Structures that no longer functioned efficiently were removed to make room for healthier replacements. Rather than breaking down, the body was quietly restoring order.

Mitochondria, the engines of cellular energy, appeared to benefit especially from this cycle. Scarcity signaled that efficiency mattered. Under these conditions, poorly functioning mitochondria were removed, while healthier ones were preserved and encouraged to multiply. Energy production did not simply continue; it improved. What looked like deprivation from the outside functioned as renewal from within.

This reframed scarcity once again. It was no longer just a survival response, but a biological reset, a period during which the body recalibrated, repaired, and prepared for what came next. Adaptation was not accidental. It was intentional.

Yet just as science was uncovering the protective and restorative nature of scarcity, modern life was eliminating the conditions under which it occurred. Constant food availability replaced natural rhythms of feeding and fasting. Hunger became something to suppress rather than interpret. The quiet metabolic space where the body transitions between fuels, the very space where repair begins, was increasingly avoided. Ketone metabolism, once a normal and expected response to environmental variability, came to be seen as unnecessary, extreme, or even dangerous. Scarcity had taught the body resilience. Abundance taught society to forget it.

Richard Veech and the Rediscovery of Efficiency

Decades later, the understanding of ketones took a decisive turn through the work of Richard Veech, a biochemist whose research quietly but profoundly reframed what ketosis actually represents. Until then, ketones were largely viewed as a backup fuel, useful in emergencies, but inferior to glucose under normal conditions. Veech showed that this assumption was backwards.

Through careful study, Veech demonstrated that ketones are an exceptionally efficient fuel. When used by the body, they generate more usable energy from the same amount of oxygen, while producing fewer damaging byproducts along the way. In practical terms, this meant cleaner energy, less strain on the system, less oxidative stress, and more reliable power delivery to cells with high energy demands. Ketones did not burden mitochondria. They supported them.

Even more striking was the realization that ketones were not merely fuel. They acted as signals, influencing how cells respond to stress, inflammation, and scarcity. Ketosis altered the internal environment of the cell in ways that favored stability and resilience. It was not a sign that the body was failing. It was a sign that the body had shifted into a more efficient operating mode. Ketosis, in this light, was not a breakdown state. It was a high-efficiency state.

Yet this insight arrived at an awkward moment. Medicine had become increasingly organized around pharmaceuticals, symptom control, and continuous nutritional input. Health was treated as something to be managed through external intervention rather than internal adaptation. Ketones did not fit easily into this model. They were not something to prescribe. They were a *state*: one that emerged only when certain conditions were allowed to exist.

Ketones required low insulin, access to stored energy, and metabolic patience. They emerged only after an initial period of discomfort, after the first waves of transition had passed, when the body was allowed to move forward rather than retreat. They could not be packaged, dosed, or easily monetized. They did not align with a culture built around constant feeding and the immediate correction of discomfort.

So once again, the pattern repeated itself. A powerful insight into human energy metabolism was established, documented, and largely confined to academic circles. The science was there. The implications were profound. But without a cultural framework to receive them, ketones remained misunderstood, recognized in theory, yet rarely allowed to be experienced in practice.

The Brain Remembers What Culture Forgot

As neuroscience advanced, researchers uncovered another paradox— one that challenged long-standing assumptions about how the brain fuels itself. In aging and in many neurodegenerative conditions, the brain's ability to take up and use glucose gradually declines. Yet even as this primary pathway weakens, the brain's capacity to absorb and use ketones remains largely intact.

Research led by Stephen Cunnane helped bring this insight into focus. His work showed that ketones could serve as a *rescue fuel*, supplying energy to regions of the brain that were struggling to metabolize glucose. Rather than shutting down, the brain adapted: drawing on an alternative source of energy that bypassed impaired pathways and restored function where possible.

This reframed cognitive decline in a quiet but profound way. Instead of viewing it solely as a problem of chemistry, structure, or inevitable loss, it introduced the possibility that part of the decline reflected an energy crisis—a failure of fuel delivery rather than a failure of the brain itself. Where glucose could no longer flow smoothly, ketones offered a different route.

Yet outside of research settings, this capacity remained largely invisible. Public understanding continued to equate mental energy with constant carbohydrate intake, frequent stimulation, and uninterrupted feeding. Fat was rarely considered a brain fuel. Ketones were rarely mentioned at all. The idea that the brain could thrive on internally generated energy remained absent from education, clinical conversation, and cultural habit.

Biology, it turned out, remembered what culture had forgotten. The brain retained its ancient flexibility even as modern life trained us to ignore it.

Diabetes, Obesity, and Heart Disease: Diseases of Fuel Restriction

Seen through the historical lens of energy metabolism, modern chronic disease begins to look far less mysterious, and far more predictable. We live in a world of constant food availability and rising body weight, yet at the cellular level, energy is often scarce. Fatigue, hunger, inflammation, and degeneration coexist with abundance.

Type 2 diabetes is often described as a disease of excess sugar, but at its core it is better understood as a state of fuel restriction. Chronically elevated insulin keeps fat locked away in storage and forces the body to rely almost exclusively on glucose, even as glucose becomes increasingly difficult and damaging to use. The body is surrounded by energy, yet unable to access it.

This mirrors Warburg's original insight. When efficient energy pathways are lost, cells fall back on inferior ones. They survive, but at a cost. What Warburg observed in diseased cells begins to appear system-wide in modern metabolism: when flexibility disappears, dysfunction follows.

Obesity, viewed through this lens, takes on a very different meaning. Rather than representing excess nourishment, it reflects trapped energy. Fuel accumulates in storage because insulin remains high, but little of it reaches the tissues that need it most. Cells behave as though they are starving, driving hunger, slowing metabolism, and demanding constant intake—while energy continues to pile up in reserves. Obesity is not the opposite of starvation. It is often *starvation in the midst of plenty.*

Heart disease reflects the same underlying mismatch. The heart is among the most metabolically flexible organs in the body, designed to run efficiently on fatty acids and ketones: fuels that provide dense, stable energy. Yet modern metabolic conditions restrict access to these fuels. As insulin resistance and inflammation rise, the heart is pushed toward greater dependence on glucose, a less efficient option under stress. Over time, energy efficiency declines, and vulnerability increases.

In both diabetes and heart disease, the problem is not a lack of capacity. The machinery for efficient energy use is still there. What is missing are the conditions that allow it to function. We keep the system turned on all the time, until it forgets how to run any other way.

Healing, in this context, becomes difficult not because the body has forgotten how to recover, but because modern life prevents it from accessing the fuels and signals required to do so.

Why This Fuel System Keeps Disappearing

At each stage of discovery—through the work of Warburg, Krebs, Cahill, and Veech—the same pattern quietly repeats itself. Science uncovers metabolic flexibility, revealing the body's capacity to shift fuels, adapt to scarcity, and activate repair when conditions allow. At the same time, culture builds systems that steadily suppress that flexibility.

This suppression is rarely deliberate. It emerges from habits, norms, and institutions that feel benign—even supportive—but gradually reshape metabolism over time. Hunger comes to be feared rather than interpreted. Nutrition messaging is simplified for clarity and compliance, often reducing complex biology to a single preferred fuel. Constant food availability removes the natural pauses that once allowed alternative energy systems to emerge.

Few cultural shifts illustrate this more clearly than the modern introduction of breakfast as a requirement. What was once a flexible, optional meal became a moralized necessity, *the most important meal of the day*. Food marketing and institutional guidance reinforced the idea that the body should never begin the day without immediate carbohydrate intake. Overnight fasting, once normal, came to be framed as something that must be urgently broken.

This pattern intensified with the rise of fast-food breakfast, engineered for speed, convenience, and repeat consumption. Energy was no longer something the body transitioned into; it was something delivered immediately, externally, and predictably—often before natural hunger had even emerged. Insulin rose early and often, and the window for accessing stored fuel quietly narrowed.

Snacking followed the same trajectory. Where meals were once separated by hours of metabolic quiet, eating became continuous. Snacks were framed as necessary for "keeping energy up," preventing hunger, or stabilizing mood. The body was rarely allowed to finish one fuel cycle before another began. The conditions required for fat and ketone metabolism: low insulin, time, and patience, became increasingly rare.

Even holidays and celebrations, once brief periods of abundance surrounded by restraint, expanded into extended seasons of constant indulgence. The exception became the norm, recovery time shrank, and metabolic adaptation was repeatedly interrupted before it could complete.

Medicine and health education evolved alongside these patterns. Clinical focus shifted toward downstream markers: blood sugar, cholesterol, weight—rather than upstream fuel access and metabolic state. Commercial incentives aligned with continuous intake, frequent eating, and constant stimulation. States that emerge quietly when feeding pauses and insulin falls did not fit easily into this framework.

As a result, these adaptive states did not disappear because they were disproven or irrelevant. They faded because modern life removed the conditions required to experience them. What is rarely accessed becomes poorly understood, and what is poorly understood is often mistaken for unsafe.

Living Inside a Body We No Longer Trust

Most people today live inside bodies built to adapt, yet rarely get to experience that design. Hunger is treated like an emergency. The early discomfort of change is treated like failure. And the quiet moment when the body begins to shift fuels—the space where repair and stability often begin—feels unfamiliar in a world that has forgotten how to pause.

As a result, many people retreat just as adaptation is beginning. They eat at the first hint of hunger, medicate the earliest sign of fatigue, and interrupt the body's attempt to shift gears before it has time to complete the transition. The discomfort is real, but often temporary. Without context, normal physiological signals are mistaken for warning signs.

This pattern has little to do with willpower. Most people are doing exactly what they have been taught to do. From an early age, we are instructed to eat regularly, avoid hunger, stabilize energy at all costs, and treat fluctuations as problems to be solved immediately. Education rarely explains why these signals occur or what they might be leading toward. The body's capacity to adapt is assumed to be fragile rather than resilient.

Ketones exist within this educational gap. They represent a biological capacity that modern life has trained us not to access, and therefore not to trust. Because few people experience ketone-based metabolism intentionally, it feels unfamiliar. And what is unfamiliar is often labeled unsafe. The irony is that this fuel system is not new or extreme. It is deeply conserved, protective, and remarkably efficient.

When the body is allowed to adapt, the experience is often subtle rather than dramatic. Energy stabilizes instead of spiking. Hunger quiets instead of intensifying. Mental clarity emerges without stimulation. But these signals appear only after the transition has been allowed to complete—something modern habits rarely permit.

Living inside a body we no longer trust has consequences. We outsource regulation to schedules, apps, external rules, and generalized recommendations, while ignoring the signals designed to guide us internally. In the absence of trust, lived experience is discounted—treated as unreliable or subjective rather than as real-time biological feedback. Over time, confidence in our own physiology erodes. Healing becomes something done *to* the body rather than something the body actively participates in and communicates through experience.

This alternative energy system reminds us of a different relationship—one based on understanding rather than control. It reveals that the body is not constantly on the edge of failure, but quietly capable of adaptation when given the chance. Most people have simply not been given that chance, and have not been given access to the full set of tools already built into the body, tools that do not come from the outside but are already present within us. Rebuilding that trust does not require force or discipline. It requires education, patience, and the willingness to experience a capacity that has always been there—a system so well-designed it doesn't need to be dominated as much as it needs to be supported.

Returning to Energy

This is not an intervention added to the body. It is a capacity revealed within it. They emerge when the body is allowed to rest from constant intake—when energy demand is met through internal reserves rather than external supply. The shift does not announce itself dramatically. It unfolds quietly, often surprising those who enter it for the first time.

When people reach this state, they tend to describe the same pattern: energy steadies instead of spiking, hunger fades into the background, mental clarity improves, and a calm efficiency replaces urgency. *What feels new is, in fact, ancient.* Ketones were never lost. They were simply inaccessible.

As individuals adapt to fat- and ketone-based metabolism, a consistent pattern often emerges in how the body responds to physical effort. Exercise no longer requires constant refueling. Energy delivery becomes steadier, supplying muscles, heart, and brain without the peaks and crashes of glucose-dependent fueling. At comparable workloads, heart rate often declines, reflecting improved energetic efficiency. Recovery accelerates. Training sessions stack more smoothly, without the cumulative fatigue that once seemed inevitable.

What stands out most in this state is not maximal performance, but consistency. Effort feels sustainable rather than urgent. The background stress of fueling fades. The body no longer appears to negotiate for energy mid-exertion—it simply has access to it. This shift is not about pushing harder or overriding limits, but about moving more efficiently through time, effort, and recovery.

Those same principles extend well beyond movement. Mental focus holds longer across the workday. Attention becomes steadier rather than fragmented. Hunger no longer dictates timing or interrupts concentration. Energy supports meetings, problem-solving, and creative work as reliably as it supports movement. Recovery feels restorative rather than incomplete, not just after exercise, but after long days of thinking, decision-making, and engagement. Energy becomes something stable rather than something to manage, allowing daily life to be lived with greater presence and less negotiation.

This is not a rejection of modern life, nor a call to extremes. It is an invitation to understand the body more fully—to recognize that health is not built on constant supply, but on the ability to adapt when supply fluctuates. Ketones remind us that resilience does not come from excess, but from flexibility. For many, rediscovering this capacity feels less like learning something new and more like remembering something fundamental. A quiet confidence emerges in realizing the body was built with redundancy, adaptability, and self-repair already in place. Trusting that resilience, rather than overriding it, can feel like trusting the wisdom of the design itself.

Once this forgotten fuel is understood, and experienced, energy is no longer something we chase. It is something we return to.

CHAPTER FIVE

ENERGY BY DESIGN

Returning to the Biology We Were Built For

"Every disorder of the body is a disorder of energy."
GEORGES LAKHOVSKY

Many of us start life with plenty of energy. In our youth, we move freely—sports, school, social life—seemingly with no limit. But then something shifts. Maybe it's in your thirties, maybe your forties. You notice mornings take more effort. You need more coffee to stay alert. You crash on weekends.

This isn't just "getting older." It's a slow unraveling, your capacity for physical, mental, and emotional energy diminishes. For some, it's subtle. For others, it hits like a wall. And today, it's not just adults. Teenagers and young adults are facing exhaustion, metabolic issues, and emotional instability, too.

What's happening to us?

We're given diagnoses: fatigue, anxiety, depression, burnout, weight gain. Sometimes they come with lab values or prescriptions. Often they come with reassurance. But they are just names. Do they explain what's actually breaking down? Too often, we treat the label, not the system behind it.

At the time, I didn't have a clear explanation, only patterns I couldn't ignore. My thinking felt dull. My body struggled to recover. I moved through my days with the sense that I was running on fumes, even while doing everything I thought was "right." Exercise, supplements, more food—it didn't help. Weight crept up. Momentum faded.

Looking back, even my constant snacking had a purpose. I often ate to avoid the next migraine. I didn't understand it then, but my brain may have been signaling distress inside a body of excess—seeking energy in the only way it knew how. That uncertainty led to a different question: what is energy?

The Hidden Engine:
Mitochondria and the Energy System

Energy isn't just a number on a calorie label. It's how your body transforms fuel into usable power, through tightly regulated biological systems. That transformation happens inside your cells, through microscopic organelles called mitochondria.

Mitochondria are small organelles inside the cell, often present in the hundreds to thousands per cell, depending on how much energy that tissue needs. For decades, they were described as the body's "powerhouses," because they convert food into usable cellular energy (ATP). But it's become increasingly clear that they do much more than produce energy. They act as metabolic sensors and signaling hubs—helping coordinate stress response, inflammation, recovery, and adaptation. They don't operate in isolation. They work symbiotically with the rest of the cell, shaping how that cell functions, repairs, and survives. They help decide whether a cell builds, repairs, defends, or shuts down.

But like any engine, mitochondria depend on clean fuel, structure, timing, and communication. When those inputs are disrupted, output declines—even if calories remain high. We are constantly overfeeding, not just in quantity, but in type and timing. Ultra-processed foods make up much of the modern diet. They are stripped of fiber, overloaded with sugar and refined starches, destabilized by industrial fats and seed oils, and packed with chemical additives. They flood the bloodstream quickly, spike blood glucose, and slam the energy system with repeated surges it was never designed to handle.

To make it worse, we eat all day long—three meals, plus snacks, drinks, protein bars. We never give the system time to clear out byproducts, repair cellular damage, recycle broken components, or reset hormonal signaling.

This creates a chaotic metabolic environment: blood sugar swings, insulin resistance, mitochondrial overload. Cells become inefficient, inflamed, and disorganized. Over time, energy stops flowing and starts backing up. You don't "burn" fuel efficiently, you store it as fat while feeling sluggish and drained. This is not just bad nutrition. This is biological confusion.

Restoring Energy: Fuel Quality and Rhythmic Repair

When Carol and I changed how we fueled ourselves, everything shifted. We focused on real, whole foods—foods that provided steady, reliable energy. We eliminated ultra-processed snacks and chose meals that supported stability instead of triggering the familiar crash an hour later.

Over time, another change followed: strategic fasting. Not starvation. Not extremes. Simply allowing space between meals. Time for the body to breathe. That space created the conditions for recovery at the cellular level. Mitochondria had time to recover and adapt. Cells could heal, recycle damaged components, and restore function through natural processes like autophagy.

The effects became difficult to ignore. Sleep improved. Thinking cleared. Energy steadied. Physical strength began to return. Workouts felt more sustainable. Mood stabilized. Body composition shifted—without calorie counting or constant effort. This wasn't about trying to "boost energy." It was about restoring the system that manages energy in the first place.

From Symptoms to Systems

We've grown accustomed to dysfunction. Low energy is often treated as a natural phase of life or a personal shortcoming. It's neither. It reflects a system that has lost coherence—overstimulated, mistimed, and poorly fueled.

Modern culture rewards fixes that are fast and visible, even when they only address the surface. Fatigue is treated with stimulation. Weight gain is treated with restriction. Sleep is treated with sedation. Symptoms are managed, but the underlying energy system remains disrupted.

Energy isn't about doing more. It's about restoring structure, communication, and flow within the body. When the type of fuel changes and the timing of that fuel shifts the system begins to regulate itself. The constant overload eases. Efficiency returns. Energy isn't something you chase. It's something you return to—by giving the body the conditions it needs to thrive.

The Evolutionary Mismatch

For roughly 2.5 million years, humans lived as hunter-gatherers. The Paleolithic era accounts for the vast majority of human evolutionary history, during which food availability, movement, and fasting were inseparable parts of daily life.

Agriculture, cities, and structured food systems occupy only a small fraction of that timeline, roughly the last 10,000 years. Ultra-processed, packaged foods are newer still, emerging almost entirely within the past century.

Human Evolution

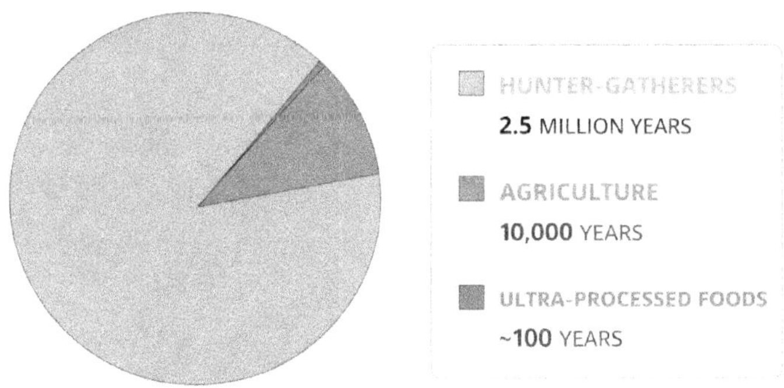

Note: This graphic is conceptual, not proportional.
Over 99% of human evolutionary history occurred prior to agriculture, with
ultra-processed foods emerging only in the most recent fraction of time.

Our ancestors evolved in an environment where food was hard-earned. They didn't eat every few hours or graze all day. They hunted animals for protein and fat and gathered wild plants, roots, seeds, and occasional fruits when in season. Carbohydrates were limited, fibrous, and seasonal, not a constant daily staple.

These conditions shaped how human metabolism functions. The body evolved to operate within a natural rhythm of feasting and fasting. When food was available, people ate. When it wasn't, the body didn't fail, it adapted. Stored fat became the primary fuel, producing an alternative energy supply for the brain when glucose was scarce. Liver glycogen provided short-term energy when needed, but fat supplied the long-term reserve.

Human biology has not changed much in the last 10,000 years. We are still built for scarcity, variation, and metabolic flexibility, not constant abundance. Agriculture is relatively new, and industrial food is newer still, but the metabolic machinery that governs fuel use and fasting was shaped over hundreds of thousands of years.

This wasn't merely survival. It was an efficient, resilient design. During periods without food, mental clarity often sharpened, alertness increased, and physical endurance was preserved. At the cellular level, mitochondria shifted toward efficiency, and the body engaged in repair processes such as autophagy—recycling damaged components and renewing tissue. For most of human history, people didn't eat to manage blood sugar or regulate numbers. They ate for nourishment and survival. Meals were centered on foods that were naturally satiating and difficult to overconsume. Fullness was the signal, and it was trusted.

Between those meals were long stretches without food—hours, sometimes days—shaped by hunting, migration, seasonality, and scarcity. Yet during those fasting windows, the body maintained function. Energy systems shifted, internal fuel was accessed, and both physical and cognitive capacity were maintained. This is the metabolic rhythm humans evolved with: flexibility, efficiency, and recovery. Food was fuel, not entertainment. Hunger was natural, not a crisis. Energy was steady, not chaotic.

Now fast forward to today. We live in a food-saturated environment—near-constant access, frequent eating, and a steady supply of ultra-processed products built around refined carbohydrates, industrial fats, and added sugars. Many people eat far more often than previous generations, and much of what is consumed bears little resemblance to the foods humans historically relied upon.

This pattern alters how the energy system operates. Repeated intake favors immediate use and storage, leaving little opportunity to draw from internal reserves. A metabolic state that once served as a routine fallback becomes unfamiliar. The body's ability to shift into stored-fuel metabolism is no longer experienced as normal, it's often treated as optional, or even extreme.

Instead of moving fluidly between fuel sources, the system becomes biased toward constant availability. Energy signals remain elevated, storage is prioritized, and release is delayed. Mitochondria are asked to process incoming fuel continuously, even when cells are already saturated, leading to inefficiency rather than flexibility. The result is one of the defining patterns of modern life: energy is abundant at the surface, yet poorly available at the cellular level—not because fuel is lacking, but because access is constrained.

This is more than a matter of diet alone. It reflects an evolutionary mismatch: a gap between the conditions under which our energy systems evolved and the environment in which they now operate.

Human biology is shaped over hundreds of thousands of years, even millions. Agriculture occupies only a brief moment in that timeline. Ultra-processed foods are newer still. Our physiology has had little time to adapt to this level of constant availability.

The biology that once kept us resilient through scarcity is now being asked to operate under continuous abundance. We're running an ancient energy system in a modern environment—and the strain shows up everywhere: chronic fatigue, weight gain, anxiety, diabetes, brain fog, and metabolic disease. The body isn't broken. It's responding exactly as designed—just operating under conditions it was never meant to face nonstop.

The Way Back

Reconnecting with energy begins with restoring the conditions the body recognizes: real food, lower glycemic load, regular movement, and enough space between meals for stored fuel to be accessed and cellular repair to occur.

Fatigue, brain fog, insulin resistance, and weight gain aren't random. They're signals—evidence that the system is overloaded and out of rhythm. When dysfunction is no longer constantly reinforced, function re-emerges. Biology doesn't need to be forced or micromanaged, it needs the right conditions.

This isn't about going backward. It's about moving forward with biological clarity. The path to performance, focus, and health isn't found in hacks or pills, but in alignment with how the human body is built to thrive.

Energy by Design

The human body was not built to fail. It was built to adapt, repair, and endure. Across evolutionary history, survival favored systems that preserved function, responded to stress, and restored balance, not systems that collapsed at the first sign of strain.

That adaptive design is visible everywhere: damaged cells are replaced, tissues are renewed, and order is maintained through constant communication within and between systems. When energy is available and signals are clear, the body tends toward repair, resilience, and stability. The system is not broken—it is responsive.

Health, then, is not determined by genes acting alone. DNA provides instructions, but instructions are expressed inside a living environment and that expression is shaped by energy availability, signaling, and metabolic context.

This is not a hierarchy as much as it is a conversation. The nucleus holds the blueprint, while mitochondria help determine what gets built, when, and under what conditions.

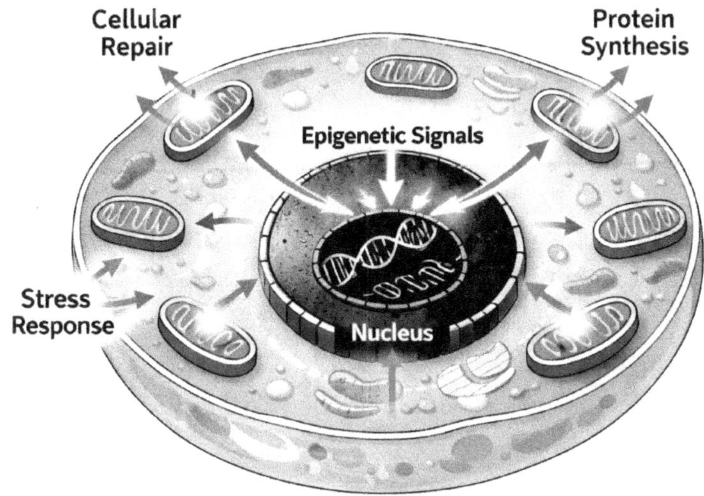

Conceptual model of cellular energy signaling.

Mitochondria do not create energy; they transform fuel into usable energy and generate signals that influence cellular repair, stress response, and gene expression. The nucleus contains genetic instructions, but how those instructions are expressed depends on energy availability and intracellular communication. This graphic is illustrative, not proportional.

Without adequate energy flow, even the most sophisticated genetic code cannot be carried out. The problem is rarely faulty instructions—it is a system that has lost the energy and communication required to execute them. And when that environment breaks down, genetic errors become more likely. Mutations do not always appear as the original cause; they often emerge as a consequence of stress, damage, and instability inside the cell.

At the center of this coordination are the mitochondria. By transforming fuel into usable energy, they help set the conditions under which repair, adaptation, and signaling can occur. Modern biology recognizes this regulatory layer as the epigenome: the interface between genes and the environment, where expression shifts in response to fuel availability, stress, and rest. Genes matter, but they do not operate in isolation. They respond to the metabolic and signaling environment around them. The body functions as an integrated whole, with mitochondria helping coordinate how cells, tissues, and organs adapt in the service of survival, repair, and long-term health.

The body heals through order, cooperation, and rhythm. Nourishment followed by rest. Effort balanced with recovery. Fuel that supports the system rather than overwhelms it. Time between meals that allows repair to occur. These rhythms are not inventions of modern science—they are written into our biology and reflected throughout creation.

We have drifted from that design. Modern life promotes constant consumption, artificial stimulation, and reactive intervention. Yet we have not been left without guidance. We were given awareness, self-regulation, and the ability to choose alignment over excess. Energy is not restored by overriding the system. It returns when the system is honored.

When we return to the conditions that support repair through rhythm, rest, and recovery, health follows. Energy flows again. Strength returns. Clarity emerges. This is not about hacks, trends, or willpower—it is about trusting the design that was there from the beginning. Healing is not something we manufacture; it is something already built into us.

We are not passive observers of our health. We are stewards of it. When we honor the design, fuel the body wisely, and give it room to work, healing is already in motion.

CHAPTER SIX

THE HORMONE WE FAILED TO MEASURE

*"Telling an obese person to eat less and move more
is like telling a depressed person to quit being sad"*
DR. JASON FUNG MD

For years we knew there had to be a better way. A better way to regain stable energy and quiet our hunger. Deep down, we sensed that constant fatigue and a constant pull to eat weren't normal, and that the traditional advice we had followed for decades wasn't giving us the stability we needed. What we didn't realize then—and what most people still don't know today—is that the most powerful determinant of energy, metabolism, and long-term health is not willpower, calories, or exercise. It's a hormone almost no one talks about, rarely measured in routine checkups, and almost never explained in a way that makes sense.

We didn't have the key. Not the key to a new diet. Not the key to discipline or willpower. The key to unlocking trapped energy. .The signal that allows the body to burn fat efficiently and run on steady, stable energy.

Most people never find this key because they're never told where to look. Routine labs don't reveal it. Most doctors never bring it up until disease is already present. And many of us spend years adjusting habits without ever understanding the single hormone that determines whether energy is burned or stored.

That hormone is insulin.

Insulin is released primarily in response to food, especially carbohydrates. But insulin isn't just a "blood sugar hormone." It is a growth and storage signal—one that tells the body what to do with incoming fuel. When insulin rises, the message is clear: store energy, build, and hold. And fat cells listen. They don't argue. They comply.

When insulin remains elevated, access to stored fat is restricted. When it falls, the body regains metabolic flexibility, the ability to shift between fuel sources as needed. But knowing the key exists isn't enough. You must know how to turn it.

What We Failed to Measure

For generations, medicine focused on what was easiest to quantify. Calories. Weight. Blood sugar. Cholesterol. These numbers became proxies for health itself, not because they explained the system, but because they were visible and familiar. What we failed to measure was the signal coordinating them all.

There's an old principle: what we do not measure, we do not improve. In metabolism, this blind spot proved costly. We measured outcomes, not control signals. We tracked the consequences of dysregulation while overlooking the hormone directing energy storage, release, hunger, and metabolic rate. Insulin wasn't absent. It was active, constantly shaping physiology, just rarely examined in context. And because it wasn't routinely measured, it wasn't routinely managed.

What remained unexplained was often labeled a lack of discipline. That realization began for us in August 2019, long before we understood the science. Both of us had spent years trying to reclaim our health—struggling with weight gain, persistent fatigue, and the cumulative effects of stress and illness. We tried all the usual approaches: walking more, tracking points, joining gyms, watching portions. Nothing held. Hunger returned. Energy dipped. Any progress quickly unraveled. At the time, it felt like failure. Now we know it was something we didn't yet understand. We weren't losing a battle with calories—we were up against a hormonal lock we didn't know existed.

Our forties were busy: parenting, working, coaching, and the routines that crowd out sleep and consistency. We assumed the gradual weight gain and low energy were simply part of getting older. What we didn't recognize was how constant stress, irregular meals, and disrupted sleep were steadily pushing insulin higher, reshaping our metabolism long before any doctor suggested there might be a problem.

We kept trying harder, unaware the body can't access trapped energy until insulin falls low enough for the lock to open. Then, in the quiet of late summer 2019, a new idea began to take shape, one that challenged decades of conventional wisdom. It wasn't a diet. It wasn't a trend. It was a shift in understanding.

Food is not math. Food is information. And insulin interprets every message. Some foods raise insulin sharply, locking the body out of fat-burning. Others keep insulin low, unlocking access to stored energy. Some choices whisper "burn." Others shout "store." Once we understood this, we finally had the key. This was the moment the door we had been pushing against finally began to open.

We've long been told that health is just math calories in, calories out. Eat less, move more, and everything should fall into place. But most people eventually discover it's not that simple.

The same number of calories can produce very different effects depending on their source. Some foods keep you full, steady, and focused for hours. Others leave you tired and hungry soon after. The body doesn't respond to calories in isolation—it responds to the signals food creates through type, timing, and frequency. Over time, some choices favor energy use, while others quietly push fuel into storage. That's why the old formula works less and less as we get older.

This was our light-bulb moment. For the first time in a long while, we felt genuine hope. We weren't searching for a new diet, we were beginning to understand a new way of living.

Insulin, Glucagon, and Fuel Access

At the core of metabolic health is a simple question: can the body access the energy it has already stored? Fat tissue is not inert mass; it is an active energy reserve designed to expand when fuel is abundant and release energy when demand rises. Whether this system supports health or dysfunction depends far less on calories than on the hormonal balance between insulin, which governs energy storage, and glucagon, a counter-regulatory hormone that governs energy release.

When insulin is elevated, the body is signaled that energy is plentiful. Incoming fuel is directed into storage, and fat cells undergo hypertrophy, expanding to accommodate that energy. At the same time, insulin suppresses lipolysis, preventing stored fat from being broken down. Fatty acids are kept out of the bloodstream, and ketone production remains minimal. This is appropriate after meals, but harmful when insulin remains chronically elevated.

Modern eating patterns: frequent meals, snacking, and refined carbohydrates—keep insulin high for much of the day. As insulin rises, glucagon signaling is suppressed and the normal contrast between fat storage and fat release disappears. Fat continues to accumulate, yet access to that energy is restricted. The body may hold substantial reserves, but it cannot reliably use them. This is the state of trapped energy.

At the cellular level, trapped energy is experienced as scarcity. With limited fatty acids entering circulation and minimal ketone production, tissues depend heavily on incoming glucose. When supply fluctuates, cells signal distress. Hunger intensifies, cravings increase, energy becomes unstable, and fatigue rises. Despite growing fat mass, the body behaves as though it is starving.

This is not a failure of discipline, but a predictable biological response. When insulin remains elevated, fat cells are instructed to retain energy and enzymes responsible for fat release are suppressed. Even with abundant stored fuel, access remains limited, leaving cells dependent on frequent glucose intake to meet ongoing demands.

Metabolic Rate, Starvation Signals, and the Failure of "Eat Less, Move More"

At the most basic level, insulin determines whether fat cells store energy or release it. When insulin is high, most often due to frequent carbohydrate intake, fat cells receive a signal of threat and are instructed to retain stored fuel. When insulin is allowed to fall, that signal reverses. Fat cells release energy, glucagon rises, and stored fuel becomes accessible again. This is the point at which energy returns.

To understand why this state is so difficult to sustain, it helps to clarify what metabolic rate actually represents. Metabolic rate is not simply how many calories a person "burns." It reflects how much energy the body is willing to expend to maintain basic functions—body temperature, organ activity, movement, and repair. In essence, metabolic rate is a signal of safety. When energy is accessible, the body maintains or even increases expenditure. When energy feels scarce, it conserves.

It is critical to distinguish between actual scarcity and perceived scarcity. In true scarcity, intake falls, insulin drops, fuel access opens, and metabolic flexibility emerges. In modern metabolic disease, however, the body does not lack energy it lacks access. Yet hormonally, the signal is the same: conserve energy, reduce expenditure, and prevent stored fat from being released.

Metabolic State: *A Signal of* **Scarcity or Safety**

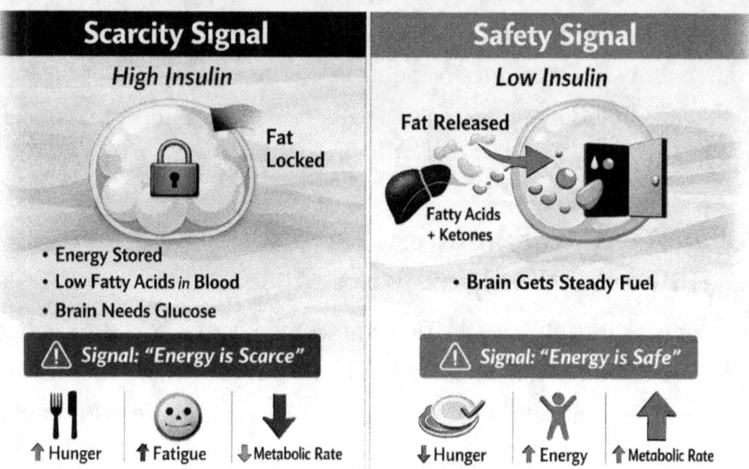

In a trapped-energy state, calorie reduction without lowering insulin intensifies the problem. Intake falls, but fat remains locked away. Lipolysis stays suppressed, fatty acids do not enter the bloodstream, glucagon cannot rise adequately, and ketone production remains minimal. From the body's perspective, energy supply has dropped while access to stored fuel has not improved. The signal is not progress—it is starvation. The response is predictable and adaptive. Hunger intensifies. Metabolic rate declines to conserve energy. Fatigue increases. Exercise tolerance decreases. Recovery slows. Adding more activity in this context raises energy demand without improving fuel availability, further reinforcing the starvation signal. Weight loss stalls or reverses. What appears to be a lack of willpower is actually normal physiological defense.

Translation: This is why people can "do everything right" and still feel worse. A late-night meal, poor sleep, and insulin resistance can keep the body in storage mode—even the next morning. So they cut calories and add movement, but hunger climbs, energy drops, and the body conserves.

It's not a character flaw. It's physiology defending against perceived scarcity.

When insulin levels are allowed to fall, the system reorganizes. Glucagon signaling rises, lipolysis increases, and stored fat is released. Fatty acids enter the bloodstream, supplying muscles, the heart, and other tissues. The liver converts some of this fuel into ketones, providing a stable energy source for the brain. Internal energy becomes reliable rather than dependent on constant intake. This is the accessible-energy state. In this state, hunger quiets rather than escalates. Energy steadies. Metabolic rate is preserved because the body no longer perceives danger. Exercise becomes sustainable instead of costly, and recovery improves. Regulation replaces compensation.

> ***Translation:*** This is what it feels like when the lock finally opens—often by changing the foods and patterns that keep insulin elevated. Hunger quiets instead of escalating. You can go longer between meals without thinking about food. Energy stays steady, workouts feel sustainable, and recovery improves—not because you forced it, but because the body finally has access to its own stored fuel. In this state, the body no longer senses danger—it senses safety.

The failure of "eat less, move more" is not a failure of effort. It is a failure to restore fuel access. Health does not come from enduring deprivation, but from restoring the hormonal conditions that allow the body to use its own energy. When energy is trapped, the body defends itself. When energy is accessible, the body regulates itself.

Insulin governs storage and growth. Glucagon enables lipolysis and release. Fatty acids and ketones reflect restored metabolic flexibility. Once access is restored, regulation follows naturally.

When Understanding Changes Behavior

Blood sugar matters because it directly drives insulin, and insulin determines whether the body can access stored energy. When blood sugar rises quickly, insulin rises sharply in response. That insulin surge pushes energy into storage and temporarily blocks fat release. As blood sugar then falls, the body is left with limited accessible fuel, triggering hunger, cravings, and a perceived energy failure that drives the urge to eat again.

A growing movement has helped people understand how different foods affect blood sugar and the body's ability to use energy efficiently. The core message is simple: by choosing foods that keep blood sugar steady, the body can become responsive again rather than trapped in a constant cycle of storing and craving. For many, learning concepts like the glycemic index and seeing how the body reacts to certain foods becomes a turning point. It opens a new world of discovery and possibility. And often, without fully realizing it at the time, that new understanding leads to bold steps toward better health.

The glycemic index (GI) is a system that ranks foods based on how quickly they raise blood sugar levels after consumption. The glycemic index ranges from 0 to 100, with pure glucose measuring 100 glucose being a type of sugar found in the blood that comes from the food we eat. Foods with a higher GI are rapidly digested and absorbed, causing a swift and significant rise in blood sugar levels. In contrast, low-GI foods are digested more slowly, leading to a gradual increase in blood sugar. Foods that are primarily fat and protein have little direct effect on blood sugar and therefore trigger a much smaller insulin response, allowing blood sugar to remain steadier and stored energy to be released from fat tissue. This slower, more stable fuel flow helps maintain steady energy levels and reduces hunger.

This was amazing. For the first time, there was a tool that could clearly show how different foods convert into sugar at completely different rates. It wasn't guesswork anymore. This chart lays out in simple numbers what's happening inside the body after every meal.

Some foods lead to a steady flow of energy and conservation, while others create sharp spikes followed by crashes and fatigue. Seeing it in this structured way changes everything—it provides a roadmap for choosing foods that work with the body instead of against it.

The body's insulin response is tightly linked to blood sugar and the glycemic index. High-GI foods cause rapid spikes in blood sugar, which trigger equally rapid releases of insulin to move that sugar out of the bloodstream. When this pattern repeats day after day, cells begin to protect themselves from constant insulin signaling by responding less strongly. This adaptive response known as insulin resistance forces the body to produce even more insulin to keep blood sugar in range. Over time, this cycle strains metabolic control and sets the stage for type 2 diabetes and related metabolic disease.

Many of our conventional strategies for weight loss revolve around meticulously counting calories. I often refer to this as the "starvation diet" because, for most people, this approach never truly satisfies hunger. Caloric intake drops, but the gnawing sensation of hunger persists throughout the day—creating a cycle of deprivation and frustration. And in many cases, it can actually worsen outcomes: energy declines, cravings intensify, and the body begins conserving. The result is often a stall, a rebound, or a pattern of progress followed by collapse—not because the person failed, but because the body is defending against perceived scarcity.

The other widely accepted pillar of weight loss is the mantra of "move more." We are constantly reminded that to achieve our goals, we must lift weights, run, walk, and engage in various forms of exercise to burn off more calories than we consume. And so, many of us dutifully join gyms, commit to regular workouts, and push our bodies to the limit. Under the wrong internal conditions, exercise doesn't feel like liberation—it feels like added demand. If insulin remains elevated and stored fuel is still difficult to access, movement increases energy requirements without improving supply. The body responds the way it was designed to respond to perceived scarcity: hunger rises, fatigue builds, recovery slows, and conservation takes over. Exercise isn't the problem. The timing and metabolic context determine whether movement becomes a powerful signal of strength—or another signal of stress.

But what if our understanding of how our bodies respond to calories is incomplete? What if we could eat nutritious whole foods and not count calories? What if the key to lasting health and weight management isn't counting every calorie, but choosing the right kinds of foods? Imagine shifting your focus from restricting calories to eliminating processed foods and embracing whole, nutrient-dense options that naturally promote fullness and satiety. Imagine enjoying food again—really enjoying it—without guilt or confusion. Could this simple yet profound change hold the answer to the struggles so many of us face in the pursuit of a healthier, more balanced life?

The glycemic index (GI) was developed in 1981 by Dr. David Jenkins and his team at the University of Toronto in Canada. They created the GI to measure how different carbohydrate-containing foods affect blood glucose levels, providing a tool to help manage conditions like diabetes and guide healthier dietary choices. Carbohydrates were chosen for this measurement because they have the most direct and immediate effect on blood sugar after eating. When we eat carbohydrates, they are broken down into glucose—the primary sugar that circulates in the blood. As blood glucose rises, the body responds by releasing insulin to move that energy into cells. Understanding this process helps explain why certain foods influence energy and hunger differently than others.

To make the glycemic index more practical, it helps to see how common foods rank. The lower the number, the slower the conversion to blood sugar and the steadier the energy that follows. The higher the number, the faster the conversion and the greater the likelihood of an energy spike and crash. Think of it like building a fire: you want a steady log fire that burns slowly through the day, not a quick burst of kindling that flares and fades.

Slow burn
energy that lasts

Fast flare
energy that fades

Here's a simple list of foods across the spectrum:

Note: The glycemic index applies to carbohydrate-containing foods. Foods made mostly of protein and fat contain little to no carbohydrate, so they have little direct effect on blood sugar and are listed as GI 0.

T-Bone Steak (GI: 0)

Source: Whole Food – Protein rich with natural fats and no carbohydrate. Beef contains no carbohydrate, making it neutral on blood sugar while supporting long-lasting satiety.

Ground Beef (GI: 0)

Source: Whole Food – Unprocessed meat containing protein and natural fat. With no carbohydrate, it has little direct effect on blood sugar and tends to be very filling.

Chicken Breast (GI: 0)

Source: Whole Food – Lean protein with natural fats and no carbohydrate. Chicken is neutral on blood sugar and pairs well with low-GI foods for stable energy.

Salmon (GI: 0)

Source: Whole Food – Pure protein – Pure protein with natural fats. Salmon contains no carbohydrate, making it neutral on blood sugar while providing long-lasting nourishment.

Eggs (GI: 0)

Source: Whole Food – Another protein-rich food that supports stable energy. Eggs have virtually no direct effect on blood sugar.

Peanuts (GI: 7)

Source: Whole Food – A legume rich in healthy fats, protein, and fiber. Peanuts digest slowly and provide sustained energy.

Broccoli (GI: 10)

Source: Whole Food – A cruciferous vegetable low in calories, high in nutrients, and digested gradually.

Apple (GI: 36)

Source: Whole Food – A fruit with natural sugars and fiber that slows absorption, producing a mild rise in blood sugar.

Banana (GI: 62)

Source: Whole Food – A fruit rich in natural sugar and potassium. Ripeness can raise its GI value.

Whole Wheat Bread (GI: 69)

Source: Processed Food – Made from whole wheat flour and still raises blood sugar moderately quickly.

White Bread (GI: 75)

Source: Processed Food – Highly refined and digested rapidly, producing sharp spikes in blood sugar.

Pancakes (GI: 80)

Source: Processed Food – Made from refined flour and digested quickly, producing a rapid rise in blood sugar and a stronger insulin response.

Cornflakes (GI: 81)

Source: Processed Food – A cereal that quickly converts to sugar, giving fast energy followed by a crash.

Mashed Potatoes (GI: 85–90)

Source: Processed Food – Cooked potatoes that are mashed digest quickly and can raise blood sugar sharply, especially when made from instant flakes or eaten without added protein/fat.

Foods high in healthy fats and protein tend to sit at the very bottom of the index, meaning they provide energy without sudden swings. In contrast, refined and processed foods rank near the top, offering fast-burning fuel that quickly fades. Recognizing this pattern helps us choose what keeps the fire burning steady—sustained energy, clear mind, and balanced appetite—rather than a cycle of peaks and crashes.

Calories are not created equal. The body responds very differently to calories from fat, protein, and refined carbohydrates—each triggering distinct hormonal and metabolic effects.

High Glycemic Index Foods

High–glycemic-index foods such as white bread, sugary drinks, and rapidly absorbed starches cause a sharp rise in blood glucose after eating. This sudden increase prompts the pancreas to release a strong surge of insulin. Insulin does more than lower blood sugar; it directs glucose into cells, supports glycogen storage in the liver and muscles, and temporarily slows the liver's own glucose production. But when glucose intake exceeds the limited space available in glycogen stores, the extra glucose is converted into fat through a process called de novo lipogenesis. Those newly formed fats are then packaged as triglycerides often increasing triglycerides in the bloodstream and eventually stored in adipose tissue, the body's fat-storage tissue. Over time, this contributes to increased fat accumulation and metabolic strain. The rapid rise and fall in blood glucose from high GI foods can lead to:

1. **Energy Crashes:** After the initial spike in energy, the quick drop in blood glucose can cause feelings of fatigue, hunger, and a craving for more high-GI foods, creating a cycle that can lead to overeating and weight gain.

2. **Increased Fat Storage:** Because high insulin levels promote the storage of glucose as fat, regular consumption of high GI foods can contribute to increased body fat, particularly if the energy intake consistently exceeds energy expenditure.

3. **Insulin Resistance:** Over time, repeated spikes in insulin can lead to insulin resistance, where cells become less responsive to insulin, making it harder for the body to regulate blood glucose levels. This can lead to metabolic disorders like type 2 diabetes.

Low Glycemic Index Foods

Low GI foods, such as berries, olives, vegetables, steak, eggs, cheese, avocados result in a slower, more gradual increase in blood glucose. This leads to a more moderate insulin response, allowing for a steady supply of energy over time without the sharp peaks and troughs associated with high GI foods. The effects of low GI foods on energy metabolism include:

1. **Stable Energy Levels:** The slow release of glucose provides a steady source of energy, which can help maintain consistent energy levels and reduce the likelihood of energy crashes.

2. **Reduced Fat Storage:** With a more moderate insulin response, less glucose is likely to be converted to fat. This can be beneficial for weight management and reducing the risk of obesity.

3. **Improved Insulin Sensitivity:** Regular consumption of low GI foods will improve insulin sensitivity, making the body more efficient at using glucose and reducing the risk of developing insulin resistance and type 2 diabetes.

4. **Enhanced Satiety:** Low GI foods are often higher in fats and proteins, which can promote feelings of fullness and reduce overall calorie intake, further aiding in weight management.

When Carbohydrates Become Fat: A Short Story

Imagine a carbohydrate entering the body. It begins in the mouth and stomach, but its real transformation happens in the small intestine. There, enzymes break it down into glucose the simplest, usable form of carbohydrate. Glucose enters the bloodstream quickly, raising blood sugar. This rise does not go unnoticed.

The pancreas responds by releasing insulin. Insulin's job is not to create energy, but to direct it. It tells cells throughout the body that fuel is available and instructs them to take glucose in and use it.

At first, this works as designed. Muscle and liver cells absorb glucose and store some of it as glycogen, a short-term energy reserve. But glycogen storage is limited. Once those tanks are full, incoming glucose has nowhere obvious to go.

That's when the liver takes over. Faced with excess glucose and persistent insulin signaling, the liver converts surplus glucose into fatty acids through a process called *de novo lipogenesis*. Those fatty acids are then packaged together with glycerol into triglycerides. The triglycerides are released into the bloodstream and delivered to fat tissue for long-term storage.

At this point, the original carbohydrate has changed form entirely. It is no longer glucose. It is no longer glycogen. It has become stored fat. This process is not a malfunction. It is a normal response to abundance. The problem arises when carbohydrate intake is frequent, insulin remains elevated, and this conversion pathway stays active day after day. Fat storage increases. Triglycerides rise. Yet because insulin remains high, access to stored fat is restricted. Energy is plentiful on paper, but unavailable in practice.

This is why elevated triglycerides are so often misunderstood. They are commonly blamed on dietary fat, when in reality they frequently reflect excess carbohydrate being converted and stored under persistent insulin signaling. From the outside, it looks like overconsumption. From the inside, it behaves like trapped energy.

If triglycerides are ultimately stored in fat tissue, why do we measure them in the blood at all? Because triglycerides don't move on their own. They must be transported. Triglycerides circulate through the bloodstream as packaged energy, carried inside particles called lipoproteins, on their way from the liver to fat tissue or other destinations. In a healthy system, this transit is brief. Triglycerides rise after meals and fall as they are cleared and stored or used. When triglycerides remain elevated in the blood, it is a sign that energy is being overproduced, overpackaged, or poorly cleared. In many cases, it reflects a system under constant instruction to store—often driven by excess carbohydrate intake and persistent insulin signaling—rather than simply the amount of fat being eaten.

Macronutrients and the Signals They Send

Not all calories send the same metabolic message. Each macronutrient: carbohydrates, protein, and fat carries its own instruction set, shaping hormones, hunger, fuel use, and the body's ability to shift between burning incoming food and stored energy. Macronutrients are the three main fuel sources in food, and the body processes each one differently. These differences explain why some meals keep us full and energized for hours, while others leave us hungry again soon after eating.

Carbohydrates have the most immediate and powerful effect on blood glucose and insulin, especially when refined or rapidly absorbed. High–glycemic-load carbohydrates cause sharp rises in blood sugar, triggering equally rapid insulin release. Insulin directs glucose into cells and stores excess as glycogen, but once glycogen capacity is reached, additional glucose is diverted toward fat storage. Because insulin suppresses fat burning, carbohydrate-heavy meals tend to lock the body into glucose dependence. Energy rises quickly, but falls just as fast, often followed by renewed hunger and cravings. Lower-GI carbohydrates digest more slowly, moderating insulin release and supporting steadier energy, but carbohydrates as a category still favor short-term fuel over long-term access to stored energy.

Protein has a more complex and stabilizing metabolic profile. It triggers a modest insulin response, but one that works in concert with other hormones to stabilize blood glucose rather than spike it. Protein strongly stimulates satiety signals—including GLP-1, peptide YY, and CCK—telling the brain that nourishment has arrived and hunger can quiet. This matters because GLP-1 is not just a medication target, it's also a **natural** satiety signal the body can produce in response to food. For many people, higher-protein meals and lower-carbohydrate patterns can deliver some of the same practical benefits associated with GLP-1 therapy: quieter hunger, fewer cravings, and improved metabolic stability. Medication amplifies these signals pharmacologically; food choices can support them physiologically. It also supports muscle repair and maintenance without readily contributing to fat storage unless consumed in large excess. By slowing digestion and dampening blood sugar swings, protein helps sustain energy between meals and reduces the urgency to refuel.

Dietary fat has the lowest direct impact on insulin and plays a key role in stabilizing energy metabolism. Because fat generates minimal insulin response, it allows fat oxidation to continue rather than shutting it down. Fat is energy-dense and burns slowly, promoting longer intervals between meals, fewer cravings, and more consistent energy. It also increases satiety hormones like CCK and delays stomach emptying, contributing to lasting fullness. When paired with protein, dietary fat supports meals that minimize glucose volatility and make the transition to burning stored fat far easier.

Why Energy Stability Is About More Than Insulin

Once you understand insulin's role in directing energy, the next layer is recognizing that your body operates on a full hormonal conversation, a coordinated dialogue between the gut, the brain, and the metabolism. This conversation is what determines hunger, satisfaction, cravings, mood, focus, decision-making, and even the motivation to move. And the way different foods influence this hormonal network explains why some meals leave you energized and clear while others leave you anxious, foggy, or hungry again an hour later.

Every bite of food triggers a cascade of signals that travel through the bloodstream and, ultimately, into the brain. These signals tell you whether you're hungry or full, energized or fatigued, calm or irritable. They influence neurotransmitters, inflammation, mitochondrial workload, and your brain's ability to maintain stable focus. These hormones: ghrelin, leptin, GLP-1, CCK, peptide YY, cortisol, dopamine, and even insulin itself form the core of this conversation. And once you know how they work, your daily choices start to make sense in a completely new way.

The Hormones That Drive Hunger

Ghrelin is released primarily from the stomach and signals the brain that it's time to eat. It rises before meals and falls when you're fed. But the type of food you eat dramatically affects the pattern. Highly processed carbohydrates cause ghrelin to fall quickly and rise again just as fast explaining why someone can eat a bowl of cereal at 7:00 AM yet feel ravenous by 9:00 AM.

- High-GI foods → fast ghrelin drop → fast ghrelin rebound → cravings.
- Lower-GI, protein-rich, and fat-rich meals suppress ghrelin more steadily, keeping hunger quiet for longer.

Dopamine isn't just a pleasure chemical; it's the "do that again" signal. Hyper-processed foods are engineered to create large dopamine spikes: sugar, refined grains, seed oils, and flavor enhancers. These foods overwhelm the natural signaling system, reinforcing habits and creating patterns of overeating—not because of weak willpower, but because the biochemical reward loop is overstimulated. Whole foods generate slower, steadier dopamine responses, supporting balanced eating without compulsive patterns.

The Hormones That Drive Satiety (Fullness)

Leptin is produced by fat cells and tells the brain that the body has enough stored energy. When leptin works properly, meals naturally feel satisfying and hunger fades. When insulin is chronically elevated, leptin's message becomes distorted—a condition often called leptin resistance, in which hunger persists not because energy is lacking, but because the brain no longer recognizes that it already has enough. The result is eating beyond fullness, frequent snacking, and persistent hunger despite abundant energy stores. High insulin → impaired leptin signaling → constant hunger.

GLP-1 (glucagon-like peptide-1) is released from the gut when you eat protein and healthy fats. It slows stomach emptying, stabilizes blood sugar, increases satiety, and reduces appetite. It's one of the body's most powerful anti-craving tools. In fact, modern GLP-1–based medications (like **Ozempic, Wegovy, Mounjaro, and Zepbound**) work by amplifying this same pathway. But you can activate GLP-1 naturally through food choice—especially meals rich in protein, fiber, and healthy fats.

CCK and Peptide YY - These hormones are released in the gut when you eat protein and fat. They travel to the brainstem and hypothalamus, reinforcing the sense of fullness and promoting calm, steady energy. Meals with balanced macronutrients stimulate these hormones far more effectively than high-carb, low-fat meals. Protein and fat signal *stop*. Refined carbohydrates signal *more*.

How Hormones Shape Brain and Mood

The brain depends on steady fuel and consistent signaling. Sharp insulin spikes after high-GI meals are often followed by crashes that disrupt clarity, focus, impulse control, sleep, and stress tolerance. This is why sugary or rapidly absorbed meals can leave people irritable, anxious, fatigued, or foggy soon after eating. The problem isn't weakness—it's unstable energy availability and a system pushed out of rhythm. Stable insulin → stable glucose → stable brain.

Why This Matters More Than Willpower

At the surface, it may look like some people have more discipline. But the truth is deeper than behavior. When the internal environment is unstable, cravings grow louder and healthy choices feel harder to sustain. When the environment is stable, those same choices begin to feel natural. Food choice becomes easier—not because you're stronger, but because your biology is finally aligned.

The Real Turning Point

Once this is understood, modern eating starts to make sense. Some foods create calm, steady energy. Others trigger a predictable loop of spikes, crashes, and urgent hunger. And the symptoms so often dismissed as "getting older" start to look less like fate and more like a pattern that can be changed. This is where real transformation begins. This is the moment you learn the difference between fueling a life and merely getting through one.

Decades of research have reinforced this same theme: hunger and craving are not simply failures of effort, but predictable outcomes of modern food patterns. Highly refined, high–glycemic-load foods can destabilize appetite signals and reward circuitry in ways that make "just eat less" a losing strategy.

When the internal environment is out of balance, eating feels harder to control. When it's aligned, the struggle changes—and for many, it quiets. And when it quiets, life opens up.

CHAPTER SEVEN

METABOLIC SYNDROME

"Disease is life under altered conditions."
RUDOLF VIRCHOW

A Modern Epidemic With Deep Roots

Metabolic syndrome did not emerge suddenly, nor is it a collection of unrelated medical problems. Its history dates back to the mid-20th century, when physicians began noticing that high blood pressure, elevated blood sugar, widening waistlines, and abnormal cholesterol often occurred together, not separately. These patterns puzzled researchers for decades until a clear picture began to form: the conditions were not random. They were interconnected.

By the 1980s and 1990s, the cluster had become unmistakable across societies. A set of problems once rare was becoming so common that it now defines modern health in industrialized nations. The causes were not genetic destiny or chronological aging. They were the predictable consequences of an environment that had drifted far from the conditions the human body evolved to handle.

Ultra-processed foods, constant snacking, engineered sweetness, chronic stress, disrupted sleep, relentless light exposure, and declining physical activity quietly pushed the metabolic system far beyond its design limits. What modern medicine labeled as "diseases" were actually the body's adaptive attempts to cope with conditions it had never encountered in human history.

Understanding this context changes everything. Metabolic syndrome is not a fate sealed by age or ancestry. It is a response, and responses can change.

When the Body Follows Instructions From the Wrong Environment

Before 2019, I was asking questions many people quietly ask themselves but rarely voice aloud: Why is it so hard to regain control of your health once it begins slipping? Is decline inevitable with age? Is there a point where the body simply gives up?

I tried countless approaches, different diets, more exercise, new routines without understanding the deeper physiology driving my results. I didn't yet grasp that metabolism is not just a system; it is a language. Food, stress, sleep, and movement all send signals, and the body responds with remarkable consistency.

The previous chapter introduced a fundamental truth: Food is information. Food does not push the body forward in a simple arithmetic of "calories in, calories out." It triggers hormonal cascades, influences inflammatory pathways, and shapes metabolic flexibility. Once you see food this way, many of the well-meaning slogans you grew up hearing dissolve:

- "Eat in moderation."
- "Never skip breakfast."
- "Snack between meals."
- "A calorie is a calorie."
- "Fat makes you fat."

These messages guided public health for decades, yet they never aligned with how biology actually works. The body wasn't built to respond to food in a mathematical fashion. It responds biochemically. Even equal calories can produce completely different hormonal footprints, hunger signals, and metabolic outcomes.

How One Root Problem Becomes Many Diagnoses

Metabolic syndrome is widely understood in the medical literature as a cluster of manifestations driven by shared upstream dysfunction, not independent diseases coincidentally occurring together. The dominant unifying mechanisms cited across endocrinology and cardiometabolic research include:

- Insulin resistance / hyperinsulinemia
- Adipose tissue dysfunction
- Ectopic fat accumulation
- Chronic low-grade inflammation
- Mitochondrial and cellular energy dysregulation

These mechanisms express themselves across multiple systems:

- Glucose regulation (fasting glucose, A1C)
- Lipid metabolism (▲ triglycerides, ▼ HDL)
- Blood pressure regulation
- Visceral adiposity
- Inflammatory signaling

Major consensus bodies explicitly describe metabolic syndrome as a single pathophysiologic condition with multiple clinical expressions, even though it is operationalized as five measurable criteria. To the outside observer, these processes appear as distinct diagnoses: hypertension, fatty liver, insulin resistance, obesity, fatigue, high triglycerides, low HDL. But inside the body, they are all part of a single story: a metabolic system overwhelmed by the pressures of modern life. The body is not failing. It is adapting to survive an environment it was never meant to endure. Seeing this connection is liberating. Instead of managing five or six separate conditions with five or six separate medications, you focus on the unifying problem—impaired metabolic signaling. Fix the signaling, and the seemingly unrelated conditions begin shifting together.

Reclaiming the Kitchen: Where Change Begins

For us, the first real step didn't involve a doctor or a clinic. It began in the kitchen. Healing requires an environment that makes healing possible—and our pantry and fridge were doing the opposite. They were stocked with foods engineered to trigger cravings, keep us snacking, and pull us right back into the cycle of hunger and fatigue. So we cleared them out, completely.

We learned something fast: keeping "healthy choices" and "problem foods" side by side is a recipe for losing. You can place an apple next to a cookie jar and tell yourself you'll choose the apple, but the brain doesn't negotiate like that. It remembers the reward. It reaches before you've even decided. Those two worlds can't share the same shelf—at least not when you're trying to change your trajectory.

Our kids were older, Megan in college and Stuart entering his senior year, which made the transition easier. We didn't throw everything away. Stuart's favorite snacks went into a separate cabinet—out of sight, out of reach. Not forbidden. Just not waiting for him at eye level.

That alone mattered, because seeing food is a signal. The brain responds instantly—pulling up memories, pulling up desire, pulling you toward what it knows will deliver a hit. Willpower is rarely a fair fight against a well-trained cue. We made success easier by removing the daily battle.

Then came the fridge and freezer. Frozen pizzas. Ice cream. Sugary condiments. Old leftovers that kept the whole system stuck. The space opened up. Stuart wasn't thrilled. He stared into the empty freezer like something sacred had been taken. But he adapted quickly. And the house began to feel different—lighter, calmer, less reactive. For the first time, our kitchen matched the future we were trying to create.

A Slow Drift Toward Dysfunction

Most people never realize they're slipping into metabolic dysfunction. It happens slowly, often over years. The Standard American Diet (SAD) normalizes highly processed foods, refined sugars, seed oils, and chemicals engineered for hyper-palatability. It trains the body to expect constant glucose availability and trains the brain to chase reward instead of nourishment.

Research published in *Metabolic Syndrome and Related Disorders* shows that only a small fraction of adults today meet the criteria for full metabolic health. Most are on the path toward metabolic syndrome long before a doctor ever diagnoses it.

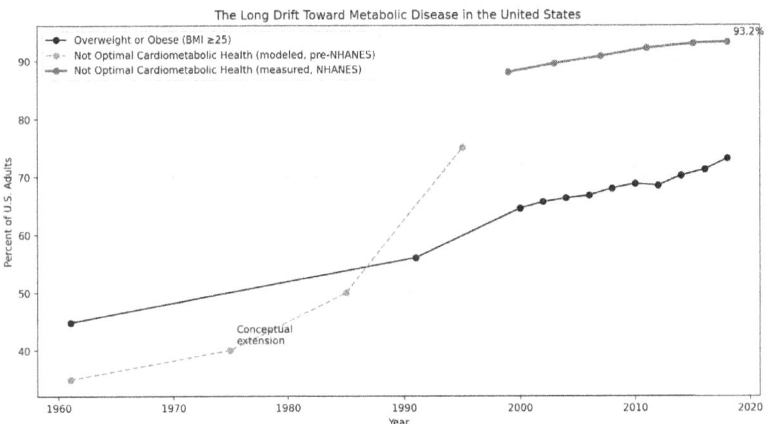

The long drift toward metabolic disease in the United States.

This figure shows two major trends: the rising prevalence of overweight or obesity (BMI ≥25) beginning with national examination surveys in 1960–1962, and the growing share of U.S. adults not in optimal cardiometabolic health using NHANES laboratory and examination data. A dashed segment indicates a conceptual extension based on historical prevalence of cardiometabolic risk factors prior to systematic measurement. By 2017–2018, more than 93% of adults failed to meet criteria for optimal cardiometabolic health.   But here's the good news: drifting into dysfunction may be slow, but the body's ability to recover is much faster than people imagine.

When Metabolism Begins to Shift

Metabolic dysfunction is not a disorder of willpower or character. It is the biological signature of a system that has lost its ability to switch fuels. This loss known as impaired metabolic flexibility—is the engine beneath most modern chronic disease. Metabolic flexibility allows the body to burn glucose after meals and switch to stored fat between meals. When this system works, energy is stable, appetite is predictable, and weight regulation feels natural.

When the diet is dominated by ultra-processed foods and frequent carbohydrate spikes, the metabolic "gear shift" becomes stuck. The body grows dependent on constant glucose. Fat burning slows. The liver fills. Insulin demand rises. Mitochondria become overwhelmed by excess fuel. The remarkable part is how quickly this process can reverse.

Metabolic flexibility often begins returning within days of meaningful change. Shifting toward lower-glycemic, whole-food meals stabilizes blood sugar, reduces insulin demand, and gives mitochondria room to recover. The response is rapid and predictable. What follows is the pattern commonly observed when someone commits fully to removing processed foods and adopting nutrient-dense meals centered on protein, healthy fats, and low-GI carbohydrates.

What Can Change: A Week-by-Week Picture

Week 1—The Reset Begins
- Blood glucose swings begin narrowing.
- Insulin starts falling between meals.
- Satiety increases with higher protein and healthy fat intake.

- Cravings loosen as dopamine-driven eating patterns begin to fade.
- Many experience their first evening without snacking in years.

Week 2—Energy Stabilizes

- Mid-afternoon energy crashes decrease.
- Hunger between meals continues to decline as fat oxidation reawakens.
- Water retention drops as insulin stabilizes.
- Sleep often improves due to more stable nighttime blood glucose.

Weeks 3–4—Metabolic Flexibility Returns

- The metabolic "gear shift" becomes active again—better switching between glucose and fat.
- Visceral fat begins to decline, improving hormonal signaling.
- Triglycerides often fall measurably in this window.
- Blood pressure may improve as insulin-driven sodium retention decreases.
- Appetite becomes calmer, more predictable, and less emotionally driven.

Weeks 5–6—Reversal Gains Momentum

- Early signs of fatty liver reversal often appear.
- Morning fasting glucose becomes more consistent.
- Many individuals with type 2 diabetes reduce medications (with medical oversight).
- Mitochondrial efficiency improves, reducing post-meal fatigue.
- Mental clarity and emotional steadiness increase.

Week 8 and Beyond—A New Baseline

- Metabolic flexibility is significantly stronger.

- Waist circumference continues decreasing as visceral fat mobilizes.
- HDL cholesterol rises, triglycerides fall, and cardiometabolic risk drops.
- Daily energy feels smoother, lighter, and more stable.
- Hunger normalizes; the relationship with food becomes calmer and more intuitive.
- The body begins operating closer to its ancestral metabolic blueprint.

Why This Matters

These changes are not miraculous. They are expected. When the pressures that caused metabolic dysfunction are lifted, the body transitions from survival mode to repair mode. Metabolic syndrome stops being a destiny and becomes a pattern—a pattern that can change.

Why the Diagnostic Markers Are Late Warnings

The diagnostic criteria for metabolic syndrome: waist circumference, blood pressure, fasting glucose, triglycerides, and HDL are downstream signs of a much earlier problem, impaired metabolic switching.

By the time someone meets the criteria, metabolic strain has been present for years. These markers are not causes; they are consequences. They are the body's attempt to compensate for an environment that forces it to store energy constantly and burn almost none. Recognizing this shifts the focus from treating numbers to restoring metabolic conditions.

How Modern Living Quietly Dismantles Metabolism

For nearly five decades, nutrition guidance shaped by selective and biased research steered people deeper into metabolic dysfunction:

- Eat six small meals per day.
- Avoid fat.

- Choose low-fat processed foods.
- Snack between meals.
- Prioritize carbohydrates.
- Fear hunger.
- Whole grains and carbohydrates are essential for health.

These messages taught people to eat constantly, spike insulin repeatedly, and avoid the very nutrients that stabilize energy. Meanwhile, modern life layered on additional pressures: sleep disruption, artificial light, high stress, minimal movement, and constant food cues. The result is a metabolic environment in which dysfunction is not the exception—it is the norm.

Why Conventional Medicine Struggles With Metabolic Syndrome

Modern healthcare treats each symptom separately:

- Statins for cholesterol
- Metformin for glucose control
- GLP-1 receptor agonists for glucose control and weight management
- ACE inhibitors or ARBs for blood pressure
- Proton pump inhibitors or H2 blockers for acid reflux
- CPAP for sleep apnea
- SSRIs for mood symptoms

These medications have value, but none rebuild the metabolic system. They adjust numbers while the underlying engine continues to struggle. This is why people often accumulate prescriptions without accumulating health.

The Metabolic Model: Fix the System, Not the Symptoms

A more effective approach focuses on restoring the environment in which human metabolism functions best:

- Lower glucose exposure
- Reduce insulin
- Increase nutrient density
- Allow true fasting windows
- Remove processed foods
- Reintroduce movement
- Improve sleep
- Reduce chronic stress signals

When these changes take hold, mitochondria repair, inflammation drops, visceral fat decreases, energy stabilizes, and appetite normalizes. The conditions that created metabolic syndrome begin to unwind together.

Why Real Change Starts With Biology

Medications can help, especially GLP-1 agonists for certain individuals but even the most effective drugs work by altering one or two pathways. They do not rebuild flexibility, mitochondrial capacity, or the body's ability to burn fat cleanly. Biology changes through daily patterns—eating, moving, sleeping, recovering, managing stress. Once people understand how insulin, hunger hormones, mitochondrial function, and nutrient quality interact, the process stops feeling like a battle of discipline and starts feeling like a partnership with physiology. That's when transformation begins.

The Path Forward Starts at Home

Healing doesn't start in the clinic. It begins in the pantry, at the dinner table, in grocery aisles, in the hours between meals. It takes shape in the rhythm of sleep, the cadence of daily movement, the reduction of daily stress signals. Medication can support the process, but lifestyle determines the direction.

Low-Carb Nutrition and Exercise: A Synergistic Reset

A low-carbohydrate, whole-food approach paired with exercise is one of the most reliable ways to restore metabolic health. But for many people, the order matters. If you don't first reduce the signal that keeps fat locked away—chronically elevated insulin—exercise becomes harder, cravings stay louder, and progress often stalls. Lowering carbohydrate intake helps restore the metabolic environment by doing what willpower alone can't:

- flatten blood-glucose peaks
- reduce insulin demand
- reopen access to stored fat
- reduce liver fat
- decrease oxidative stress
- increase metabolic flexibility

Once the signaling begins to normalize, exercise starts working the way people expect it to. Muscle contraction pulls glucose and fatty acids into cells with minimal insulin, giving the pancreas a break and improving insulin sensitivity for hours afterward. Together, low-carb nutrition and exercise don't just "burn calories." They restore access—fuel availability, stable signaling, and the flexibility to run on more than one energy source.

Why Nutrition Must Come First

Exercise becomes powerful once the body is running on steady fuel. But when the body is stuck in a cycle of hunger, cravings, and energy crashes, working out doesn't feel like progress, it feels like another thing you have to survive. And for years, that was me showing up, trying harder, pushing through, assuming the answer was more effort. I didn't yet understand nutrition. I didn't understand metabolic flexibility. I didn't realize that you can't "outwork" a body that's still trapped in the wrong fuel environment. I see it in the gym constantly.

The most discouraging stories aren't about people who don't try. They're about people who *do*. They show up on schedule. They follow the plan. They sweat. They push. And yet week after week they don't get the return they expected. The scale barely moves. Their energy doesn't rise. Their sleep doesn't improve. They leave proud of their effort but quietly confused that their body isn't responding.

That's usually the moment the advice gets louder. A trainer says, "You've got to fuel your workouts." A dietitian says, "You need balance." A doctor says, "Everything in moderation." A friend says, "Don't skip breakfast." A wellness influencer says, "Start your day with a smoothie." Most of it is well-intended. Most of it sounds healthy. But for people already stuck in metabolic dysfunction, a lot of these "healthy" routines don't stabilize anything. They keep the cycle alive.

A smoothie for breakfast can feel clean and energizing—until it doesn't. A granola-and-yogurt breakfast can look like health on a plate and still leave you searching for food again by mid-morning. And once that pattern starts, it doesn't just affect the next meal. It affects your entire day. Some foods don't just satisfy hunger. They create more of it. They keep you chasing energy instead of building it. That's why nutrition has to come first—not because exercise doesn't matter, but because exercise isn't the foundation. It's the amplifier. And if I had to put numbers on what mattered most at the beginning, I would say this: nutrition and time-restricted eating were about 95% of the solution. Not because workouts are useless—but because you can't out-train a body that never gets a chance to stabilize.

For many people, exercise becomes truly effective *after* energy is reawakened—after hunger quiets down, cravings loosen their grip, and the body stops acting like it's running on emergency power. Then the gym becomes what it was always meant to be: a place where strength builds naturally, endurance returns, and progress finally matches effort. Get healthy first. Then performance, endurance, and strength become possible.

Returning to Energy

As metabolism heals, everything changes. Energy stabilizes. Hunger normalizes. Brain fog clears. Recovery improves. You regain access to the best version of your physiology. Metabolic syndrome is not a life sentence. It is a pattern—and patterns can change.

This chapter marks the point where understanding becomes empowerment. Now you know the biology. Now you understand the environment. Now you've seen how rapidly the body can heal. In the chapters ahead, I'll draw from both science and my own journey back to energy—showing how to sustain these shifts, deepen metabolic flexibility, and cultivate lasting resilience.

CHAPTER EIGHT

THE PATH TO METABOLIC HEALING

*"The body heals itself when the
obstacles are removed."*
ANDREW TAYLOR

Still A New Beginning After Understanding the Problem

By the time most people learn the term *metabolic syndrome*, they've already experienced the symptoms for years: fatigue, weight gain, creeping blood pressure, elevated fasting glucose, stubborn belly fat, or that persistent sense that something is "off" inside the body. Understanding the condition is important. But understanding how the body heals from it is life-changing.

Metabolic syndrome isn't simply reversed through willpower, strict discipline, or a handful of medications. It heals by returning the body to the biological conditions under which human physiology evolved. When insulin drops, when mitochondria breathe freely, when the liver no longer drowns in excess fuel, and when cells reacquire their ability to switch between glucose and fat, the entire metabolic system reawakens.

The path forward begins with one central premise: Healing doesn't require fighting the body. It requires supporting what the body is already designed to do. To understand this healing process, we need to look deeply at the metabolic blueprint—how a healthy metabolism actually works, what created the dysfunction in the first place, and what conditions allow the body to reverse the damage with surprising speed.

Metabolic Healing Begins With Returning to Biological First Principles

The body is built on rhythms: fed and fasted states, periods of rest and exertion, cycles of repair and renewal. For most of human history, those rhythms were unavoidable. Food wasn't available every hour. Night brought darkness and quiet. Movement wasn't scheduled—it was woven into daily life. Stress came in short bursts, not constant background pressure. Food was real, unprocessed, nutrient-dense, and scarce compared to modern abundance.

In that environment, insulin rose briefly after meals and dropped soon after. The body burned glucose when food was available and accessed stored fat when it wasn't. This metabolic switching wasn't optional, it was essential for survival.

Today, the rhythm is reversed. Most people spend their entire day in the "fed state." Insulin remains elevated from breakfast through late-night snacks. Glucose is constantly available. Fat storage increases while fat burning shuts down. Sleep is compromised. Stress hormones remain elevated. Mitochondria are forced to run at full throttle without rest. Over months and years, this pattern erodes metabolic flexibility and builds the foundation for metabolic syndrome.

Healing begins when we reverse that pattern. Restoring health is not about depriving the body; it's about giving the body back the operating conditions it evolved to expect.

Lowering the Glucose Burden:
The First Lever of Healing

One of the fastest ways to change metabolic health is to lower the daily glucose burden. Metabolic syndrome thrives in a state of glucose oversupply, when carbohydrates dominate the diet, when the eating window stretches from morning until night, and when food choices are engineered to digest rapidly and spike blood sugar. Reducing the glucose burden accomplishes two critical things: Insulin falls, allowing fat burning to resume. Metabolic switching returns, restoring flexibility and stability. You don't need extreme diets or starvation. You need foods that stabilize blood sugar rather than destabilize it. You need spacing between meals. You need nutrient density instead of nutrient dilution. When glucose exposure declines, healing begins.

Food Should Provide Energy, Satiety,
and Stability—Not Chaos

A healthy metabolism is one in which meals provide lasting energy and natural satiety. To achieve that, protein and healthy fats must anchor the diet. Protein triggers powerful satiety hormones (like peptide YY) that quiet hunger between meals. Healthy fats provide slow, steady fuel that doesn't spike insulin. Low-glycemic carbohydrates, especially non-starchy vegetables and small amounts of berries provide micro-nutrients without overwhelming the metabolic system.

When the macronutrient balance supports metabolic health, the results are profound: fewer cravings, fewer swings in mood or energy, and an intuitive connection to hunger cues rather than emotional or impulsive eating. In contrast, highly processed carbohydrates and sugars disrupt these natural signals. They generate fast-burning energy that fades quickly, triggering more hunger, more insulin, and more storage. The metabolic system becomes chaotic, and the cycle repeats itself daily. Healing requires shifting the balance toward foods that calm the system instead of inflaming it.

Restoring Insulin Sensitivity: Reversing the Core Dysfunction

Insulin resistance is the central engine of metabolic syndrome. It evolves quietly but steadily as the body becomes less responsive to insulin's signals. To compensate, the pancreas produces more insulin to maintain normal blood sugar. This keeps glucose in range, but the cost is steep—rising insulin drives fat storage, amplifies inflammation, strains mitochondria, and increases the risk of nearly every component of metabolic syndrome.

When insulin falls, the story changes:

- Fat becomes available as fuel again.
- The liver begins releasing stored fat.
- Triglycerides decline.
- Blood pressure improves.
- Inflammation drops.
- Hunger stabilizes.

This isn't theory. It's physiology. It's what happens when the core dysfunction is addressed rather than managed.

The Real Role of Exercise After Energy Stability Returns

Exercise is often misunderstood as a tool for burning calories. It isn't. Its real power lies in how it transforms metabolic signaling. Whenever muscle contracts whether through walking, strength training, cycling, or resistance work it acts like a sponge for glucose and fatty acids. It pulls fuel out of the bloodstream without requiring insulin, giving the pancreas a break and dramatically improving insulin sensitivity.

This effect lasts well beyond the workout itself, often 24 to 48 hours. Consistent movement, even simple movement, becomes a metabolic amplifier. It increases mitochondrial number, improves energy efficiency, reduces oxidative stress, and enhances the body's ability to switch fuels. Exercise does not heal metabolic syndrome alone. But combined with a nutrient-dense, lower-glycemic diet, it becomes one of the most powerful therapies available.

The Metabolic Roots of Common Conditions

Fatty Liver (NAFLD)

Non-alcoholic fatty liver disease is a visible marker of metabolic overflow—when the liver becomes a storage site for excess incoming energy. As fat accumulates inside liver tissue, triglycerides rise, signaling becomes distorted, and the whole system becomes harder to regulate. When the metabolic load is reduced and fuel handling improves, liver fat often declines surprisingly fast—sometimes within weeks— restoring energy, improving lipid markers, and relieving downstream strain on the entire cardiometabolic system.

Type 2 Diabetes

Type 2 diabetes is not a sudden diagnosis. It is the late-stage expression of a long process of impaired fuel regulation. The system can maintain "normal" numbers for years while compensation quietly intensifies in the background. When the metabolic pressure is lowered, blood sugar frequently improves quickly—often before major weight loss occurs—because the body is no longer trying to manage constant oversupply. In many cases, the condition can move toward remission as regulation returns.

Hypertension

High blood pressure is often treated as a plumbing problem, but it commonly reflects a system operating under constant internal load. Fluid retention, vascular stiffness, sympathetic activation, and impaired vessel relaxation create a state where the cardiovascular system must run "tight" all day. As metabolic signaling stabilizes, the body often releases unnecessary volume, vascular tone becomes more responsive, and pressure improves earlier than people expect, sometimes long before dramatic changes appear on the scale.

Acid Reflux (GERD)

Reflux is often blamed on "too much acid," but for many people the core issue is pressure, timing, and impaired digestion, not simple acid excess. Frequent eating, late-night meals, and highly processed foods can increase stomach volume, delay gastric emptying, and raise abdominal pressure, especially when the system is already inflamed and overloaded. As meals become simpler, less frequent, and more physiologically compatible, reflux often improves at the source rather than requiring indefinite suppression.

Gut Symptoms, Digestion, and the Microbiome

Many gut symptoms are not isolated gut problems—they are downstream effects of a system under chronic strain. Constant grazing, high fermentable loads, and processed carbohydrates can increase bloating, gas, urgency, and unpredictable stool patterns, especially in reactive or inflamed intestines. Over time, the microbiome may shift toward dysbiosis, reinforcing symptoms through altered fermentation and signaling. When meals become more spaced, simpler, and nutrient-dense, digestion often becomes calmer and more predictable without constant intervention.

PCOS (Polycystic Ovary Syndrome)

PCOS is one of the clearest examples of metabolic dysfunction presenting as a reproductive disorder. The ovaries respond to the metabolic environment, and when signaling is distorted, androgen production rises, ovulation becomes irregular, and symptoms intensify. As metabolic stability returns, cycles often become more regular, fertility may improve, and the clinical picture frequently softens—because the root environment is no longer pushing the system toward disruption.

Brain Fog, Mood, and Cognition

Brain fog is often experienced as a mental problem, but it frequently reflects unstable energy delivery and impaired cellular signaling. When the brain is forced to operate on inconsistent fuel availability, cognition becomes slower, mood becomes more reactive, and stress tolerance shrinks. As metabolic stability returns, many people notice clearer thinking, steadier mood, fewer crashes, and improved emotional resilience—because the brain is no longer operating in a fluctuating energy environment.

Joint Pain and Osteoarthritis

Joint pain is commonly treated as purely mechanical wear-and-tear, yet metabolic dysfunction can amplify pain through inflammatory signaling, tissue swelling, and impaired recovery capacity. When the internal environment becomes calmer, joints often feel less stiff and less reactive, sometimes early in the process, even before major weight loss occurs. For some, improved metabolic stability reduces symptoms enough to delay surgery—or improves outcomes if surgery is still needed.

Sleep Apnea

Sleep apnea is often framed as an airway problem, but it frequently reflects deeper metabolic strain that alters anatomy, fluid distribution, and respiratory stability overnight. Visceral fat and fluid shifts can narrow the airway, while repeated breathing interruptions disrupt oxygen delivery and recovery physiology. As central fat decreases and the metabolic load lightens, airway stability often improves, breathing disruptions decline, and sleep becomes more restorative—allowing repair systems to function again.

Chronic Kidney Disease

Chronic kidney disease is often treated as a kidney-only condition, yet it frequently develops in the context of systemic metabolic strain. Over time, elevated pressure within the kidney's filtering units, altered vascular dynamics, and persistent volume stress accelerate decline. When metabolic signaling stabilizes, fluid handling improves and vascular resistance may soften, reducing chronic stress on vulnerable kidney tissue and supporting better long-term regulation.

Escaping the Illusion of Moderation

"Everything in moderation" is usually offered as reassurance. It's meant to lower anxiety and make change feel sustainable. And for some people, it works. But for many people struggling with metabolic dysfunction, moderation doesn't feel like freedom. It feels like constant negotiation. Moderation assumes a stable internal system, one where hunger is proportional, appetite is reliable, and food choices stay within reasonable boundaries without much effort. It assumes the body gives clear signals and the brain can calmly weigh options without being pulled into urgency. In a modern food environment filled with ultra-processed products designed to be irresistible, those assumptions often fall apart. When appetite has been trained by years of stimulation, "a little" doesn't stay little. Not because someone is weak, but because the system has learned what to crave, when to crave it, and how to demand more. For a person in that state, moderation can become less of a strategy and more of a trap: the very thing that keeps the cycle alive.

What the Conversation Should Sound Like in the Exam Room

If "everything in moderation" isn't the answer, what should a clinician say instead? The problem is not that patients haven't tried. Most people with metabolic syndrome have already practiced moderation.

They've tried portion control. They've tried counting calories. Many have done it for years while quietly watching their energy decline and their cravings intensify.

The conversation needs to start in a different place. The patient should not feel blamed. They should feel understood. Your struggle is not a character flaw. It is a predictable response to your environment and your biology. When certain foods repeatedly trigger hunger, instability, and loss of control, continuing to recommend moderation is not helpful. In the beginning, some foods need to be removed to restore stability. Some foods do not behave like normal food in your body. They behave like a trigger. That does not mean the patient is weak. It means the nervous system, reward system, and metabolic system have been conditioned over time. Many people are not fighting bad habits. They are responding to learned, automatic patterns tied to stress, fatigue, comfort, and reward. The goal is not moral perfection. The goal is stabilization.

This is what that conversation should sound like:

> **Doctor:** Let's start here—you're not failing. Your body is responding exactly the way it's been trained to respond.
>
> **Patient:** Then why does it feel like I can't control it?
>
> **Doctor:** Because certain foods are driving that response. They're not neutral in your system—they're triggering hunger and keeping your metabolism unstable.
>
> **Patient:** So I just need more discipline?
>
> **Doctor:** No. This isn't about discipline. It's about changing the conditions. Right now, moderation with those foods isn't working because they keep the signal active.
>
> **Patient:** What do we do instead?
>
> **Doctor:** We remove the triggers for a period of time. Not forever—but long enough to let your system settle

down. We're not aiming for moderation right now—we're aiming for calm.

Patient: So this isn't about willpower?

Doctor: No. It's about giving your body a chance to stabilize so hunger, energy, and control can return.

This is the shift: Change the conditions that create the symptoms.

Healing Starts Where Food Is Chosen

Metabolic healing does not begin with prescriptions. It begins in the grocery store.

Most of the foods that disrupt metabolism are found in the same place: the aisles. This is where refined carbohydrates, added sugars, industrial seed oils, and engineered food combinations dominate. These products are designed for shelf life, hyper-palatability, and convenience, not metabolic stability. They deliver rapid glucose exposure, repeated insulin demand, and persistent reward signaling that keeps the system locked in dysfunction.

The perimeter of the store tells a different story. Fresh proteins, eggs, vegetables, whole fruits, dairy, and unprocessed fats live there because they require refrigeration, spoil with time, and resemble food as it has existed throughout human history. These foods provide complete nutrition without overwhelming hormonal systems. They support stable energy, predictable appetite, and access to stored fuel.

This single shift, shopping the perimeter and minimizing time in the aisles—quietly resolves many of the obstacles people struggle to manage consciously. Protein intake rises without calculation. Hidden sugars disappear. Meal frequency naturally decreases. Cravings soften. Blood sugar stabilizes. The metabolic environment begins to change without micromanagement. Medication may play a role, but no medication can fully overcome a food environment that continually pushes the body back into dysregulation. When food choices support biology, healing becomes possible. When they do not, even the best medical tools are forced to work uphill.

Returning to the Metabolic Blueprint

As the body heals, everything changes. Energy stabilizes. Hunger calms. Inflammation falls. Mental clarity increases. Weight loss becomes natural, not forced. Movement becomes energizing rather than draining. This is the restoration of metabolic flexibility. This is the reclaiming of the body's ancestral blueprint. This is the path out of metabolic syndrome.

The body is not broken. It has been overwhelmed by an environment it never evolved to handle. Remove the pressure, restore the rhythm, rebuild the foundation—and the body responds with remarkable speed. Healing is not impossible. It is biological. It is expected. It is the next step in your journey back to energy, clarity, and health.

CHAPTER NINE

FASTING: RECLAIMING THE BODY'S PRIMARY FUEL SYSTEM

*"Human beings evolved to move for long
periods without constant feeding."*
DANIEL LIEBERMAN, PHD

asting is not new. It is not a trend, a modern weight-loss strategy, or a biohacking experiment. It is one of the oldest forms of healing on record. The Bible is filled with references to fasting—not as punishment or deprivation, but as a way to renew the mind, discipline the body, and draw closer to clarity. Moses fasted for forty days. Daniel abstained from rich foods and became stronger than those who feasted. Jesus fasted in the wilderness before entering his ministry. Isaiah described fasting as a means to "loosen the chains," a phrase that echoes liberation in every sense—spiritual, emotional, and even physical.

Across ancient civilizations: Greek, Jewish, Christian, Islamic, Ayurvedic, and Chinese fasting played a central role in maintaining health. Long before microscopes revealed mitochondria or molecular pathways, fasting was recognized as a way to restore balance. Its benefits were observed, repeated, recorded, and passed down through culture and medicine. What Hippocrates hinted at has since been confirmed through modern metabolic research: the human body possesses a built-in healing system, one that awakens when digestion stops.

The 19th and early 20th centuries produced several physicians who continued this tradition at a time when laboratory science was still emerging. They observed improvements in patients with digestive disorders, pain, mood disturbances, and metabolic dysfunction and they credited fasting for those outcomes. Although many of these doctors lacked the biochemical tools we have today, their clinical observations align remarkably well with modern metabolic science.

A Brief History of Therapeutic Fasting in Western Medicine

One of the earliest Western figures to reintroduce fasting into medical care was Dr. Henry S. Tanner. In 1880, Tanner undertook a supervised 40-day fast that shocked the medical community. Instead of wasting away, he reported improvements in chronic pain, digestion, and mental clarity. His results challenged the belief that prolonged fasting was inherently dangerous. His published accounts laid the groundwork for fasting's re-emergence in clinical settings.

Around the same time, Dr. Edward Hooker Dewey began using fasting to treat what we might now call metabolic syndrome, depression, and digestive disorders. His book *The No-Breakfast Plan* argued that the human body often needed rest, not food. He believed that persistent digestion placed a burden on the system, preventing the body from repairing itself. His ideas viewed skeptically at the time now align with what we know about autophagy, reduced nutrient signaling, and hormonal recalibration.

The early 20th century introduced one of the most important fasting clinicians: Dr. Hugh W. Conklin, an American physician who used fasting to treat epilepsy. Conklin believed that seizures were triggered by toxins in the body and that fasting allowed the system to clear them. His patients underwent water fasts lasting up to three or four weeks, and his clinic recorded substantial improvements, especially among children. Remarkably, his seizure-reduction outcomes prompted Mayo Clinic physicians to investigate fasting more rigorously. It was Conklin's work that inspired Dr. Russell Morse Wilder at Mayo to develop a diet that mimicked the metabolic effects of fasting, the ketogenic diet, which has since become a cornerstone therapy for refractory epilepsy.

Wilder understood something that modern medicine is only beginning to appreciate: epilepsy is fundamentally metabolic. When the brain is deprived of stable, efficient energy, electrical regulation falters. Fasting and keto-nutrition provide an alternative fuel source ketones which stabilize neuronal function. Wilder's ketogenic diet was, in essence, fasting without complete abstinence.

During the same period, in Germany, Dr. Otto Buchinger Sr. used fasting to treat chronic pain, arthritis, metabolic disease, and mood disorders. After recovering from a debilitating rheumatic illness through supervised fasting, Buchinger devoted his career to refining therapeutic fasting protocols. His fasting clinics, founded nearly a century ago, continue to operate today and are supported by peer-reviewed research demonstrating improvements in inflammation, metabolic health, digestive function, and mental well-being.

Mid-century America saw the rise of Dr. Herbert Shelton, who supervised thousands of fasts. Although some of his methods lacked the rigorous medical oversight that would be required today, Shelton meticulously documented clinical improvements in hypertension, chronic inflammatory diseases, digestive disorders, and obesity. His writings preserved one of the largest bodies of anecdotal evidence supporting the therapeutic effects of fasting.

A more scientifically foundational figure emerged in the latter half of the 20th century: Dr. George Cahill at Harvard Medical School. Through controlled fasting studies, Cahill demonstrated that the human brain runs exceptionally well on ketones, that fasting preserves muscle far more effectively than previously believed, and that insulin sensitivity improves rapidly during fasting. Cahill's work helped decode the metabolic shifts that earlier physicians had observed but could not explain. His research provided the biochemical framework for understanding why fasting enhances mental clarity, stabilizes mood, and supports neurological function.

Across centuries and continents, these physicians saw what we can now measure in labs: when the body stops receiving constant incoming fuel, it switches into a different mode, a mode of repair, clarity, and metabolic restoration.

Fasting and Muscle Preservation: Debunking a Persistent Myth

One of the most enduring fears surrounding fasting is the loss of muscle. The concern feels intuitive: if food is absent, the body must begin consuming itself. For decades, that assumption shaped both medical guidance and athletic nutrition. Yet it rests on a misunderstanding, but because human metabolism does not break down at random. It responds with hierarchy, selectivity, and purpose.

Foundational work by George Cahill revealed that fasting is not a metabolic free fall, but a carefully orchestrated shift. As insulin recedes, other signals rise to take its place. Growth hormone increases to protect lean tissue. Glucagon steps in to stabilize blood glucose without sacrificing muscle. Norepinephrine mobilizes stored fat, opening access to long-term fuel. Even cortisol, often misunderstood, remains tightly regulated in short and moderate fasts. The system pivots smoothly toward fat oxidation and ketone production. Muscle is not dismantled. It is preserved.

A central reason lies in the brain's fuel requirements. Under fed conditions, the brain depends heavily on glucose, creating pressure to generate glucose through muscle-derived amino acids. During fasting, however, this constraint is rapidly relieved. As ketone production rises, the brain transitions to ketones as its primary fuel, reducing its glucose demand by as much as two-thirds. With that shift, the metabolic need to break down muscle largely disappears.

This insight reframed starvation physiology. What was once interpreted as muscle wasting was, in fact, a misunderstanding of *context*: prolonged famine, illness, or malnutrition versus hormonally regulated fasting in a metabolically intact individual.

Decades later, this work was extended and clarified by Stephen Phinney and Jeff Volek, who demonstrated that individuals adapted to low-carbohydrate, ketogenic metabolism maintain and often improve lean mass, even under conditions of reduced caloric intake. Their research showed that fat-adapted muscle preferentially burns fatty acids and ketones, sparing both glycogen and structural protein. Muscle, in this context, is not expendable tissue. It is functional capital.

Clinical observations from Jason Fung further reinforce this reality. In fasting patients, muscle loss is not the rule; it is the exception, typically associated with underlying illness, advanced age with immobility, or inadequate protein refeeding. In metabolically healthy individuals, fasting triggers conservation, not collapse.

The deeper truth is this: the human body did not evolve to sacrifice muscle during short or moderate periods without food. Muscle is essential for survival, for movement, thermogenesis, and eventual food acquisition. Adipose tissue, not muscle, is the intended buffer against periods of scarcity. When fasting is entered from a state of metabolic flexibility, the body sheds what is *excess or damaged*, not what is vital.

These benefits are amplified in a low-insulin, fat-adapted state, where fasting represents continuity rather than metabolic disruption. This is why fasting can be profoundly therapeutic. It is not a process of breakdown, but of prioritization. The body repairs, reallocates, and renews—preserving strength while restoring metabolic order.

Our First Encounter With Fasting and Fat Adaptation

When we first shifted to a low-carbohydrate, nutrient-dense diet, the changes were immediate. Within days, hunger softened. The cravings we had lived with for years evaporated almost overnight. A week later, five to ten pounds of water weight had melted off, not because we exercised more, but because lower insulin allowed the body to release retained fluid.

The kitchen itself became a symbol of transformation. Packaged foods disappeared. There were no more boxes of crackers, no more cereal, no more frozen desserts hiding in the freezer. Without processed carbohydrates dominating our meals, we found ourselves naturally gravitating toward real food—rich, flavorful, satisfying. Steak with butter. Bacon and eggs. A crisp romaine leaf wrapped around avocado and cheese. Everything tasted better because our palate, no longer hijacked by sugar, began to reset.

Most importantly, our metabolism began to reset. Once insulin dropped and our bodies began burning fat more consistently, fasting emerged organically. Breakfast felt unnecessary. Lunch became optional. We didn't force ourselves to fast; our physiology simply stopped demanding constant feeding. This, we would later learn, is the hallmark of fat adaptation—a metabolic state in which the body prefers fat as its primary fuel.

Dr. Stephen Phinney originally coined the term "keto-adaptation" in 1980 to describe this shift. Though Phinney didn't invent the process—fasting practitioners had observed it for centuries—he was the first to articulate its metabolic mechanisms. Fat adaptation changes everything. Hunger stabilizes. Energy becomes reliable. Mental clarity sharpens. Exercise feels easier. Mood becomes more consistent. And once this shift occurs, fasting becomes not a test of willpower but a normal, effortless part of daily life.

Around this time, research surfaced that finally gave language to what we were experiencing. In 2016, the Nobel Prize in Physiology or Medicine was awarded to Yoshinori Ohsumi for discoveries of the mechanisms of autophagy: the cell's intrinsic system for clearing damaged proteins and organelles. What felt like a new breakthrough was, in truth, a renewed understanding of something the body has always been equipped to do especially when nutrients are temporarily scarce.

One day in the gym locker room in 2019, the topic surfaced almost casually, until it came from my mouth. Mentioning autophagy changed the atmosphere immediately: the idea that periods without food can activate internal renewal systems, now recognized at the highest levels of science, was entirely new to everyone there. The usual locker-room chatter stopped mid-sentence. Several of the older guys paused mid-routine, turned back, and started asking questions. Autophagy? None of them knew the word, but they immediately grasped the concept. They wanted to know how it worked, how long it took, whether it was safe, and why no one had ever explained it this way before. Not because it sounded extreme, but because it made intuitive sense.

That moment stayed with me. It underscored how often modern science ends up validating what the body already knows—and why fasting, once understood through the lens of cellular renewal, feels less like deprivation and more like restoration.

Autophagy: Going Deeper Into the Biology of Fasting

Autophagy is one of the most important biological discoveries of the 21st century. The term means "self-eating," but in reality, it refers to cleansing, repairing, and renewing. When nutrients are scarce, the body shifts away from growth and storage and instead focuses on dismantling damaged proteins, dysfunctional mitochondria, and intracellular waste. Put simply, autophagy allows the body to take out its own cellular trash.

Ohsumi's work revealed the machinery behind this process: the genes, vesicles, and lysosomal pathways that identify and break down damaged cell components. Prior to his research, autophagy was poorly understood and often overlooked. His discoveries showed that autophagy is essential for preventing the accumulation of toxic cellular debris implicated in aging, neurodegeneration, immune dysfunction, and cancer.

This discovery should have revolutionized healthcare. It provided a molecular explanation for why fasting has been beneficial across cultures and centuries. Yet despite the significance of Ohsumi's Nobel Prize, autophagy remains rarely mentioned in clinical practice. Most physicians receive little training in nutrition or metabolic repair, and fasting is often dismissed as a fringe or alternative therapy.

The irony is stunning: We found the mechanism behind fasting's healing power—and the system barely acknowledged it.

Autophagy typically begins after 18–20 hours of fasting and intensifies significantly after 24 hours. By 36 to 72 hours, autophagy peaks, sweeping through the body and removing damaged cellular components. This includes mitophagy, the targeted removal of dysfunctional mitochondria. Because mitochondria convert stored energy into usable ATP, their renewal is crucial for preventing chronic disease.

When dysfunctional mitochondria accumulate, oxidative stress rises, inflammation escalates, and metabolic efficiency declines. Autophagy reverses this trajectory. It rejuvenates the internal architecture of the cell, enhances metabolic flexibility, and supports long-term health.

It is no exaggeration to say that autophagy is one of the body's most powerful defense systems—and fasting is one of the most reliable ways to activate it.

Why Fasting Works: The Physiology of Metabolic Healing

What many historical physicians witnessed clinically, we can now explain biochemically. When fasting begins, insulin falls. This single shift triggers a cascade of beneficial changes. Low insulin unlocks the door to stored fat, allowing the body to burn what it has accumulated. Hunger hormones—ghrelin and leptin—begin to regulate properly, reducing cravings and stabilizing appetite. Blood sugar swings diminish. Inflammation lowers. Mitochondria work more efficiently.

Fasting also restores metabolic flexibility, the ability to switch between burning glucose and burning fat. Modern diets rich in refined carbohydrates keep insulin constantly elevated, preventing this switch from occurring. Fasting reawakens the ancient metabolic machinery that allowed humans to thrive during long periods without food.

The progression unfolds in predictable stages. As insulin falls, the body first draws on glycogen stored in the liver and skeletal muscle—using liver glycogen to help stabilize blood glucose while muscle glycogen supports local energy needs. Then it enters nutritional ketosis, producing ketones that can serve as clean, stable fuel for the brain. With consistent daily fasting windows over the following weeks, tissues gradually adapt to fat oxidation. Mitochondria remodel. Enzymes involved in fat metabolism increase in number and activity. Eventually, the body becomes fully fat-adapted—a state many people have not experienced in decades.

Ketones: The Body's Secondary Energy Network

When liver glycogen finally thins, another system comes online, one that is older, quieter, and far more stable. Fatty acids begin arriving at the liver in volume, and the liver does something remarkable: it converts part of that fat into ketone bodies. These are not emergency fuels or metabolic accidents. They are deliberate, portable energy molecules designed to move through the bloodstream and deliver energy where glucose cannot always go.

Three ketone bodies are produced: acetoacetate, beta-hydroxybutyrate, and acetone. Of these, acetone plays only a minor role and is largely exhaled. The real story belongs to the other two. Acetoacetate is the first to appear, but it is beta-hydroxybutyrate—more stable, more abundant, and longer-lived that dominates circulation and carries most of the signal.

Once released, ketones travel freely through blood and cross cell membranes without the need for insulin. They pass through the blood–brain barrier with ease, supplying neurons directly at a time when glucose availability is intentionally reduced. This is not a backup plan—it is a priority pathway. The brain, which cannot burn fatty acids directly, readily accepts ketones and often functions more steadily with them than with fluctuating glucose.

The heart is another major destination. Under normal conditions, cardiac muscle already prefers fat-based fuels. In ketosis, it draws heavily on ketones, which yield more energy per unit of oxygen consumed than glucose. In a tissue that never rests, efficiency matters. Ketones provide a dense, clean-burning fuel that supports constant output with less oxidative strain.

Skeletal muscle adapts more slowly, but over time it, too, becomes proficient at oxidizing ketones and fatty acids. As this shift unfolds, the demand for glucose drops, sparing it for tissues that truly require it. What emerges is a quiet redistribution of energy traffic—less frantic signaling, fewer spikes, and a steadier flow.

Ketones are not just fuel; they are also messengers. Beta-hydroxybutyrate interacts with cellular signaling pathways involved in inflammation, oxidative stress, and gene expression. It communicates abundance in the midst of restraint, signaling that energy is available even without constant intake. This signaling role helps explain why ketosis often coincides with reductions in inflammatory markers and improvements in metabolic resilience.

There is also evidence that ketones influence mitochondrial efficiency through mild uncoupling—a controlled reduction in energy loss that lowers reactive oxygen species production while maintaining output. This is not metabolic inefficiency, but protection. The mitochondria trade a small amount of raw efficiency for long-term stability and reduced damage. We will return to this mechanism in more detail later, because its implications extend far beyond nutrition.

What matters most is this: ketones are not a niche adaptation or a dietary trend. They are a response to urgency. They emerge precisely when the body senses that constant intake is no longer guaranteed and that survival, repair, and clarity must be preserved. In a world where so many people live in a state of resistance—cells flooded with fuel but unable to use it—the ketone system represents access restored. It delivers energy to the brain at a time when many adults notice something slipping: focus dulling, memory slowing, resilience fading compared to earlier years. The brain, once richly fueled, is often left underpowered without anyone recognizing why.

Ketones exist to meet that moment. They provide the brain with a stable, uninterrupted supply of energy when glucose can no longer do so reliably. They support tissues that cannot pause—heart, neurons, repair systems—without demanding constant stimulation from the outside world. This is not an override of biology but its fulfillment. A system designed to activate under constraint, to preserve function, to protect the most vital organs while the body restores balance elsewhere. For much of human history, this pathway was entered and exited naturally. Today, many people live their entire adult lives without ever engaging it—unaware that this energy system exists, or that it was designed not merely for survival, but for healing and repair.

The Return to Metabolic Rhythm

Fasting is less about restriction and more about restoration. It realigns metabolism with the rhythms of human evolution. It allows insulin to fall, mitochondria to repair, hormones to rebalance, and cells to renew themselves through autophagy. It recalibrates appetite and restores metabolic flexibility.

Most importantly, fasting has the power to stabilize energy. not just physical energy, but mental clarity, emotional resilience, and cognitive performance. Many people describe a subtle euphoria, a sense of lightness and focus that emerges once fasting becomes part of life. This is the metabolic signature of a body that is burning clean fuel and operating efficiently.

Fasting is not new. It is not radical. It is one of the oldest and most scientifically grounded interventions available. It is a bridge connecting ancient healing practices with modern cellular biology, uniting scripture, observation, and Nobel-winning research. It is a therapy that asks little and offers much. A therapy rooted not in addition, but in subtraction. A therapy that reveals what the body can do when given the chance. Fasting is, in many ways, the most natural medicine we have.

CHAPTER TEN

OUR N-OF-1 CLINICAL TRIALS

*"The patient is the most important
part of the experiment."*
CLINICAL MEDICINE

There is a moment in every scientific journey when theory must meet lived experience—when ideas move out of textbooks and into real lives. From the beginning, we understood that the path we were taking was not the standard approach to health. Nothing about what we were doing aligned with conventional guidelines or dominant nutrition frameworks. And because of that, we made a deliberate choice: if we were going to depart from established models, we would document everything.

We tracked our daily glucose and ketone biomarkers. We reviewed our medical charts and gathered our lab results. We monitored patterns in hunger, mood, sleep, and energy. We followed the changes step by step, from beginning to end.

Very quickly we saw what many researchers already know: metabolism does not unfold in averages. It unfolds in individuals. Population-wide recommendations can describe what tends to happen, but they cannot predict what will happen in a single body with a single history. Every person carries a unique combination of physiology, hormone patterns, stress load, genetics, sleep quality, and dietary response, variables that shape results and cannot be reduced to a one-size-fits-all model. This is why N-of-1 clinical trials matter.

In research, an N-of-1 study is the smallest possible clinical experiment: one person, one intervention, one measurable before-and-after outcome. The participant becomes both the subject and the control. Instead of burying meaningful changes inside the noise of hundreds or thousands of people, an N-of-1 trial reveals what truly happens in a single, real human body when the fuel source is changed.

These trials are respected across personalized medicine, nutrition science, and metabolic therapy because they often show what large population studies obscure: the magnitude of improvement possible when the right intervention meets the right biology. They expose metabolic flexibility—or rigidity. They reveal whether blood pressure improves when insulin falls. Whether triglycerides drop when carbohydrate intake is reduced. Whether hunger stabilizes when the body shifts from glucose dependence to fat oxidation.

In this chapter, we present two such N-of-1 trials. . Two structured self-experiments documented as case reports. They are not stories of willpower or inspirational anecdotes. They are structured case reports documenting measured physiological change. They show what happened when we moved from a chronically elevated-insulin lifestyle to a metabolically aligned one. When we shifted from constant feeding to strategic fasting. When we replaced ultra-processed carbohydrates with whole-food, higher-fat nutrition. These are our results, presented in a case-report format.

CASE REPORT A—CAROL

Patient Background

Adult female; age at baseline: 48 Age at follow-up: 51 (Phase One, 3 years), 54 (Phase Two, 6 years)

- **Past medical history:** Early-onset hypertension requiring medication beginning at age 34; history of breast cancer diagnosed at age 43, treated prior to metabolic intervention.
- **Presenting symptoms:** Unstable energy levels, carbohydrate-driven hunger, and difficulty maintaining metabolic stability.
- **Pre-intervention diet:** Higher-carbohydrate, lower-fat intake with multiple meals per day.
- **Presenting problem:** Clinical findings were consistent with longstanding insulin resistance, reduced metabolic flexibility, and features of metabolic syndrome.

Intervention

- Low-carbohydrate, whole-food diet
- Increased dietary fat + adequate protein
- Morning fasting; time-restricted eating
- Removal of ultra-processed foods
- Daily walking; improved sleep routine

Clinical Outcomes—Three- and Six-Year Follow-Up

- Blood pressure normalized → antihypertensive medication discontinued
- Significant reductions in triglycerides and VLDL
- HDL increased
- Weight and BMI normalized
- Stable energy and mood, reduced hunger variability

Baseline and Phase One–Phase Two Biomarkers

Marker	Baseline (2019)	Phase One (2022)	Phase Two (2025)	Units
Age	48	51	54	years
Weight	184	125	124	lbs
BMI	33.7	22.9	21.54	kg/m2
Glucose	100	85	80	mg/dL
Cholesterol	228	255	204	mg/dL
Triglycerides	185	40	30	mg/dL
HDL	42	85	67	mg/dL
TG/HDL Ratio	4.4	0.45	0.47	-
LDL	149	128	128	mg/dL
VLDL	37	6	6	mg/dL

Clinical Interpretation

Carol's biomarker improvements reflect metabolic realignment, with improved insulin sensitivity, normalized lipid metabolism, reduced vascular resistance, and stable fuel utilization. The concurrent return to a healthy body weight supports a shift away from chronic energy-storage signaling toward improved metabolic flexibility.

CASE REPORT B—MICHAEL

Patient Background

Adult male; age at baseline: 49 Age at follow-up: 52 (Phase One, 3 years), 55 (Phase Two, 6 years)

- **Past medical history:** Episodic reactive hypoglycemia, chronic fatigue, migraines, cognitive fog, declining exercise tolerance, and features of insulin resistance.
- **Presenting symptoms:** Reliance on frequent carbohydrate intake to prevent headaches and energy crashes.
- **Pre-intervention diet:** Moderate-to-high carbohydrate intake with regular between-meal snacking.
- **Presenting problem:** Despite regular physical activity, the patient exhibited impaired metabolic flexibility, characterized

by elevated fasting glucose, reactive hypoglycemia, unstable energy availability, and limited access to stored fuel.

Intervention

- Whole-food ketogenic nutrition
- Intermittent fasting
- Fasted endurance training
- Reduction of refined carbohydrates

Clinical Outcomes—Three- and Six-Year Follow-Up

- Sustained reduction in triglycerides
- Increase in HDL cholesterol
- Reduction and maintenance of lower body weight and BMI
- Improved endurance performance during periods of fasted training
- Stabilization of energy levels and mood
- Modest circulating ketone levels consistent with long-term metabolic adaptation

Baseline and Phase One–Phase Two Biomarkers

Marker	Baseline (2019)	Phase One (2022)	Phase Two (2025)	Units
Age	49	52	55	years
Weight	232	170	168	lbs
BMI	31.9	23.4	23.1	kg/m2
Glucose	108	100	103	mg/dL
Cholesterol	162	216	251	mg/dL
Triglycerides	51	52	42	mg/dL
HDL	51	96	91	mg/dL
TG/HDL Ratio	1.00	0.54	0.46	-
LDL	101	107	147	mg/dL
VLDL	10	10	8	mg/dL

Clinical Interpretation

This case shows restored metabolic flexibility, enhanced mitochondrial efficiency, and improved cardiometabolic risk markers. Performance, mood, and daily energy became stable without reliance on frequent fueling.

Clinical Insight—LDL Interpreted in Metabolic Context

In lipid assessment, individual markers must be interpreted within the broader metabolic pattern. In follow-up labs, LDL and total cholesterol increased alongside two well-established indicators of improved cardiometabolic health: markedly lower triglycerides and higher HDL. When triglycerides are low and HDL is high, cardiovascular risk is generally lower, and LDL alone becomes a less reliable standalone marker of risk than when viewed in isolation. In this context, the overall profile—low triglycerides, higher HDL, improved glucose regulation, and normalized weight—reflects metabolic improvement rather than increased cardiovascular risk.

Key References: Miller (2011), Gaziano (1997), Ravnskov (2016)

What These Changes Mean in Real Life

Numbers can tell a powerful story, but lived experience often reveals something the labs alone cannot: the felt experience of a body returning to metabolic stability. When we reviewed our data side by side, one pattern was immediately clear. Both of us began this journey meeting criteria for Class I obesity, with weights and body composition associated with rising insulin levels, increasing metabolic strain, and early warning signs of chronic disease. Within a relatively short period, however, both of us moved into the normal weight range—not through forced restriction, but as appetite quieted and fat-adapted metabolism restored access to stored fuel. The shift carried implications far beyond the scale.

Carol: Physiological and Cognitive Recovery

For Carol, the transformation extended into nearly every aspect of daily life. Weight declined steadily, but the deeper story was metabolic recovery. Blood pressure normalized, allowing antihypertensive medication to be discontinued as ongoing pharmacologic support was no longer required. Her lipid profile shifted in a clinically meaningful way: triglycerides and VLDL fell markedly, while HDL increased. This pattern—lower triglyceride-rich lipoproteins with rising HDL—is strongly associated with improved insulin sensitivity, reduced cardiometabolic strain, and a substantially lower atherogenic burden. Hunger quieted. Energy stabilized. Mood leveled. The small daily struggles she once assumed were "normal" revealed themselves to be signs of impaired energy access rather than inevitable consequences of aging or stress.

The physical changes were only part of the story. Shortly after nutrition changed, fasting became routine, and daily movement stabilized, mental and emotional improvements emerged. Optimism increased in ways neither of us anticipated. She described feeling lighter in every way—physically, cognitively, and emotionally. Mental clarity sharpened. The fog that had lingered throughout the day lifted, and mood stability emerged with a consistency she had not experienced in years.

At work, cognitive focus improved noticeably. Tasks that once demanded sustained effort became smoother and more manageable. These changes preceded major shifts in laboratory markers, suggesting that the brain may respond to improved metabolic conditions faster than the body as a whole. In many ways, this early cognitive transformation became the clearest signal that her physiology was moving back toward health.

Michael: Cognitive, Metabolic, and Endurance Adaptation

As the metabolic changes took hold, improvements in cognition and mental clarity were just as striking. Focus sharpened in ways I had not experienced since my early twenties. The persistent cognitive fog, mid-afternoon crashes, and effort required to sustain attention began to lift. Energy stabilized across the entire day, creating a steadiness I had never achieved through willpower or discipline alone. My work improved not because I pushed harder, but because my thinking was simply clearer.

Alongside these cognitive changes, my emotional and spiritual life deepened. My outlook softened and expanded; gratitude came more easily, and my connection with God grew more grounded and consistent. Even my love for reading and research returned—something I had lost for years because I was too exhausted after work and too depleted on weekends to sustain interest. What had once felt like a daily battle to remain alert and engaged gradually transformed into a sense of alignment: physically, mentally, and spiritually. These were not isolated cognitive gains, but signs that my internal environment had shifted toward clarity, stability, and purpose.

What surprised me most was how athletic performance changed. Once I was no longer dependent on glucose to get through a workout, endurance expanded in ways I had never experienced, not even in earlier years. Fasted runs that once felt impossible became routine. I completed long training sessions, extended swims, and eventually ultra-endurance efforts without the familiar cycle of carbohydrate loading, crashing, and refueling. My body was no longer fighting for energy; it had reliable access to it.

Interestingly, circulating ketone levels often remained modest—an unexpected finding that fits long-term metabolic adaptation. A well-adapted system does not require chronically high ketones in the bloodstream; it becomes efficient at producing and using fuel as needed. Even during a prolonged reduction in running due to injury, metabolic markers and mental clarity remained stable, reinforcing that the foundation was metabolic—not dependent on high training volume.

Shared Interpretation: Beyond Weight and Biomarkers

Taken together, these changes told a story far beyond weight loss or improved cholesterol values. They reflected a deeper metabolic realignment—one that reversed every hallmark of metabolic syndrome. What emerged was a shared pattern: when the body transitions from glucose dependence to fat-based fuel utilization, improvement is not linear. It is exponential. Blood pressure normalizes. Lipids reorganize. Hunger softens. Energy steadies. Performance rises. Systems that once strained to maintain balance begin working in harmony again.

These personal transformations—measured, documented, and lived—revealed something essential. What unfolded was neither exceptional nor accidental. It was the predictable physiological response to restoring proper fuel signaling and metabolic flexibility. When metabolic dysfunction is addressed at its root, the body does not merely improve, it reorganizes. What followed was not a new intervention, but a natural consequence of regained health: movement became possible again in ways it had not been before.

PART III

Movement, Performance, and Practice

CHAPTER ELEVEN

HEALTH LED BACK TO RUNNING

"When the body heals, it wants to move."
REHABILITATION PRINCIPLES

Turning Point: The Path to Healing

My health began improving fast—faster than people realize. To understand what made the difference, I had to look back. My forties were marked by a long struggle to reclaim any lasting sense of well-being. Each year followed the same pattern: a slow decline through fall and winter, a partial rebound in spring and summer—but never without resistance. The body was always fighting me. Energy was low. Recovery was slow. Nothing felt sustainable.

Everything changed when I finally understood nutrition not as calories or willpower, but as metabolic signaling. The shift began with adopting a low-carbohydrate approach. Almost immediately, my trajectory reversed. Energy returned. Weight dropped not just to baseline, but well beyond it. At my worst, I had been diagnosed obese: 234 pounds at 5'11¾". Even in my best summers eating "in moderation," counting calories, and staying consistent with the gym and movement I would settle around 210 to 220, what I called baseline.

But in 2019, something clicked. That summer didn't end in the usual backslide. Instead fall and winter brought even greater progress.

My weight dropped from 224 to 187—down 37 pounds in just over seven months. And this time, the change wasn't just physical. My energy surged. Movement felt natural again. Exercise made sense again. By late fall, I was signing up for 5K community runs. Running was back.

Rediscovering Movement: A New Engine

Walking, running, and resistance training returned to my life not as chores, but as things I actually enjoyed again something that felt familiar, like it used to when I was younger. For the first time since adolescence, movement felt easy again. My body wasn't fighting me, it was responding, adapting, and recovering. This wasn't just about weight loss. The transformation ran deeper. Something fundamental had shifted in how my body produced and used energy.

At the time, none of this language was familiar to me. I wasn't thinking in terms of metabolic pathways or fuel partitioning—I was responding to how my body felt. But as my performance improved in ways that contradicted everything I had been taught about endurance nutrition, curiosity took over. I wanted to understand *why* this was happening. What I was experiencing felt paradoxical: fewer carbohydrates, yet more endurance; less frequent eating, yet steadier energy. That question: how this new way of fueling could outperform the old one—pulled me into the research. That shift, I came to understand, was the result of becoming fat-adapted.

Research in ketogenic metabolism has shown that when the body transitions to using fat as its primary fuel—rather than carbohydrates—profound changes occur in metabolic efficiency, endurance, and recovery. In tightly controlled laboratory studies of elite ultra-endurance athletes, long-term fat adaptation was associated with fat-oxidation rates once thought physiologically unattainable—exceeding 1.5 grams per minute—while preserving glycogen availability and maintaining performance (FASTER Study, 2016). For decades, conventional endurance models assumed fat oxidation peaked at roughly half that level, with carbohydrates required to sustain higher workloads. These findings challenged that assumption, demonstrating that the human body is capable of far greater reliance on stored fat than previously believed.

This was groundbreaking because it challenged the long-standing belief that carbohydrates were essential for peak endurance performance. In my own experience, these principles proved not just valid, but liberating. Once I was fully fat-adapted, I no longer had to chase food throughout the day. My energy was stable. Long walks, hours of physical labor, and eventually multi-hour training runs could be completed without crashing or constant refueling.

A ketogenic metabolism enhances mitochondrial efficiency: meaning the organelles that convert stored fuel into usable ATP do so more effectively, especially in muscle tissue. Beta-oxidation (fat metabolism) generates more ATP per gram of fuel than glycolysis, the glucose-dependent pathway that rapidly breaks down carbohydrate for short-term energy, and it does so with far less oxidative stress. This isn't just theory—I felt it. Recovery from workouts came quicker. Inflammation markers fell. The soreness that used to linger for days began resolving within hours. I could train more often, sleep better, and think clearer.

Resistance training returned to my weekly rhythm, not with the goal of bulking up, but of rebuilding stability and metabolic resilience. Strength came back gradually, but with less joint pain and fewer setbacks than in years past. Running, too, returned with a new feel. I wasn't chasing split times or worried about crashing halfway through a workout. I was running sustained—fueled by a clean-burning, steady engine.

The beauty of fat adaptation is metabolic flexibility. Beneath the surface my body had become more efficient at matching fuel use to demand. At low to moderate intensities trail running, hiking, steady movement, or lifting fat provided the dominant fuel. During brief surges of intensity, such as a steep climb, a short acceleration, or a heavy lift, the body could still draw on glycogen when needed even though total glycogen stores were lower than in a carbohydrate-dependent state.

Importantly, this glycogen was preserved and strategically deployed rather than rapidly depleted or dependent on constant carbohydrate intake. This balance reflects a metabolically resilient system—one capable of supporting both fat- and glucose-based energy pathways, and therefore both performance and long-term health.

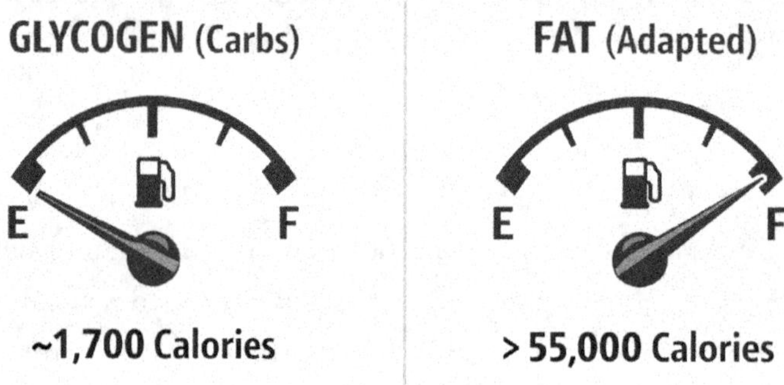

Energy availability in carbohydrate-dependent versus fat-adapted metabolism.

Glycogen stores are small and finite, providing roughly 1,500–1,700 kcal of readily accessible energy. Stored fat represents a vastly larger fuel reserve often >55,000 kcal even in metabolically healthy individuals. Fat adaptation shifts the primary fuel source toward fat oxidation while preserving glycogen for brief, high-intensity demands, increasing metabolic resilience and endurance capacity.

So this wasn't just a return to exercise. It was the discovery of a whole new paradigm of human performance—one that rejected the standard fueling model and embraced what our physiology was always capable of: long, sustainable movement powered by the body's own stored energy. And I had tens of thousands of calories of energy available at all times. That changed everything.

Running, walking, lifting—these were no longer separate activities. They became expressions of health. My body had re-learned how to move, not in short bursts fueled by sugar, but in sustained efforts powered by fat. I didn't just lose weight. I gained back metabolic power.

The Shift: From Pavement to Dirt

Everyone should find a passion for movement. Mine became running—not just the act, but the steady motion and the mental clarity it brought. That flow state where thought falls away, and the body leads. I rediscovered it as my energy returned, but something unexpected changed everything: trail running.

Before 2019, the woods were something I walked through—the occasional hike, nothing more. That year, I began running them. It opened a door to movement I hadn't known I needed. I wasn't logging road miles anymore. I was weaving through trees, dodging roots, and climbing hills. Trail running demanded more—balance, strength, attention—but it was easier on my body. The constant variation spread the work across muscles and joints instead of loading the same tissues again and again on flat pavement. It was more demanding, yet more enjoyable. And that was the surprise.

Trail running was counterintuitive. The terrain was wild, uneven, constantly shifting. Hills burned the lungs. Technical sections tested balance and focus. But despite the added challenge, my body held up better than it had on pavement. My joints didn't ache like they used to. There was less pounding, less repetitive stress. The varied surfaces— dirt, grass, leaves, sand—absorbed impact in a way concrete never could. The constant variation forced micro-adjustments, recruiting stabilizing muscles throughout the feet, ankles, hips, and core, and spreading the load rather than concentrating strain in the same places.

I found myself running longer with fewer issues. No shin splints, no road fatigue, no familiar aches that used to follow even short road sessions. I was pushing harder, yet recovering faster. It made no sense until it did—the variability was the medicine.

Just outside Mt. Pleasant, Michigan, I found Deerfield Nature Park. The trails loop through old woods, tracing the quiet bends of the Chippewa River. Each loop runs three to four miles—perfect training distances—wrapped in beauty and solitude. No cars. No sidewalks. Just forest and trail. Every visit reset something in me.

Trail running was also gaining traction elsewhere. I discovered that many runners in midlife were ditching pavement for dirt—not just for scenery, but for sustainability. Out there, no one cared about pace. The goal wasn't speed. It was presence, endurance, and joy.

That fall, I picked up my first pair of trail shoes—not for fashion, but for function. I chose the Salomon Speedcross 5, and they were a game changer. The aggressive lugs bit into dirt and leaves instead of skidding across them. The snug, locked-in fit stabilized my foot on uneven ground, and the more responsive feel brought me closer to the trail itself. Loose gravel, roots, wet grass—none of it felt unpredictable anymore. The terrain stopped fighting me, and my body began learning how to move across it. Not long after, we picked up a pair for Carol as well, and she felt the difference immediately. I wasn't just jogging anymore. I was training for something different.

Day after day, I returned to Deerfield Park. I ran alone. I ran without music. I ran without goals, and yet each run brought gains— physically, mentally, metabolically. Trail running didn't just build strength; it sharpened awareness. Every step was different, so every step required attention. At times you walk and enjoy the calm. That kept me engaged and grounded. In a world built for distraction, this was sacred space.

The dirt became my teacher. The terrain demanded respect but offered so much in return—resilience, adaptability, and a kind of quiet fulfillment no treadmill or road ever gave me. I wasn't escaping life. I was stepping deeper into it, one stride at a time.

This shift—from smooth pavement to unpredictable dirt—didn't just make me a better runner. It made me more durable, more present, more alive. It reminded me that sometimes the harder path is actually the one that heals you.

Faith, Energy, and the Sacred Trail

Trail running didn't just change my body it changed my spirit. As I spent more time in the woods, on quiet paths far from the noise of everyday life, movement turned into communion. The deeper into the forest I ran the more I felt the presence of God. Not in any forced or scripted way—but in stillness, in breath, in the footsteps that felt like an answer to prayer.

Modern medicine doesn't have a framework for this kind of energy. But it's real. I experienced it. Once my body adapted to using fat for fuel, there was a calm, steady energy always available—not just physical stamina, but clarity. That energy made room for awareness. And in that awareness, I found God walking with me.

There's something ancient about trail running. You leave behind the straight lines and concrete grids. You move through creation the way it was meant to be experienced—unpaved, imperfect, alive. Each run became a conversation. Not always with words, but with gratitude, awe, and sometimes hard questions. I wasn't alone out there.

This was a kind of healing that no lab test could measure. Yes, my weight dropped. Yes, my endurance returned. But more importantly, my faith deepened. Nature gave me perspective, but it was God who gave it meaning. The silence wasn't empty, it was full of presence, reassurance, and reminders that God has a greater plan for us when we're shown a new framework aligned with His design. That connection turned out to be essential. Because what came next would test the world—and every individual in it.

By early 2020, the COVID-19 pandemic arrived, and fear and uncertainty spread quickly alongside it. Isolation was prescribed as protection. Stillness became a virtue. But there is a difference between *solitude* and *isolation*. Solitude can be grounding: chosen, meaningful, and full of presence. Isolation is something else entirely: imposed, uncertain, and often saturated with fear.

And this response never fully reconciled with what humans have long understood about health and resilience. Long before modern medicine, survival depended on movement, sunlight, social connection, and meaning. These were not fringe ideas. They were foundational.

In a surprisingly short time, the trails had given me more than cardiovascular health. They had begun to build resilience—physical, emotional, and spiritual. I was learning how the body settles when it moves, how the mind clears in nature, how light restores after long winters. I was learning that strength grows through engagement, not withdrawal.

So when the world shut down, I kept moving. Not recklessly, but deliberately. Faith was not separate from my training—it was woven into it. Each step through the woods was grounding. A way to process what was unfolding. A way to stay aligned with something older than fear. And that steadiness carried beyond me, shaping how I showed up for the people around me.

Running became a form of reconciliation. Private. Humble. Honest. Some runs were filled with joy. Others with questions or frustration. But every one of them was a way to return to the truth: I am not in control, but I am not alone. This connection—to God, to the land, to the renewed energy flowing through my body made life better in ways I couldn't have planned. It gave me strength. And that strength carried me straight into the unknowns of the pandemic with steadiness, focus, and faith.

Breaking Thirty and Breaking Old Rules

My first race was a local 5K in Fall 2019. I was only three months into my new life without carbohydrates and sugar. Everything I had been taught about fueling before a road race told me to eat pasta, bread, and more bread. Carbohydrates were mandatory before a race. Yet there I was the night before eating my steak and salad. It felt reckless, but something in me knew it was right.

I woke up on the morning of CMU's homecoming 5K ready and excited. Everything about this race felt different. I wasn't heavy from the steak dinner the night before and all I needed that morning was coffee and water—nothing else. From that moment forward, every training run, gym workout, and race start line began the same way: coffee and water.

I felt ready and steady. My goal was to break 30 minutes. By the two-mile mark, my heart was racing, but my mind and stomach were calm. The push to the finish was stronger than I expected, and I crossed the line in under 28 minutes. It wasn't the time that mattered—it was what it represented. That day changed everything. Running ketogenic became my new normal. No carb loading. No breakfast. No stomach trouble on the course. I ran fasted, with only coffee and water fully fat-adapted.

I registered for my next race, the Dickens Christmas 5K wanting to understand why my body felt so different. Why energy stayed level. Why recovery was so quick. Why my gut felt silent and reliable.

The science behind it was straightforward. When carbohydrate intake remains very low, the body shifts its overall fuel strategy. Glycogen, once the primary driver of endurance, becomes a secondary reserve. Fat metabolism takes the lead. Muscle cells increase the machinery required to oxidize fat efficiently, and mitochondria adapt to produce a steady sustained supply of energy from stored fat. For endurance athletes this means access to a vastly larger fuel reservoir without the volatility or constant need for external fueling.

It explained exactly what I was feeling in races. Those early 5Ks weren't long enough to truly test endurance limits but they revealed something important: the spikes and crashes were gone. My energy curve flattened, which made pacing easier and effort steadier. Running in a fasted state no longer felt risky. It felt natural.

There was another layer I hadn't expected. With the body relying less on digestion for quick carbohydrates, the gut stayed quiet. Blood flow remained where I needed it in my legs, not shuttling back and forth between muscles and my stomach. Runners often have to deal with stomach issues or porta-john management. In a fat-adapted state those problems faded for me. My stomach no longer required constant timing and planning.

And then there was recovery. High carbohydrate intake during training can drive large swings in glucose and insulin, which are associated with increased inflammation and oxidative stress in muscle tissue. When the body shifts to fat and ketones as primary fuels those swings diminish. Many athletes in a fat-adapted state report less soreness, reduced inflammation, and a faster return to training after hard efforts. I felt that change week after week. I could race, sleep well, and wake up ready to train the next day.

By the time the Dickens Christmas 5K arrived, I wasn't questioning it anymore. I wasn't wondering whether the science held up or whether my experience had been a fluke. I was operating in a different metabolic gear and it was reshaping the way I ran.

That morning began like most of my training days had by then—no breakfast at all. I bundled up with a winter cap and gloves, stepped into the cold, and felt the same quiet confidence that had been building all fall. Running without carbohydrates didn't weaken me. It stabilized me.

That morning the streets were busy with families, volunteers, and runners shaking out their legs before the start. I was still new to this but something in me felt settled. I knew my body would deliver. The horn blew and we moved through town, and by the final stretch I was running strong down the center of Broadway Street. Another personal best. Another moment when everything felt lighter.

Winter Warriors

Winter 2020 came fast, but the weather did not stop me from running and training. At times I trained through snowstorms on the trail and frigid wind, and each run pulled me deeper into that same fat-adapted performance. I can remember my first snowstorm run. I was deep on the trail miles from my car, with no worry about energy, pace, or whether I had enough in the tank to get back. The world had slowed into something calm and untouched. Snow drifted down in heavy flakes that clung to the trees, turning every branch into a white sculpture. The trail disappeared under a soft ankle-deep blanket that cushioned each step. My strides made almost no sound, just a gentle crush of powder beneath my trail shoes.

Again, no music, no headphones, no noise, no interruptions. Only the quiet sound of trees and the steady fall of snow around me. The air felt cold and clean. The woods carried a stillness you can only find in winter, as if the whole world paused and waited for you to move through it.

It was pure joy. Beauty surrounded me, from the frosted branches overhead to the path unfolding ahead. I wasn't running away from anything or chasing anything. I was simply there, alone with it all. That day sealed my love for winter running.

Three more races followed, the 2020 Winter Warrior Race Series in January, February, and March. These were often run in single-digit temperatures, with wind blowing hard off open farm fields as we left town for the countryside and turned back again. Each 5K and 10K reinforced what I was feeling. More fitness. More fun. More effortless recovery. Every race that winter confirmed the same truth. My engine had changed, and it was finally working the way it was meant to work.

What this winter showed me was simple. When the energy system I'd always relied on stopped working the way it used to, another one was already there. By shifting fuels, I didn't force performance—I uncovered it. This approach is not for everyone, but for athletes struggling with energy or recovery, it may offer another path forward. For me, it wasn't about doing more. It was about finally using what had been there all along.

When the World Closed, the Mountains Called

When the world shut down, every race in Michigan vanished almost overnight. Spring events I had marked on the calendar were quietly removed, replaced with the same message everywhere: stay home and stay safe. The momentum built through the winter—the miles logged in snow and wind—suddenly had nowhere to go. The race community went still. Start lines disappeared.

The summer of 2020 opened with a sense of absence. After a long winter, people were ready to move again, to aim toward something tangible. Instead, there were no crowds. No organized events. No shared markers on the calendar. Training continued, but without direction, and the question lingered: how do you carry momentum forward when the usual paths close?

For me, the answer wasn't resistance—it was redirection. If Michigan was shutting down, preserving health and purpose would require finding spaces where movement was still possible—quiet, wide open, and free of crowds. Places that offered more distance, more perspective, and more challenge. Colorado kept coming to mind. High trails. Long climbs. Open wilderness. The kind of effort that tested not just fitness, but resolve. Ridgelines that demanded attention. Miles that asked for commitment rather than speed. Not an escape—but another way forward.

The idea of adventure—far from uncertainty, far from closed calendars—began to take shape. A few questions surfaced. Where was this still possible? What kind of challenge was big enough to focus the months ahead? If I was going to put something on the calendar, it needed to be substantial—something that justified the training, the travel, and the risk of hope. Something ambitious. Something definitive.

As I started looking—scrolling race calendars, scanning trail sites, seeing what still existed—the same distance kept appearing. A standard marker in the ultra world. Long enough to demand respect. Big enough to anchor a season. A 100K in the mountains. Big miles. Big elevation. A challenge worthy of everything that had been rebuilt from the inside out.

Michigan shut it down. No more community road races. No more registrations. So the search moved beyond Michigan—toward the race that demanded the most.

Never Summer 100K. Held in Rocky Mountain National Park near Gould, Colorado. A 100-kilometer race, just over 64 miles that year, with 14,000 feet of elevation gain.

The race describes itself as *"a mountain race in the truest sense of the term,"* with long stretches of high alpine ridge running, two alpine peaks, and four alpine lakes along the route. Much of the course sits above timberline, offering wide-open views of the Never Summer Mountains, expansive western vistas, and even glimpses north toward Wyoming's Snowy Range.

The terrain is deliberately unforgiving—meadows and streams, mud deep enough to wade through, boulder fields, alpine trail, and stretches of road that offer little relief. Wildlife is part of the experience: moose, elk, and occasionally bighorn sheep.

The organizers are clear about what this race demands. *"This is a true mountain wilderness experience,"* much of it between 10,000 and 12,000 feet. Come prepared.

Plane tickets booked. Colorado, July 2020. Then the email from the race director:

You are registered.

CHAPTER TWELVE

WE'RE GOING TO COLORADO

"The moment one definitely commits
oneself, Providence moves too."
W. H. MURRAY

You are registered.

Those three words changed everything. Once the decision was made, the abstract became concrete. The training had a destination. The miles had a reason. What had been an idea was now a commitment.

Registration was complete for the Never Summer 100K—a 64-mile mountain ultra with over 14,000 feet of elevation gain, run over terrain known to humble even seasoned runners. High altitude. Long climbs. Miles above timberline where preparation matters and intention is tested.

But before I could stand at that start line, I needed to learn how to breathe, move, and think in thin air. And I wasn't going alone.

My eighteen-year-old son, Stuart, and I boarded an American Airlines flight to Aspen with only our travel essentials. All the gear we needed—trekking poles, backpacks, a bear canister, and equipment for days of altitude work—had already been rented and scheduled to arrive at a staging location near the trailhead. The logistics were handled in advance: a ten-day block of hiking, high-elevation back-packing, mountain biking, swimming in alpine lakes, and long acclimation climbs, all leading toward the moment I would step onto the start line of the Never Summer 100K at 3:45 a.m.

Stuart agreed to come not just as my crew, but as my partner in the adventure. He was ready for all of it—the training, the travel, the unknowns. And he understood that his role was to support, observe, adapt, and help me reach that start line prepared.

As the plane approached Aspen—our first time out West—the world outside our window transformed. Mountain peaks rose out of the landscape like cathedral spires—green slopes sharpening into rocky ridges, with pockets of snow still clinging stubbornly to the summits despite the July heat. But as we admired the beauty below, it became clear we were not descending at all. The plane began circling back toward the east, and I remember thinking, *This is strange—shouldn't we be landing soon?* It felt like we were hovering on the edge of another world, close enough to see it, but not yet allowed to enter.

A few minutes later, the pilot's voice came over the intercom—steady, professional, and carrying the kind of calm that tells you something unexpected has happened. "Ladies and gentlemen, our brake warning indicator has illuminated. Because Aspen's runway is too short for a safe landing under those conditions, we're diverting to Denver."

We touched down in Denver without issue. No frustration. No disappointment. Just a quick change of plans. But once we reached the rental counters, every company seemed to give the same answer: no cars available.

We scrambled from kiosk to kiosk until finally a manager stepped forward and said there was one vehicle left on the entire lot. Not exactly what we planned but exactly what we needed. A Ram 1500 Big Horn, a deluxe full-size truck built for mountain terrain.

It felt almost intentional, like Colorado wanted us to experience it properly. We signed the paperwork, grabbed the keys, tossed our bags into the cab, and climbed in. With the bigger engine beneath us, we pulled out of the airport and pointed the truck west toward Crested Butte—our actual destination from the start.

Funny enough, even though we had booked flights to Aspen, the drive from Aspen to Crested Butte would have taken about four hours. The mountains between the two towns are so rugged and impassable by road that you have to loop all the way around them. In the end, Denver wasn't much farther. It felt like the mountains had simply rearranged our route, not our plans. The Front Range rose ahead like a massive stone wall, daring us to keep going. The adventure had already started. We both understood this was only the beginning—we had to be ready for every obstacle ahead.

Stuart wasn't just along for the ride. At eighteen, entering his senior year of high school, he stepped into the role of crew with a seriousness and maturity that felt far beyond his age. He was my support, my partner in the mountains. To me, he was simply Stu. He didn't realize it at the time, but he ended up doing everything a crew member does—managing strategy, fueling, gear changes, aid-station timing, navigation, on-the-spot problem-solving, and the most important ingredient of all: emotional support when fatigue, doubt, or pain set in. And as a former cross-country runner with three to four years under his belt, he already understood what it meant to dig deep when it mattered.

In ultra-distance trail running, crews are standard and essential; they serve as the runner's anchor through every stage of the race. And because this was my first ultra, the stakes were even higher. These races have strict aid-station cutoffs: if you don't reach each checkpoint by a designated time, your event ends—no appeals, no exceptions. Simply put: your race is over.

In this race, the crew role carried even more weight. The final stretch—nearly fourteen miles through the night—allowed runners to be paced by a crew member on the trail, a safety measure born from the reality that deep into these events, runners can face fatigue, sleep deprivation, and mental fog. Having someone you trust beside you isn't optional—it's essential. Stuart would be that person for me, the steady presence in the dark.

Truthfully, this trip became as much his adventure as mine. Long before he became my crew on race day, I was the one handling the planning. Stuart didn't know the daily itinerary, but he trusted that whatever I mapped out would lead to a good experience. All he really knew was that we were heading into the mountains together.

What he didn't know was the full plan:

- Days of backpacking, each of us carrying 40–60 pounds of gear,
- Climbing five ridgelines, each rising above 12,500 feet,
- Trails covering 40 miles through rugged Colorado backcountry,
- Backpacking overnight in bear country,
- Keeping the food canister far away from our tent at night,
- Timing our climbs around the unpredictable midday thunderstorms that roll suddenly across alpine ridges.

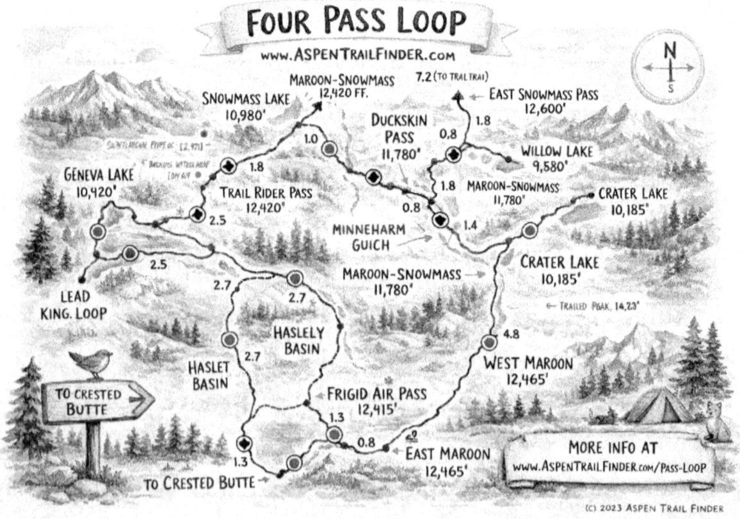

This was pre-race preparation in its rawest form—long days at altitude moving with intention, keeping our effort low and steady, staying aerobic so we could build efficiency and resilience. For two flat-landers from Michigan, this was a different world entirely. It was the kind of work that could either build us, creating memories we'd carry for a lifetime, or break us, forcing us to confront challenges we had never experienced before. Every step, every ascent, every night on the mountain was designed to strengthen our lungs, legs, and minds for what lay ahead.

At altitude, adaptation doesn't respond to intensity—it responds to time, exposure, and conditioning. With less than ten days before race morning, we had a narrow window to get this right. Normally, you taper before an ultra—rest, reduce volume, protect your legs. We didn't have that luxury.

Stu was all in. He wanted to see Colorado's raw beauty, breathe the thin alpine air, and test himself alongside me. In truth, we were stepping into something unknown and rare—we had never done any of this before. We trusted each other to support each other, and that was the plan: a father and son sharing a once-in-a-lifetime journey together.

In the months leading up to the trip, we trained together a few times not knowing we would need that training in more ways than one. This was on top of my regular trail runs, gym workouts, and the road races that had ended due to the pandemic, leaving us to create our own training at home. Our garage and a one-mile loop around the block became our shared training grounds: a concrete floor, a few scattered weights, and a chin-up bar suspended from the rafters. That was enough.

This training was different from the low, steady aerobic work I relied on for endurance. Stu introduced me to the Murph Challenge—a timed workout built on intensity and grit rather than restraint:

- 1-mile run
- 100 pull-ups
- 200 push-ups
- 300 air squats
- 1-mile run

A deceptively simple test of grit, one that teaches you how to manage discomfort, how to steady your breathing when muscles are burning, and how to keep moving when everything in you wants to stop. It turned out to be a perfect counterbalance to our mountain preparation, though we had no idea it would be tested in the most surreal way possible. The next morning came early. We woke at 5 a.m. No time to waste.

Day 1 (10 days to 100K)

We geared up for Day One—a point-to-point hike of just over eleven miles, heading north across one of the most beautiful valleys either of us had ever seen. Wildflowers were in full bloom, splashing the meadow with color. Above us rose the Maroon Bells, their rocky ridgeline cutting a jagged line against the sky as we climbed toward 12,490 feet.

The trail carried us over the top and down the far side into a wide, open valley. We crossed snowpack and icy runoff streams, stepping from rock to rock as meltwater rushed beneath our boots. Every so often we passed another small group of hikers, sharing a smile or a few words about the view—because how could you not? The weather was perfect: partly sunny, temperatures in the 60s, and clear mountain air filling our lungs with every breath. All of it pulled us gently toward Aspen.

This hike covered the first section of what would become a multi-day backpacking route over the days ahead—a four-pass loop around the Maroon Bells. But for this day, we traveled light. Stu and I carried only what we needed: light packs, trekking poles, a water canister, food for the day, a pair of dry socks, and sunscreen. That was it. And it was all we needed for one of the most unforgettable day hikes of our lives.

Snow-capped peaks. A lush green valley stretching for miles. A sky so blue it seemed unreal. Not a man-made structure in sight. Just nature—untouched and enormous—and the two of us hiking together, one in front of the other, almost speechless. No headphones. No music. Just views, sounds, and the steady rhythm of movement.

We reached the Maroon Bells trailhead around 5 p.m., nearly ten hours after we began. A shuttle carried us—and a few equally exhausted hikers—on the long, four-hour ride back around the mountain range to Crested Butte.

By the time we reached our place for the night, darkness had settled in. We grabbed a quick dinner, boots still dusty, legs buzzing from the descent. Day One was complete. Day Two began with another early start.

Before gearing up for a full day of mountain biking in Crested Butte, we went out for a short morning run to shake out our legs. Stu and I finished that run by climbing up along the ski lodge and past the private homes perched above town. From up there, the entire ski village spread out below us—buildings painted in classic Colorado colors, glowing in the early light.

That afternoon, we headed into town to rent mountain bikes. Our goal was ambitious, we had imagined tackling advanced terrain, but it didn't take long to realize that what Crested Butte calls *moderate* felt plenty advanced for us. Still, we were committed. We biked out of town toward a trail system that climbed steadily upward, until the incline finally forced us off the bikes and onto our feet.

By midday, the sun was high and the heat settled in. We drank more water, knowing our rentals were only for the day and that more work lay ahead before reaching the peak—the point where gravity, rather than muscle, would take over.

We had our helmets, of course, but this wasn't a moment to relax. At the top, looking out over the valley, it was clear why we'd made the climb—this was the reward. And now there was no turning back. Gravity waited. A thin line of singletrack disappeared into the aspens. We trusted our brakes more than our nerves. Rocks, drops, and surprises hid in the shadows. We repeated the basics to ourselves: stay loose, stand on the pedals, brake smooth. Stay upright. Just get down in one piece.

Stu led the way. I stayed back just far enough that if he wiped out, I wouldn't plow into him—because I was still the parent here, trying to keep him safe, even while doing something that felt a little insane for both of us.

Not far into the descent, the trail narrowed along a right-hand edge where there was nothing but air. We let gravity pull us forward, tapping the brakes, building confidence with each turn of the wheels— until we saw what waited ahead. A mistake there would've ended the trip on the spot. In that moment, Stu nearly lost it. The trail turned left, but his bike wanted to go straight—straight off the side. His front wheel snapped left at the last second, the kind of turn you don't see until you're already in it. From farther back, I had just enough warning to steady my balance and keep my line.

By the time we reached the bottom, our hearts were pounding and our hands glued to the handlebars, every muscle flooded with lactic acid. It was the kind of fear that arrives late, blooming only after the danger has passed. One more inch. One wobble. One wrong shift of weight. I looked over at Stu, his face drenched in sweat, eyes wide with disbelief. We both knew how close it had been. We survived. And we had no intention of doing that descent again.

We hurried back, pedaling the last few miles into town to return the bikes, legs burning all over again. Afterward, we wandered through the shops before dinner. I ordered my usual—salad with protein—and with Day Two nearly complete, allowed myself a tall glass of cold light ale. Stu devoured what looked like an entire pizza. We sat outside, surrounded by locals, hikers, bikers, and vacationers.

Trail dogs wandered by—lean, tireless, with marbled coats and piercing blue eyes. I couldn't place the breed, but if I had to guess, Australian Shepherds or Border Collies, dogs built to move all day. Stu and I laughed about our beagles back home in Michigan—sweet, loyal, and not exactly built for the mountains and how someday we'd want a real trail dog of our own. As the sun dropped behind the peaks, the day settled into our bones.

Day Three passed quickly, a short training run, a shower, and then hours on the road. We drove toward Snowmass to pick up our backpacking gear and prepare for what we knew would be our biggest challenge yet: the Four Pass Loop.

We arrived late that night, checked into our room, and picked up our gear at the front desk. The rest of the evening was spent sorting and reshuffling our backpacks, trying not to pack too heavy while knowing we had little choice. The essentials added up fast—tent, food canister, warm layers, emergency kit, water pouches, trekking poles. Every time we traded one item for another, the weight only seemed to grow.

Day 4 (6 days to 100K)

Twenty-six miles. Four passes. Back on the trail, we began at Maroon Lake Trailhead, the start of our Four Pass Loop—a circuit that climbs over four high alpine passes, each rising to nearly 12,500 feet. Over the next few days we would crest West Maroon Pass (12,490 ft), Frigid Air Pass (12,415 ft), Trail Rider Pass (12,420 ft), and Buckskin Pass (12,500 ft), climbing above treeline again and again as the Elk Mountains opened up around us.

From each pass, the world seemed to open all at once. Jagged ridgelines stretched in every direction, sharp enough to look unreal, while the valleys below rolled out like a patchwork of meadows, streams, and thin ribbons of waterfalls. The sun felt different up there—brighter, closer, almost too much at times—but it lit the peaks in a way you don't forget. The red walls of the Maroon Bells and Pyramid Peak glowed against pockets of lingering snow, and beyond them the mountains faded layer by layer into blue. It was the kind of view that left us quiet for a moment—partly from the altitude, partly from simply taking it all in.

Stu and I reached our first pass in good time. We dropped our packs, snapped a few photos, and chatted with other backpackers while sitting on the edge, looking down over the valley. Then it hit us—we still had to descend the far side, climb back up, and cross the next pass before reaching our campsite for the night. I'd planned to do in three days what most people stretch into four. It didn't take long to see why they do.

I can laugh about it now, but in the moment there was nothing funny about it. As we dropped off West Maroon Pass and began the long approach toward Frigid Air Pass, I looked up at the climb ahead—more switchbacks etched into a wall of rock—and felt my heart start to race. It was mid-afternoon. Clouds rolled in without warning. The temperature dropped with every step. I squinted up at Frigid Air Pass and thought, *Now I know.* It was going to take every intentional step just to get through that stretch.

My back was screaming under the weight of my pack. My skin and mouth still felt the sun from the morning climb. We knew we had to get over the pass before any thunderstorms hit—there was no safe place to stop, no flat ground to pitch a tent. Stu could see I was struggling—my pace slipping, each climb feeling steeper than the last. To my surprise, he reached the top, dropped his pack, and came back down the trail to meet me. He didn't say much. He didn't have to. He lifted the heavy pack off my shoulders and swung it onto his own. I followed behind him, suddenly able to move again. We had to clear that ridge before the sky opened up.

We finally reached our campsite for the night, a small clearing just off the trail where backpackers can pitch a tent without disturbing the surrounding habitat. The light was already fading as we hurried to set up the tent and open the food canister. It was too late to fire up the stove, and darkness was closing in fast. All we wanted was a quick bite, a safe place to stash the canister, and to crawl into the tent.

What happened next was unexpected—and by the following night, I would learn just how quickly things can unravel on the trail.

A few hours later, we lay stretched out in our sleeping bags, finally settled in. It felt different out there. No white-noise fan. No background hum. Just the raw quiet of the mountains. Every sound carried. Something rustled outside the tent—probably small critters, though we couldn't be sure. The bear warnings from the ranger looped through my mind as we lay awake, listening.

Even with all of that, I was grateful to be resting. But Stu wasn't doing well. I could sense it. The elevation, the sun, the brutal schedule, the miles carrying heavy packs—it all hit him at once. Then I heard his breathing, fast and shallow. Stu was having a panic attack. For a moment, I didn't even understand what was happening. It felt like half the night passed as I talked him through it, trying to steady his nerves and slow his breath. We were a full day's travel from anywhere, tucked deep in the Elk Mountains, two ridgelines from the nearest trailhead, wrapped in a darkness that made the valley feel impossibly far from help. It was as alone as either of us had ever been.

The next day, we climbed toward Trail Rider Pass, our third high point of the loop. By then, something had shifted. We'd settled into a rhythm—alternating who carried the water pouches and the food canister, moving steadily with whatever the day asked of us. Stu didn't need to haul my gear up this pass. Everything felt a little less urgent than the day before, but still intentional. We had a schedule to keep.

When we crested the pass, everything opened wide. Below us lay one of the most breathtaking sights of the entire trip: Snowmass Lake, sitting at 10,980 feet, a deep turquoise bowl of water tucked beneath the towering faces of Snowmass Peak and Hagerman Peak. Waterfalls spilled down the surrounding cliffs, catching the light like silver threads. Broad valleys fanned out in every direction, with lingering snowfields clinging to the ridgelines above.

There we stood, knowing that night we would pitch our tent beside that lake. The long descent ahead didn't bother us. With a view like that leading the way, every step felt like walking toward something sacred.

That night, we camped beside the glassy blue of Snowmass Lake. This time, it was my turn. A couple of hours after crawling into the tent, the same panic that had gripped Stu the night before found me—different in its trigger but identical in its hold. My breathing and heart rate drifted apart, accelerating in different directions, and for a moment I couldn't bring either one back under control. Stu talked me through it, calmly and steadily, reminding me to breathe, to slow down, to stay present. Sleep came in fragments, but morning eventually broke, bringing with it the resolve to take on our final pass.

Those two nights became another chapter in how we carried each other through. Strange in the moment—humbling and unforgettable. Looking back, we may have simply been adjusting to the altitude—our bodies catching up to where we'd taken them.

We climbed up and over Buckskin Pass, the final high point of the loop, then carried ourselves down the far side past day hikers who were fresh, clean, and spring-legged. We were sun-scorched, worn thin, and carrying the kind of fatigue that only comes from days at altitude with heavy packs. But the memories were the kind that last—etched into our legs, our lungs, and our stories.

By mid-afternoon, we returned to the same trailhead we had stepped away from three days earlier. A shuttle bus carried us down the valley to our truck, and just like that, the wilderness gave way to pavement. We drove forty miles straight to our hotel and spa in Glenwood Springs—finally able to shower, eat a real meal, and sleep on something that wasn't solid rock. For the next couple of days, we regrouped, ran a few easy miles, and spent long stretches in the hot spring pools overlooking the canyon walls.

By the time we were back on the road again—heading north to Gould for race check-in—the days in the mountains had done their work. We were rested, prepared, and carrying something deeper than fatigue.

Race morning came early.

My start time was 3:45 a.m.

Because of the pandemic, the race director had overhauled the usual format. Instead of a 5:00 a.m. start with hundreds of runners, we were assigned staggered waves—ten runners at a time, released into the dark.

The race director called us to the line. He raised his hand and began the countdown.

10... 9... 8...

My heart was pounding.

7... 6... 5...

Everything we'd done led to this: the Winter Race Series, the Four Pass Loop, Murph training sessions with Stu—my lungs ready for altitude.

4... 3... 2...

Pitch darkness surrounded us. Headlamps carved thin tunnels of light through the high-country night.

1... Go.

And just like that, I was off, running my first ultramarathon in Colorado.

Within minutes, something was wrong. The cone of light ahead of me dimmed. Flickered. Then went dark.

CHAPTER THIRTEEN

THE START LINE

*"The start line teaches us what
the finish line never can."*
ENDURANCE SPORT PHILOSOPHY

My headlamp died.

This wasn't my hiking headlamp—it was my trail-running lamp, fully charged back in Michigan. But after ten cold nights stuffed in my pack, I hadn't thought to refresh the charge. The lamp turned on at the start of the race, but only a half mile in it began to flicker—and then died completely.

Pitch-black trail. Two hours until sunrise. And the faint silhouettes of the faster runners ahead of me were the only hope of staying on course. I had no choice. I had to catch them.

Their pace was quicker than mine, not too fast to catch, but fast enough that staying with them for two hours until sunrise would take everything I had. Losing them meant running blind through rugged backcountry—rocks, uneven ground, and only the occasional pink course flag every quarter mile, invisible without a headlamp. So I surged forward, trailing close behind, stumbling at times, letting their narrow cones of light pull me along.

The first two miles of the course were relatively flat and runnable, more like a gravel logging road. I kept the runners in sight, knowing what waited ahead. At roughly mile three, the climbing began. Not a small climb—the long ascent toward Seven Utes Mountain, the first major test of the Never Summer 100K.

From here to mile six, we were off-road, climbing relentlessly. No switchbacks, no gentle grades, just a straight punishing ascent up the mountainside. Just up. This was a true mountain ultra: hands on the ground at times, scrambling on all fours, and other moments driving my trekking poles deep into the slope to keep from sliding backward.

Seven Utes stands near 11,400 feet, and from the base to the summit the climb gains roughly 3,000 feet. It felt like every inch of it. In the darkness, the grade seemed even steeper, each step a guess, each breath a reminder that altitude has a personality of its own. Still, staying with the pack was my only light source—my only tether. And so I climbed.

We reached the top of Seven Utes just as the first light touched the sky. I finally stopped to catch my breath and let the other runners go. For the first time all morning, I wasn't chasing anyone—I could actually look around. The sky shifted to soft pink, shadows lifting off the mountains, the whole valley below slowly revealing itself.

After taking in some fluids, miles seven through eighteen carried me past Lake Agnes, a quiet, glassy bowl tucked beneath the ridgeline. From there, the course rolled through forest and open meadows until Diamond Aid Station at mile eighteen, where Stu would be waiting. Beyond that point loomed the climb up Diamond Peak—about 2,400 feet of gain in barely two miles. The tallest, steepest climb of the entire day.

Stu was happy to see me. It's not unusual for a crew member to arrive late—or miss an aid station entirely—when the only access roads are unmarked logging routes. I was relieved he'd made it, but I immediately dropped to the ground and tore into a banana and a handful of granola mix. All I kept saying was, "It was a bad start. An absolute disaster."

Diamond Peak towered in front of me as I sat there, assessing the damage, calculating the aid-station cutoffs, and wondering where the last five or six hours had gone. My tank was already leaking from the headlamp failure and the stress of running and climbing in the dark. Stu stayed calm, offered the right words, and reminded me I could still pull this together. I caught my breath, pulled myself upright, and asked him which direction I needed to go.

Diamond was unlike anything I'd ever climbed. Two miles of relentless elevation, steep enough to feel vertical at times—the kind of slope that pulls you toward the earth. There were no switchbacks, no relief, just a straight wall of loose dirt and rock. I spent much of it on all fours, quads shaking, calves firing, lungs working at their limit. Progress was painfully slow—two hours to cover two miles—and fear sat quietly beside me the whole way. Every step felt like a negotiation with gravity, but turning back wasn't an option. I just kept climbing.

Descending off Diamond Peak, I was staring at the reality ahead: forty-plus miles left if I stayed on the 100K course, or the newly created 60K pandemic route. In the months leading up to race day, the organizers had made a rare allowance, an optional 60K alternative for runners whose training had been disrupted. It was a choice you could make before the race or, uniquely, during it if your fitness or strategy began to unravel.

And that's where I found myself. Cutoff times were closing fast—one missed checkpoint and my race would end on the spot. Around mile twenty-eight, the course split: right for the traditional 100K, left for the 60K.

In that moment, I didn't want our day to end without a finish. We had worked too hard to get here. This was our race, and I had to make a decision for both of us—risk missing a cutoff and leave Stu wondering where I'd gone on the course, or take the 60K alternate route and finish strong as a team. Switching routes wasn't part of the plan, but with the cutoffs closing in, it felt like the responsible choice.

Stu was waiting for me at the last aid station before the final stretch. It was wild to think that back at Diamond, somewhere in my incoherent rambling, I'd muttered that the 60K was "in sight" if things didn't go well. And here we were, reunited, and suddenly the whole day had shifted. The pressure was off. We'd still share the ending, and we'd be back at the hotel in time for dinner. An unbelievable turn of events.

That aid station with Stu was exactly what I needed. My nutrition was refilled, my spirits shot through the roof, and I was suddenly excited—genuinely excited—that I had just over 10K left to run. The choice to take the alternative route felt completely okay.

As it turned out, I finished faster than Stu expected. I crossed the line on my own, finisher photo and an ice-cold beer in hand, daylight still to spare. Stu joined me shortly after—an ending neither of us could have imagined after the chaos of the morning.

It wasn't the route we'd planned, but it was the finish we'd earned. We adapted when the day demanded it, stayed together in the ways that mattered, and crossed the line proud.

Leadville was next on our list. We were back on the road early, heading south as the elevation rose steadily and the old mining town came into view, tucked beneath blue-black ridgelines and a thin summer sky. Sitting at 10,152 feet, Leadville carries an energy all its own, a mix of history, grit, altitude, and stories whispered from buildings older than most states.

Still sore from the race but wide awake with that strange post-ultra clarity, we decided to take the Leadville Railroad train up into the mountains. The open-air cars rattled along the old mining route, carrying us higher above the valley as the wind snapped against our jackets. It felt good to sit and watch the mountains move past instead of having to climb them. For the first time in days, our legs could rest while our eyes did the work.

After the train returned to town, hunger took over. That evening, we walked down Harrison Avenue to Quincy's Steak & Spirits, the kind of no-nonsense Colorado steakhouse where the menu is simple, the portions are huge, and the atmosphere is easy. Thick-cut steak, baked potatoes, butter melting everywhere—calories we had more than earned. It was the kind of meal that makes you instantly warm and quiet for a few minutes, just grateful to be eating something good.

We had reservations at the historic Delaware Hotel for the night, a beautifully preserved nineteenth-century building wrapped in Victorian charm. What we didn't know—or maybe tried not to think about—was its reputation. The Delaware is one of Leadville's most famous haunted hotels, and the stories are not subtle. A woman in a blue dress is said to wander the hallways, waiting for her husband, a miner who never returned. Guests talk about footsteps echoing in empty corridors, cold spots in otherwise warm rooms, and the sense of old mining-era spirits moving through the upper floors.

Let's just say Stu did not sleep well. Every sound became a question. Every creak in the floorboards a possibility. Meanwhile, I slept like someone who had just finished climbing mountains—out cold, undisturbed, oblivious.

It was a fitting ending to a trip of a lifetime. We rode a mountain train, ate steak in a century-old saloon, and spent the night in one of the oldest hotels in Colorado—haunted, if the stories are true.

By morning, we were on a flight back to Michigan, tired but grateful, carrying with us the possibility of coming back. The return home was quiet. The trip had given us challenge, altitude, confidence and sharing it together made every moment more meaningful. We created memories I'll always be grateful for.

Soon life in Michigan settled just enough for the next goal to come into focus. It wasn't an ultramarathon this time. It was something older, something rooted in my own childhood.

CHAPTER FOURTEEN

FULL CIRCLE

"What we call the beginning is often the end.
And to make an end is to make a beginning."
T. S. ELIOT

Michigan's racing calendar went quiet in 2020, including the one I'd been waiting years to run: the Bobby Crim 10-Mile Run.

So the story moves ahead to 2021 and 2022—years when racing finally returned, and I could test myself again on the streets where I first learned to love running.

The race I spoke about in my opening chapter—the one I dreamed about as a kid, the one I had to walk away from when my health fell apart—was finally here. For years I tried to make it back to that starting line. For years, I couldn't.

The Crim wasn't just another race on the calendar. It was the race I promised myself I'd return to when I finally turned things around—the line between the life I was living then and the one I'm living now. And this time, I wasn't going to miss it.

The Crim 10-Mile isn't just another Michigan road race. It's a tradition—a rite of passage for many runners, and a lifelong benchmark. Run through Flint's neighborhoods with crowds close enough to feel personal, it has its own personality: the Bradley Hills, the grit, the noise, the pride. And the finish—onto the bricks of Saginaw Street—feels like stepping onto something older than you.

For Michigan runners, the Crim has always been more than a race. It's a gathering, a reunion, a moment when the city lifts you up and carries you home. This was a personal journey for me. After years of missing the race I'd dreamed about as a kid, I finally registered in 2021—and again in 2022. These weren't complicated trips. No crew. No logistics. Just me, the road, and a chance to test the health I'd fought to reclaim.

My training was simple, mostly trail miles around Michigan, especially west of Mount Pleasant, but I was ready for the bricks of Flint. A quick ninety-minute drive downstate, and suddenly I was standing at the start line with thousands of runners, music echoing off the buildings, spectators lining the streets just like the old days.

In 2021, the heat was brutal—mid-80s and rising fast. I felt it in every mile. I averaged 11:14 per mile, finishing in 1:52:20, grateful just to be back on that course.

In 2022, the morning was cooler—60s—and something in me had changed. My stride came back. My breathing stayed steady. I ran 9:34 per mile for a 1:35:39 finish, and for the first time in a long time, it didn't feel like survival. It felt like running.

It was one of the best experiences I'd had in years—another unmistakable sign that my health was returning, mile by mile. Even so, while the Crim became my late-summer tradition in 2021 and 2022, the truth is that another race had begun calling me back long before I ever toed the line in Flint. On December 15, 2020, an email flashed across my phone:

NEVER SUMMER 100K—REGISTRATION NOW OPEN.

I stared at it, feeling that familiar mix of nerves and excitement only a mountain race can trigger. Without hesitation, I sent a message to Stu: "You want to go to the Grand Tetons for elevation—and finally finish what we started in Gould?" This time, I meant the full 100K. Stu replied almost immediately. The spark was lit. Another Colorado summer was officially on the calendar. But this time, we decided to do things differently.

Instead of flying and bouncing between hotels, we rented a four-door Toyota Tacoma with off-road capability and packed all our gear ourselves—no more shipping boxes across the country. We mapped a route west that would double as adventure and preparation: Mount Rushmore, Devil's Tower, and then the Grand Tetons, where we'd spend four days backpacking deep in the high country—four days of thin air and honest effort. Only after that would we drop south into Colorado—a few shakeout runs in Steamboat Springs, then on to Gould for race prep, course scouting, and whatever the mountains had in store for us next.

What made this comeback possible wasn't a magic workout or a shortcut. It was energy—steady, reliable energy. Running improved because the foundation underneath it finally stabilized. Once my nutrition changed, everything else followed: better sleep, faster recovery, calmer hunger signals, and a body that stopped fighting me.

And Stuart kept me honest. He wanted me to build the return with real mileage—half-marathon and marathon-distance runs that added durability, not just confidence. I listened. I put in the work. And for the first time, the return didn't feel reckless or desperate. It felt earned.

Looking back, what stands out most is the one constant that carried through every mile and every season: my life no longer revolved around constant fueling. The old cycle of snacks, gels, carb-loading, and blood sugar swings simply disappeared. And it didn't take years— it changed within weeks of adopting a ketogenic lifestyle.

My mornings were the first to transform. Fasted workouts became fasted long runs, and eventually full race-morning efforts without a single bite of food. By August 2019, I stopped "fueling" before exercise altogether. No stomach issues. No crashing halfway through, just smooth, even output.

Recovery shifted too—not dramatically, but unmistakably, like someone quietly removed a ceiling that had been limiting me for years. I could train consistently without feeling wrecked. Over time, ultra-endurance stopped being a question of how many calories I could carry and became something else entirely: managing electrolytes and trusting my own internal fuel supply.

With insulin low and fat-burning high, my brain, heart, and muscles had access to a dependable reservoir—ketones and fatty acids—a built-in endurance system I never knew I had. Travel became easier. Long drives, early mornings, and hours on the trail no longer required planning around meals. Even packing for hiking and back- packing changed; I wasn't worried about running out of fuel because I could finally access what I'd been carrying all along. That confidence carried us into what came next.

That first trip west was our introduction. Not just to altitude, mountains, and backcountry travel, but to who we were as a father–son team in unfamiliar territory. We carried more unknowns than gear. We didn't know how our bodies would hold up, how the elevation would treat us, or whether we had the right equipment for what we were about to attempt. Just as important, we didn't yet know how we would operate together in the mountains.

What were our strengths? Where were our limitations? Did we travel well together? Did we handle stress the same way? Could we navigate in sync, or would our instincts clash?

Those answers hadn't been written yet. We discovered trust moment by moment: packing the truck, choosing trails, setting up the tent, learning who excelled at what, and recognizing each other's patterns under fatigue, hunger, cold, and uncertainty. That first trip tested not only our physical endurance, but our ability to move as a cohesive pair through unfamiliar terrain.

And this return trip was designed to go deeper—to stay outside longer, carry more of what we needed, and spend more of our time where we felt most alive: out in the wild. This time we weren't guessing; we were building on experience. We understood our individual strengths and weaknesses—and more importantly, we understood them as a team. Stu's natural pace. My climbing rhythm. Who carried what. Who navigated. Who stayed calm when conditions shifted. We had learned all of that the hard way, together.

Now we were intentional—in our planning, in our travel, and in the challenge ahead. The Return Trip wasn't just another mountain adventure; it had a purpose stitched into every mile: finishing the 100K together. Not just finishing, but crossing the line with Stu beside me. That goal reshaped everything. Training, expectations, and confidence moved in the same direction.

We weren't avoiding discomfort anymore; we were ready for it. Tent nights, KOA stays, dispersed camping, long days in the truck, unpredictable weather, none of it felt intimidating. It felt like the natural landscape for the work we'd done. The Return Trip wasn't about proving we could survive another adventure. It was about showing up as who we'd become—two people who trusted each other, had a plan, and intended to cross the finish line together.

We started differently this time. The first leg wasn't west at all—it was north, to Big Bay in Michigan's Upper Peninsula. It had been a couple of years since the family last gathered at the cabin on Lake Superior, and returning there felt like easing into the adventure.

We spent our days outdoors hiking, relaxing, and even braving the cold water for a swim or two at the family's lakeside cabin, its windows facing north across the open expanse of Lake Superior. When it came time to leave, we didn't even return home. We pulled away from the bay and turned west, the truck fully loaded with backpacking gear, food rations, and everything we'd need for long days in the mountains. The plan was simple but deliberate: KOA stays on the nights we weren't on the trail, and dispersed camping whenever we wanted to be closer to silence and stars.

Day 1 (14 days to 100K)

The opening stretch was open highway and steady progress toward the American West. Our first major stop was Mount Rushmore, a long haul from northern Michigan made easier by the excitement of finally heading west. After a couple of nights at KOA campsites, we moved on to the hiking trails surrounding Devils Tower, circling its massive volcanic column and feeling the early July heat settle into our shoulders. When the hike was done, we pushed farther west, driving across Wyoming's high plains toward Jackson Hole.

By the time we reached the valley near the Grand Tetons, the light was fading. Instead of a hotel, we turned off onto a stretch of public land designated for free dispersed camping on the east side of the valley, a rugged hillscape facing the Tetons across the distance.

It wasn't glamorous, but it was exactly the kind of night we wanted: quiet, remote, intentional. The Return Trip had officially begun.

The next morning marked the start of the backpacking leg of that journey. We began early with a stop at Mormon Row, one of the most photographed corners of Grand Teton National Park—old barns and homesteads set against the skyline. The Moulton Barn stood in the foreground, weathered and quiet, with the Tetons rising straight behind it. In the early light the boards glowed softly and the peaks turned blue as the day came on. We didn't stay long, but it gave the morning a sense of history. People had built lives here long before this valley became a destination, standing under the same mountains with far less comfort than we carried.

From Mormon Row we drove south along the Moose–Wilson corridor, a narrow road winding through stands of aspen and pine. Every few minutes the trees opened just enough to reveal the steep eastern faces of the Tetons—close now, and unmistakably serious.

The road to the Death Canyon Trailhead turned to gravel and grew rough, but the Tacoma handled it without trouble. When we parked and stepped out, the air felt different—cooler, thinner, touched by that dry alpine scent you only find in the high West.

We loaded our packs and began the steady climb toward the Phelps Lake Overlook, gaining a few hundred feet almost immediately. Before setting off, Stu and I did our usual gear check in the parking area—food, water, tent pieces, all the basics. What he didn't know, and what I kept to myself, was that I had slipped a small battery-powered fan into my pack. It was mostly for me, something I thought might help at night with a little white noise. I knew the four D-cell batteries were heavier than ideal, which is exactly why I didn't mention them.

We dropped down toward the lake and continued into Death Canyon, and the walls tightened around us—tall cliffs of ancient rock carved and weathered over unimaginable spans of time. The creek ran beside us as we worked our way toward the old patrol cabin.

Past the cabin, the trail climbed another 1,500 feet toward Fox Creek Pass, where we connected with the Teton Crest Trail. From the trailhead to the Crest is roughly five to six miles, depending on the route, and you feel nearly every step. But the moment you reach the top, the whole landscape opens wide.

Turning north onto the Crest Trail, we walked along the broad, open bench known as the Death Canyon Shelf. It was our first night on the trail. Wildflowers dotted the meadows, wind-carved pines stood against the ridge, and the views stretched east toward the Tetons and west toward Idaho. We set up our tent along the shelf with a deep valley dropping away beneath us—a campsite that felt earned after roughly 7–8 miles and about 2,000 feet of elevation gain.

As we unpacked for the night, Stu found the fan I'd quietly carried. He lifted it out of the gear pile—batteries and all—and gave me a look that said everything. Who brings a fan on the trail? I told him I figured he'd find it eventually, and I also figured he wouldn't let me bring it if I asked. He laughed, shook his head, and set it down like it was confiscated. And the truth is, we used the fan every night. It didn't do much—the mountains have their own kind of quiet—but it became part of our nightly routine. A small, unnecessary piece of gear that somehow made the whole trip feel more grounded.

Day 5 (10 days to 100K)

The next morning, we moved north as the Crest Trail left the national park boundary and passed briefly into the wild openness of Alaska Basin. The terrain felt different here—more exposed rock, wide bowls carved by ancient glaciers, and fewer trees. The trail dipped and rose in long, gentle curves, carrying us steadily toward one of the most dramatic points on the entire route: Hurricane Pass.

The climb to Hurricane Pass added another 1,000 feet, bringing us above 10,000 feet. From the top, the view is unforgettable—the back side of the Teton peaks rising straight in front of you, jagged and close, with glaciers tucked into the folds of the mountains. It's the kind of view that makes you stop talking for a while.

After the pass, the trail dropped steadily toward the South Fork of Cascade Canyon. We didn't go all the way to the junction that day. Instead, we found a flat spot about three miles before the South Fork/Cascade Canyon intersection and set up our second night. By then we had covered roughly 12–14 miles total for the trip, and we were deep inside the range surrounded by tall peaks, wide basins, and a long evening light that seemed to stretch on forever.

The following morning, we continued toward the junction, and both of us knew what was coming. Before packing up, I stepped into the glacial creek beside our campsite for a quick cold plunge, then shouldered my pack and followed the trail.

The trail split ahead at the South Fork Cascade Canyon intersection—a simple wooden sign pointing in two directions that represented two very different endings to our trip. If we turned right, the trail would drop steadily for about five miles back to the main trailheads and parking areas. Civilization. Food. Comfort. A quick finish. If we went straight, we were committing to the biggest climb of the entire route: the push toward Paintbrush Divide, rising above 10,700 feet with views that stretch across the entire Teton Range. Epic, yes—but also steep, exposed, and followed by a long descent over rock faces and lingering snow.

As we got closer to the junction, we started talking about it openly. We were low on food, our packs felt heavier than the day before, and three days on the trail had settled into our legs. That right turn—the easy one—was starting to look pretty good. But we both knew the truth. We hadn't come all this way to take the shortcut home.

So when we reached the sign, we paused only long enough to look at each other and smile. Stu kicked a small rock off the trail and said something like, "Well… we didn't come here to be comfortable." He was right. We walked straight.

Choosing discomfort paid off. The climb was hard, but each switchback opened into bigger and better views—basin after basin, ridges stacked behind one another, and the Tetons rising like a wall beside us. And farther ahead, Paintbrush Divide waited with the kind of scenery you remember for the rest of your life.

We descended about five miles into Paintbrush Canyon before stopping for our third night—our last night under the stars. This stretch of trail drops quickly, passing through rocky shoulders, alpine streams, and eventually the first small pockets of trees that mark your return to lower elevation.

As we made our way down, we started passing day hikers coming up from the valley below. Almost every group mentioned the same thing: bears feeding on berry bushes farther down the canyon. Some hikers had seen them earlier that morning. Others hadn't seen the bears but had heard about them from people just ahead. Either way, the message was clear—be cautious.

That information sat in the back of my mind the entire descent. Stu and I rounded each stretch of high brush or thick foliage making plenty of noise—talking louder than normal, tapping trekking poles together, anything to avoid surprising a bear at close range. It felt a little silly at times, but every ranger and every guidebook says the same thing: noise is your friend.

As if my nerves weren't already on edge, we found a good water stream coming down from the upper canyon, but had to make another tough decision. We were tired, daylight was slipping away, and staying close to water made sense for the night. The only problem was that this was clearly bear country. After stopping to talk it through, we realized we didn't have the daylight or the energy to push farther down the canyon. So we found a small, workable spot near the stream and set up camp, doing our best to settle in despite knowing the bears had likely been there earlier that day.

Neither of us said it out loud, but we were both thinking the same thing: stay alert. That night, every sound carried through the trees a little louder than usual. But fatigue won out. After a long day of climbing, descending, and managing nerves, we eventually settled in and trusted that we had done everything right—clean camp, food stored properly, and the quiet acceptance that in the Tetons sometimes you simply share the landscape. It wasn't the most relaxing campsite of the trip, but it became one of the most eye-opening.

The next morning, we hiked the final easy miles to Jenny Lake, finishing near the visitor center. From there, we called an Uber back to the Death Canyon Trailhead to retrieve the Tacoma.

Three nights. Roughly 30–35 miles. Nearly 6,000 feet of climbing—a full crossing of the Tetons from southwest to northeast. Not rushed. Not complicated. Just steady miles, a simple plan, and the two of us moving through one of the most beautiful mountain ranges in the country, our next goal already in mind: the Never Summer 100K.

After finishing at Jenny Lake, we drove back into Jackson Hole and treated ourselves to a night at a Motel 8. After three nights in a tent, we figured we deserved it. A quick shower, a little shopping, and a simple dinner felt like luxury before an early lights-out to bank some rest.

At first light we headed south to Steamboat Springs, checked into our KOA, and went out for a short run on the local trails. It felt good to move without a heavy pack, to let the legs settle, and to start shifting our focus from backpacking mode to race mode. The Never Summer course is rugged, high, and unforgiving, and I wanted to give it the respect it deserved.

From Steamboat, we drove east into the quiet forests near Gould, Colorado, settling into a small cabin for the next three nights, just down the road from the race start.

Stu took one look at the cabin and joked that it reminded him of the Unabomber's hideout. It was pretty rough—but only $50 a night it was perfect for what we needed to do. Just a small cabin tucked beneath tall pines, with a simple porch, two chairs, and a view that opened straight toward Diamond Peak and the mountains that would define race day. The joke stuck, and for the rest of the trip we called it "Ted's Cabin," laughing at how far we'd come from our Michigan routines.

We settled in quickly. The quiet was its own kind of preparation. We used those days to hike parts of the race course, figure out the remote gravel roads to the aid stations, and plan how Stu would join me for the final fourteen miles as my crew and pacer. The Never Summer course is spread across a wild patch of the Colorado State Forest—where moose outnumber people, cell service barely exists, and every ridge is either climbing or descending.

But sitting on that porch, reading our books, letting the mountains settle into our minds, we felt ready. Wyoming had sharpened us, and Colorado was about to test everything we'd built. By race morning, that test had arrived.

It was 2021, long past the era of ten-runner wave starts at 3:45 a.m. For the first time since the pandemic, we all gathered together at the start line, each of us excited to see one another again. Headlamps lit the darkness—it was 5:00 a.m., and in just an hour the sun would rise. The air was still cold, and standing shoulder-to-shoulder felt like a return to something we'd all been missing.

Even in the dark, the opening miles felt familiar. We climbed straight up toward Seven Utes Mountain, slipping as we dug in our trekking poles—no trail, just a long, patient climb through the trees. As we emerged above treeline, the sun was there to greet us, and when we reached the ridge near the peak, the view was nothing short of spectacular. From there, the descent toward Lake Agnes became a controlled scramble—rocky, narrow, and technical enough to keep every step honest. But I felt steady, locked into a pace that surprised even me.

By the time I reached the Diamond aid station, Stu was waiting—and nearly startled by how quickly I arrived. My times were sharper than any of our projections, each segment coming in well ahead of the cutoffs. I wasn't moving fast so much as moving consistently. No long stops. No wasted transitions. Just a steady forward momentum carrying me from one aid station to the next.

The weather leading into race weekend had threatened thunderstorms, so the course was rerouted around Diamond Peak instead of over the summit. Avoiding lightning at high elevation made sense, and the reroute kept the mileage the same. As the day wore on, my times kept improving, mile after mile slipping by with a quiet ease.

The funny thing was that the better I moved, the harder it became for Stu to keep up with the plan. He was tracking my splits and estimating when I'd reach each aid station, but I wasn't following any of our projections. I was ahead—sometimes far ahead. And as the race stretched into evening, then deep into the night, that pattern didn't change.

By the time I approached mile-50 aid station, Stu and I had agreed he would meet me there at 11:00 p.m. That had seemed realistic based on every chart and projection we'd made. But somewhere between the steady climbing, the unplanned surges, and the rerouted miles, I arrived at 10:00 p.m.—an hour early, moving well, ready for him.

Except... Stu wasn't there.

I rolled into the aid station alone, expecting to see him step out from the crowd with his headlamp and gear. But for me, the place went quiet. Volunteers were busy helping runners, lights moving everywhere—and still, no sign of him. After fifty miles on my feet, the one person I needed at that moment was nowhere in sight.

We were two years into this journey together—training, acclimating, braving storms, bears, and long climbs—and this aid-station mix-up wasn't going to derail what we'd built. Still, the volunteers saw me standing there alone and urged me to keep going. They warned me that waiting could tighten everything up or cost valuable time. "You might want to leave without him," one of them said.

But I knew I wasn't doing that. Not after everything we had done to get here.

Ten minutes passed. Then thirty. Then forty-five. I stretched, walked small circles, stayed loose, and kept watching the darkness beyond the aid-station lights. The volunteers tried again. "It's time to finish it!" But we had agreed on 11:00 p.m. That was the plan. And I was sticking to it. Right at 11:00, almost to the minute, a single headlamp appeared far down the side trail leading from the parking area. I said out loud, "That's got to be Stu." And it was. He came jogging in, completely geared up, unaware that I'd been waiting for nearly an hour. It didn't matter. He was there.

We headed back onto the trail together for the final fourteen miles, our headlamps carving two narrow beams through the darkness. Sometime after 3:00 a.m., we crossed the finish line—two full hours before the cutoff. We finished exactly the way we had trained: side by side.

We had our finish-line photo taken together—Stu as my crew and pacer, two headlamps and two tired faces grinning back at the camera.

The next morning we packed up the cabin and pulled away from Gould, beginning the long drive home to Michigan in the Tacoma.

Trip Two was complete. We had crossed the Tetons, raced through the Never Summer mountains, and reached every goal we set—not because we were perfect, not because of performance, but because we moved through it all together, side by side.

CHAPTER FIFTEEN

A RETURN TO WATER

*"Endurance is the art of energy
management, not energy expenditure."*
TIM NOAKES, MD, PHD

After Colorado, life settled into a new kind of normal. I kept training and signing up for local trail races across Michigan, but the pace felt different—steadier, more grounded. Carol and I began spending our weekends walking the trails together. That time became our shared routine, a way to reconnect and move through familiar places side by side.

In the spring of 2023, a training injury sidelined my running. Suddenly, summer running didn't feel certain anymore. For a moment, it seemed like the momentum I had rebuilt might slip away.

But something unexpected happened. I came across a charity event, only in its third year, an open-water swim around Mackinac Island. 8.2 miles. The idea pulled me in immediately. Swimming had always been part of my life, but open water—especially in a wetsuit—was unfamiliar territory. My longest swim at that point was one mile.

Swimming didn't aggravate the injury. In fact, it felt therapeutic, almost restorative. Maybe, for the moment, endurance didn't have to come from mountains or trail races. I remember thinking, *Maybe I could do this… Could I actually do this?* The idea brought a rush of excitement, followed quickly by a flood of questions. I would need a wetsuit, real open-water practice, and a willingness to face the fact that the event was only a couple of months away. Was there enough time to train? How would I handle what I couldn't control?

But my foundation was already strong. My training and nutrition were dialed in, and I trusted the work I had built. Just as important was the experience I carried from the mountains of Colorado—long days with Stuart where uncertainty and adversity were simply part of the terrain. I had learned how to stay calm, patient, and steady when conditions weren't ideal. I knew those same skills would matter in the water.

The event would take place in the Straits near the Upper Peninsula, in the waters around Mackinac Island—a place deeply connected to Carol's childhood. This was her part of the state, where she was born and raised, where everything still felt familiar in a way only home can. When I told her about the swim, we both saw the same picture immediately: this could be a perfect weekend getaway.

I could swim the island while she spent the day walking it with her friend—through town, along the shoreline, stopping for lunch—as I worked my way around the 8.2 miles of open water. It felt right. It felt possible. And for the first time since the injury, I felt excitement build toward a new kind of endurance.

Once I committed to the idea, I wanted to understand exactly what the 8.2-mile swim involved. Mackinac Island sits in the middle of the Straits, surrounded by deep water and shifting currents that don't behave the same from one day to the next. The island is a ring of limestone bluffs, rocky beaches, and quiet forested stretches, and from the water it appears far larger than it does from shore.

Swimming around it isn't a casual loop. In August, the water is usually in the low 60s, cold enough to require a wetsuit, steady breathing, and a willingness to be uncomfortable for a long time. The currents wrap around the island in different directions, sometimes pushing from the side, sometimes pulling slightly forward, and at times working directly against you. One mile can feel calm and glassy; the next can turn choppy without warning.

That first year would turn out to be the coldest.

As we entered the water, the shock hit immediately. Cold poured down the back of my wetsuit and stole my breath. The water was in the 50s—borderline brutal even with a wetsuit—and mine offered little protection. Sleeveless, ordered at the last minute, and never tested. A classic first-year move. I'd tested the waters in another inland lake— Lake Superior—during a trip to the family cabin. Early-morning miles. I thought it would be good practice. But this was different. Colder.

Once the swim settled in, something shifted. Stroke by stroke, my body began generating its own heat. The cold didn't disappear, but it became manageable. Nearly 400 swimmers stretched across open water, kayaks and safety boats surrounding us at the start. Then the pack slowly spread out, and each swimmer found their own pace.

After that, swimming became an underwater sport. Once technique and breathing stabilized, an ancient world opened beneath the surface—schools of small fish slipping between boulders, wide slabs of limestone layered in shades of gray and tan. In some sections, the water dropped suddenly into deep blue, falling away into cold depth. In others, it was crystal clear and shallow, close enough to see every detail and feel the temperature shift instantly. It felt like watching an oceanography film in real time—except I was inside it.

The views were spectacular on both sides of the surface: sky and shoreline above, stone and moving life below. Moving through water like that demands a different kind of attention. You can't rush it or force it. Every breath and stroke matters. To swim efficiently, you have to stay fully present—aware enough to feel what the water is giving back, yet relaxed enough to let your body settle into a steady pattern.

This was my first swim longer than a mile, by a wide margin. The distance and unfamiliarity were overwhelming at moments. What carried me through was the same thing that had carried me through Colorado: the ability to adapt. Stroke by stroke, mile after mile, I stayed patient and steady.

I finished as the last swimmer in the water, with a safety kayak beside me for the final stretch. By the time I reached shore, everything was already packed away. The tables were folded. The music was gone. The celebration long over. There was only one bag left at swim-bag pickup, mine. None of it mattered. I had finished, and that was enough.

Carol was waiting, exactly the person I needed to see. As we drove home, heat blasting, my body slowly warmed. Somewhere along the way, she asked if I would come back the next year and do it again.

My finish time was nine and a half hours. "No way."

It was August 2024, my second year registered for the Mack Swim. I remember praying for good weather, knowing how unpredictable conditions around Mackinac Island can be. This year brought a different kind of test. The morning arrived under a high wind advisory.

The water temperature was manageable, mid-60s, not warm, but nothing like the shock of the year before. The wind, however, changed everything. Steady chop, shifting currents, and rolling waves demanded focus from the first stroke.

I was prepared this year. Better gear. A wetsuit with sleeves that kept me warm. The course felt familiar now. I knew the currents, the long stretches where the shoreline barely seems to move, and the patience required when progress feels slow.

The swim begins in front of the Grand Hotel. We complete the first mile around a cone before starting the clockwise loop around the island, a workaround for the ferry lane that can't be crossed.

The first four miles felt like a washing machine. Wind and waves were strong enough to push some swimmers backward, and there was nothing smooth about it. The water churned constantly, disrupting forward movement just as consistency began to settle. Still, I stayed persistent. Stroke by stroke, forward progress came if I kept moving.

Since it's a charity swim, fins are allowed, and this second year I chose to use them. The added propulsion mattered. The fins helped me power through the waves, recover after each surge, and keep moving without burning unnecessary energy. Once conditions eased, my pace stabilized and the miles began to stack again.

Mission Point Resort marked the finish line. This time there was music, food, medals, and people celebrating along the shoreline. I had made it in time for the festivities, something I'd quietly hoped for after the year before.

My finish time was six and a half hours. A three-hour improvement—not from speed alone, but from experience, preparation, and adapting to the conditions in front of me.

By August 2025, everything came together. Conditions were near perfect: the water warmer, the morning calm, and the surface finally offering cooperation. For the first time, I felt like I truly belonged out there.

My goal was to finish under six hours—something that had felt unthinkable during that first frigid year. I finished in five hours and forty minutes. It was my best swim yet, and still there were things I knew I could improve.

Carol was waiting at the finish again—our ritual, our way of closing another summer, another season of movement. Three years. Three swims. Three very different experiences. Each one pulled me closer to the water, closer to Carol, and closer to the steady energy that had carried me through every chapter of this journey.

Only after these swims did I fully understand what had been happening. Each open-water swim lasted between six and nine hours—well beyond what stored glycogen alone can reliably support at a steady endurance pace. In carbohydrate-dependent athletes, energy is largely limited to liver and muscle glycogen, typically enough for only a few hours at sustained effort before depletion leads to fatigue, hunger, and cognitive decline. And yet my energy remained steady. There was no wall. No bonk. No progressive fade. Stroke remained consistent. Breathing stayed calm. Mental clarity held from start to finish. What carried me across open water was not sugar—it was access.

Fasted endurance reveals a fundamental distinction in human physiology: not how much energy the body stores, but how much of it is accessible. When insulin is low, the body is no longer confined to limited glycogen reserves. Fat-based fuels become available, expanding usable energy not by increasing supply, but by removing the barrier to it.

Even a lean athlete carries tens of thousands of calories of stored fat. The limitation has never been quantity. It has always been hormonal permission.

Low insulin—supported by fasting and sustained carbohydrate restriction—removes the block on lipolysis and fat oxidation. Mitochondria are no longer pushed into glucose dependence. These fuels become primary rather than emergency backups.

In environments like open-water swimming where digestion is impractical, thermoregulation is demanding, and steady output matters more than peak power—this distinction becomes decisive. Effort feels different. Heart rate steadies. Breathing becomes controlled rather than urgent. Hunger fades. The body stops asking for fuel and simply uses what it already has.

What stood out most was not endurance, but stability. The same metabolic state that sustained movement also supported focus and calm. Energy was no longer fragmented across systems, it supported the body as a whole.

This experience is not unique. A growing number of endurance athletes—particularly in ultra-running and long-distance cycling—have demonstrated that fat adaptation can support sustained performance without continuous carbohydrate intake. Research by Jeff Volek and Stephen Phinney has shown that well-adapted athletes can oxidize fat at rates once thought impossible, while preserving glycogen for moments when it is truly needed.

From a biological perspective, this is expected. Fat-based metabolism is a foundational human endurance system. It engages more slowly, but it is vastly more stable. Glucose excels at bursts. Fat excels at duration. Fasting simply removes the interference that blocks access to the most appropriate fuel for prolonged effort.

In the water, this translated into something unexpected: freedom. Freedom from feeding schedules. Freedom from fear of depletion. Freedom from constant mental negotiation. Energy became background rather than concern. What remained was water, motion, and presence.

I was doing it because it felt sacred, and because in the back of my mind, I wanted to keep testing what I'd started to believe back in 2019. The same foundational energy pathway that sustained long swims exists in everyday life. My swims were a way of testing whether stable energy can last for hours without breakfast, without constant calories, and without continuous input—not by pushing harder, but by restoring access.

If that's true in cold open water, then it may also be true in ordinary life. Perhaps many people are not running out of fuel—they are running out of access. Just as chronic disease reflects disrupted fuel access, endurance athletes, and ordinary people living with fatigue and metabolic instability, may benefit not from more fuel, but from restored access to what is already there.

These swims were not fueled by willpower. They were fueled by physiology. That shift gave me confidence I could trust. My health had changed at a foundational level, and endurance was no longer something I had to chase, it emerged naturally from a return to energy.

PART IV

Industrial Food and Human Physiology

CHAPTER SIXTEEN

THE COLLAPSE OF REAL FOOD

"Food is no longer food when it must be invented before it can be eaten."
MICHAEL MACDONALD

The Invention of Fake Food and the Disappearance of the Real

Across every supermarket aisle, brightly packaged foods line the shelves—technicolor, uniform, shelf-stable, and designed for convenience. They promise nourishment, satisfaction, and sometimes even "health." Yet beneath their aesthetic of abundance lies an emptiness: what we call food today is often not food at all. It is an engineered product, built from a narrow palette of industrial ingredients designed for efficiency and profit rather than nourishment. Our species, evolved to thrive on the complexity of ecosystems, now survives on commodities extracted from soil depleted of life and recombined in laboratories.

This is the paradox of modern civilization: in the age of apparent plenty, we are nutritionally starved. The human body, an orchestra of cellular energy systems evolved over millions of years, is now forced to metabolize substances never encountered in nature. What once came from soil, sun, and season now comes from factories, vats, and refineries. The line between food and industrial product has blurred so completely that even our vocabulary fails to distinguish them.

The transformation did not happen overnight. It emerged from the convergence of technology, policy, and economics—an alignment of convenience, profit, and a financial system that increasingly rewarded scale over resilience. As economic pressures favored constant growth, agriculture adapted. The directive was clear: more output, more yield, more calories per acre. To deliver that volume, farming became more dependent on chemistry—synthetic fertilizers, pesticides, and genetic manipulation—to push nature beyond its normal limits. The goal was no longer to grow food but to grow commodities: corn, soy, wheat, and sugar—the four pillars of modern malnutrition.

The Metamorphosis of Soil into Factory

The story begins in the soil, the original source of all food and all life. For most of human history, soil was treated as a living community—a complex microbiome teeming with fungi, bacteria, worms, and minerals that formed the basis of health for both plants and people. But under the pressures of industrialization, this living system was reduced to a medium for inputs and outputs. Nitrogen, phosphorus, and potassium replaced humus, mycorrhizae, and earthworms. Crops were no longer grown within a living soil ecosystem but sustained by direct chemical inputs that bypassed the biological processes responsible for nutrient density and resilience. Chemical fertilizers made crops grow faster but stripped the soil of biological intelligence—the self-organizing web that cycles nutrients, detoxifies pathogens, and generates the plant's micronutrient profile.

When soil dies, plants lose their vitality. They grow quickly but shallowly, producing calories but not nutrients. Studies comparing modern produce to its pre-industrial ancestors reveal shocking declines: calcium down by 20–40%, iron by 30%, magnesium by 25%. The carrots and spinach of today are ghosts of their former selves—bright in color, empty in essence. The farmer, forced by debt and subsidy to produce volume rather than vitality, has little choice. The economics of survival reward yield per acre, not nutrient density.

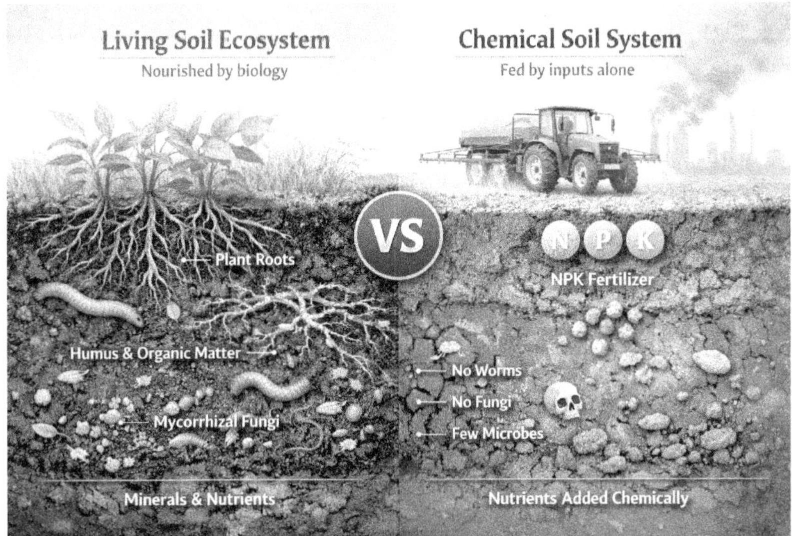

To offset this decline, food scientists invented fortification and supplementation—adding back synthetic vitamins to what had been stripped away. This cycle created the illusion of abundance while masking the underlying decay. A cereal box can carry a health claim, list ten added vitamins, and look like nutrition on paper—yet still deliver sugar, refined starch, and very little that resembles real food.

Chemical Agriculture and the Reprogramming of Soil

The industrial transformation of agriculture was not driven by fertilizers alone. It accelerated with the widespread adoption of synthetic herbicides, pesticides, and genetically engineered seeds designed to tolerate them. Among these, glyphosate became foundational—not because it nourished crops, but because it simplified ecosystems. By suppressing weeds and competing plant life, glyphosate-enabled systems allowed vast monocultures to replace landscapes once governed by ecological diversity and living soil processes.

Genetically engineered seeds were developed to withstand repeated chemical exposure, binding crop genetics to routine herbicide use. This pairing redefined farming from a biological relationship between plants and soil into a controlled industrial process. Fields increasingly functioned as managed input systems rather than living environments sustained by microbial exchange.

While these technologies increased short-term yields, they carried biological tradeoffs. Glyphosate and related chemicals alter soil microbial communities, disrupt mycorrhizal fungal networks essential for mineral uptake, and interfere with the cycling of trace elements such as manganese, zinc, and iron. Plants grown under these conditions may appear productive, yet develop within soils that no longer support the complexity once fundamental to real food.

Why This System Dominates

The dominance of chemical-dependent agriculture is not accidental, nor driven solely by ideology. It is the product of economic pressures that reward volume, uniformity, and scalability over biological function. In the United States, more than 90 percent of corn and soybeans are genetically engineered to tolerate herbicides such as glyphosate, making chemical compatibility a prerequisite rather than an option.

Glyphosate has become the most widely applied herbicide in history, used across hundreds of millions of acres each year. Its adoption rose in parallel with herbicide-tolerant seed systems, enabling large-scale monoculture by simplifying weed management and reducing short-term labor demands. For farmers operating under thin margins, debt obligations, and competitive yield expectations, participation in this system is often a matter of survival rather than preference.

While not all food is grown this way, the crops that dominate modern diets overwhelmingly are. The result is an agricultural landscape optimized for efficiency and output, but increasingly detached from the biological foundations that once governed soil health, resilience, and renewal.

The Industrialization of the Body

Once the soil was industrialized, the human body followed. The metabolic wisdom of evolution—designed to operate on whole foods, animal fats, fibrous plants, and minerals—was thrust into an alien environment dominated by refined carbohydrates, seed oils, and additives. Industrial food is not metabolically neutral. It hijacks the brain's reward systems, spikes insulin and cortisol, and floods the bloodstream with inflammatory signals. The mitochondria—our internal power plants—lose their flexibility, burning sugar inefficiently while accumulating oxidative damage.

The rise of the modern chronic disease burden—obesity, diabetes, heart disease, cancer, autoimmune disorders, depression—is not a mystery of genetics or bad luck. It is the predictable consequence of feeding an ancient biological system with artificial fuel. Each meal of processed food is a micro-dose of inflammation. Each snack made from refined flour and seed oil drives the body further from metabolic homeostasis.

This transformation of food into commodity required a linguistic and cultural deception. "Vegetable oil" sounds wholesome, but the term hides industrial lubricants extracted from soybeans, corn, and rapeseed through high heat, solvents, and bleaching agents. "Heart healthy" cereals are loaded with sugar, glyphosate residues, and emulsifiers that damage the gut lining. "Low fat" yogurts are sweetened with artificial sugars that confuse insulin signaling. Even the word "plant-based," once signifying natural wholesomeness, now serves as a marketing cloak for ultra-processed laboratory proteins extruded from genetically modified soy and peas. The irony is profound: we call these products progress, yet they represent the collapse of real food culture. Humanity has been exiled from the ecosystems that once sustained it.

The Replacement of the Real

At the heart of this transformation lies a simple pattern: every genuine thing in nature has been replaced by a counterfeit. Real fats—tallow, butter, ghee, lard—were replaced by industrial seed oils; real bread fermented with wild yeasts and whole grains was replaced by refined flour and chemical leavening; raw milk was replaced by homogenized, denatured milk—structurally altered by heat and processing, no longer biologically intact; meat from pastured animals was replaced by confined, grain-fed livestock raised on antibiotics.

The replacement is not only chemical but philosophical. Real food requires patience, seasonality, and connection to the land. Fake food offers immediacy and uniformity. Real food spoils—fake food never does. Real food nourishes quietly—fake food stimulates loudly. The human nervous system, tuned for survival in natural environments, now navigates a barrage of hyper-stimulating tastes that confuse satiety signals and override ancient hunger cues.

The fake food system functions like an addiction economy. Products are designed for bliss points—the precise combinations of salt, sugar, and fat that maximize dopamine release. The goal is not to feed the body but to capture the consumer. The population becomes both market and patient. Processed food drives metabolic disease; metabolic disease fuels pharmaceutical profits. What we call the healthcare system is largely a maintenance program for the consequences of fake food.

The Mirage of Choice

Modern consumers believe they exercise choice when selecting among thousands of brands, flavors, and products. Yet nearly all these options originate from the same raw materials: corn, soy, wheat, and sugar produced by a handful of corporations. The global diet has been homogenized to an astonishing degree: over 70% of the world's calories now come from just twelve plant and five animal species. Diversity of taste masks the monotony of substance.

This illusion of variety is sustained by marketing rather than agriculture. Advertisements equate processed food with freedom, identity, and progress. A soda becomes "happiness," a fast-food meal becomes "belonging." Behind the imagery lies the stark biology of addiction: sugar, caffeine, and artificial flavors combine to create psychological dependency and physical inertia.

As communities lose local farms, bakeries, and butchers, food sovereignty collapses. The human relationship with land becomes abstract. Children grow up thinking milk comes from cartons, not animals; vegetables from stores, not soil. This amnesia serves the system perfectly. A population disconnected from its food is easier to manipulate.

The Economics of Substitution

The more tangible story is agricultural economics itself. Industrial agriculture operates as a system of substitution: natural processes that once required time, care, and biodiversity are replaced by faster, cheaper alternatives. Sunlight becomes fertilizer. Fermentation becomes flavor chemistry. Pasture becomes feedlot. Labor becomes machinery.

These substitutions are only possible because the cost of true depletion—soil erosion, water pollution, chronic disease is externalized. The supermarket price hides the ecological and biological debt accumulating downstream. A $3 loaf of bread or a $1 hamburger appears affordable, but its true cost—measured in health deterioration, antibiotic resistance, and loss of soil carbon—is catastrophic.

Inflation magnifies this substitution by eroding purchasing power. As real food becomes expensive, populations turn to cheaper, ultra-processed alternatives. Governments, under pressure to prevent hunger, subsidize the very commodities that perpetuate disease. In this sense, fake food is a policy instrument: a method to stabilize societies by providing calories without nutrition, energy without vitality.

The Myth of Volume

The argument that industrial monoculture is necessary to "feed the world" and particularly to feed Africa is one of the most persistent justifications for chemically dependent, high-yield agriculture. It sounds compassionate, scientific, and urgent. Yet it rests on a flawed assumption: that hunger is primarily a problem of insufficient volume. In reality, the world already produces more than enough calories to nourish the global population. What is missing is not food, but access, resilience, and nutritional integrity.

According to the Food and Agriculture Organization of the United Nations, global agriculture produces sufficient calories to feed more than ten billion people—well above the current world population. Hunger persists not because food is scarce, but because it is poorly distributed, wasted, exported, or converted into industrial commodities. Roughly one-third of all food produced globally is lost or wasted, while the remaining supply is increasingly funneled into ultra-processed products, biofuels, and livestock feed rather than local nourishment.

High-volume monoculture systems excel at producing export crops, corn, soy, wheat, and sugar, but these are not the foods that historically sustained healthy human populations. They are calorie crops, not nutrient crops. When such systems are exported to developing regions, they often displace traditional farming practices that were locally adapted, nutritionally diverse, and ecologically resilient. The result is dependence rather than food security.

Across sub-Saharan Africa, for example, the expansion of industrial agriculture and imported processed foods has coincided with a rapid rise in non-communicable diseases. Rates of obesity, type 2 diabetes, and hypertension, once rare, are increasing fastest in urban and semi-urban populations undergoing dietary transition. This shift has occurred not because people are eating too little, but because they are eating more industrial calories and fewer real foods.

The promise of volume also obscures a deeper vulnerability: loss of control and independence. Industrial agriculture requires purchased seeds, chemical inputs, fuel, and external financing. Farmers become price-takers in a global commodity market, forced to compete on yield rather than nourishment. Seed saving is restricted, crop diversity declines, and resilience to drought, pests, and economic shocks erodes. When supply chains break—as they routinely do during conflict, climate disruption, or financial crises—food systems optimized for volume fail quickly.

By contrast, the most effective long-term solutions to hunger have little to do with yield maximization. Clean water access, small-scale irrigation, integrated livestock, soil regeneration, and the preservation of indigenous crops consistently improve food security and nutritional outcomes. Animal-source foods, fermented staples, and locally adapted plants provide dense nutrition that calories alone cannot replace. Where these systems are supported, communities become less dependent, not more productive on paper but fragile in reality.

The tragedy is not that the world lacks food, but that it has confused abundance with nourishment. Volume has delivered cheap calories while quietly dismantling food sovereignty, ecological stability, and metabolic health. Feeding people is not a question of how much can be produced, but of who controls food systems, what is grown, and whether those systems can sustain life without constant external intervention.

The Consequences: Inflammation as a Way of Life

The human body is now inflamed by design. Chronic low-grade inflammation has become a defining feature of modern physiology. The causes—seed oils, sugar, stress, circadian disruption—are embedded in the architecture of daily life. Mitochondria, overloaded by constant nutrient excess and oxidative stress, shift from energy production to defense. Cells become locked in the biological equivalent of fear: the cell danger response.

This metabolic imprisonment mirrors our societal condition. Just as cells cannot return to repair while danger signals persist, populations cannot return to health while fed toxic food and false information. Inflammation becomes both a biological and cultural state—an ongoing alarm that never resolves. It affects not only the body but cognition. Inflammatory diets impair memory, attention, and emotional regulation. The fog of fake food becomes the fog of modern consciousness.

The outcome is visible everywhere: fatigue, anxiety, loss of focus, and dependence on stimulants. Caffeine, sugar, and processed snacks become the fuel of a tired civilization. We eat to feel awake and take drugs to sleep. Food no longer restores energy; it depletes it further.

Reclaiming the Real

The path back begins with re-education: remembering what food truly is. Real food is not a brand or a label but a relationship. It comes from living systems that require care, reciprocity, and humility. It is grown in soil that smells alive, raised in sunlight, harvested by hand, and prepared with time. Its vitality cannot be replicated in factories because it carries the energy of ecosystems—the complexity of life itself.

Eating real food is an act of rebellion in a world built on convenience and control. It is a vote for biological truth over industrial fiction. It reconnects the eater to cycles of season, land, and body. Real food does not just feed cells; it teaches them to communicate again. It tells the mitochondria that the danger is over, that healing can resume.

This return to food integrity parallels the restoration of honest money. Both are acts of grounding—anchoring human systems back to reality. Sound money protects value; real food protects vitality. When both are corrupted, society drifts toward artificiality and decay. When both are restored, energy returns—metabolic, moral, and civilizational.

Local Food and the Restoration of Living Systems

Local and regenerative farming is not simply a nostalgic return to small-scale agriculture; it is a biological correction. When food is grown close to where it is eaten, farming shifts from an extractive model to a relational one. The farmer is no longer managing commodities for distant markets, but stewarding soil, animals, water, and people within a shared ecosystem. This proximity restores accountability—between land and grower, grower and community, and food and health.

Regenerative farms rebuild soil by prioritizing life below ground: diverse plant cover, microbial richness, deep root systems, and the integration of animals. Instead of stripping nutrients and replacing them synthetically, these systems regenerate fertility through natural nutrient cycling. The result is soil that holds minerals, retains water, and produces food with greater micronutrient density. This is not theory; it is observed repeatedly in soils managed with cover crops, rotational grazing, composting, and minimal disturbance.

The health benefits extend beyond the field. Communities supported by local farms gain access to fresher food harvested at peak ripeness, often consumed within days rather than weeks. Nutrients degrade with time, transport, and storage; local food shortens that gap dramatically. Just as important, local food systems rebuild cultural knowledge: how to cook, preserve, ferment, and share food, skills that anchor metabolic health across generations.

There is also a social metabolism at work. Local farms keep food dollars circulating within the community, supporting farmers, tradespeople, processors, and local markets rather than distant corporations. This economic recirculation strengthens resilience, reduces dependency on fragile supply chains, and reconnects people to the land that sustains them. Health improves not only because the food is better, but because communities regain agency over nourishment itself.

The Environmental Opportunity We Keep Missing

Modern environmental discourse often focuses on abstract metrics—carbon offsets, global targets, and distant timelines—while overlooking the most immediate and effective lever available: soil regeneration. Healthy soil is not just an agricultural concern; it is an environmental infrastructure system. When soil is alive, it becomes a sponge, a filter, and a stabilizer all at once.

Regenerative soils absorb and retain significantly more water than degraded soils, reducing runoff during heavy rains and mitigating both floods and droughts. Instead of water rushing across compacted fields into streams and rivers—carrying sediment, fertilizer, and chemical residues—living soil slows water, filters it, and allows it to recharge groundwater. This directly improves local water quality, protects rivers and lakes, and reduces downstream flooding.

Soil structure also determines resilience to extreme weather. Deep root systems anchor soil during storms, preventing erosion and preserving topsoil—the thin living layer upon which all terrestrial life depends. Where industrial agriculture leaves fields bare and vulnerable, regenerative systems keep the ground covered year-round, protecting it from wind, rain, and heat.

These benefits extend far beyond farms. Healthier watersheds mean cleaner drinking water, lower municipal treatment costs, restored fisheries, and more stable ecosystems. Regenerative practices reduce reliance on synthetic fertilizers and pesticides, cutting nutrient runoff that fuels algal blooms and dead zones. In this sense, soil regeneration is not a niche agricultural practice, it is a public health and environmental intervention.

The irony is that many climate narratives overlook these tangible solutions in favor of distant, centralized strategies. Yet rebuilding soil carbon, restoring water cycles, and supporting local regenerative farms produce immediate, measurable benefits—now, not decades from now. They heal landscapes and communities simultaneously, without requiring new technologies or complex global agreements.

The Future of Food and the Memory of the Earth

The question "What are we eating?" is ultimately philosophical. We are eating the byproducts of an ideology that values control over cooperation, quantity over quality, and appearance over essence. We are consuming the residues of systems that have forgotten the language of life. Every processed bite contains a hidden story of soil depletion, corporate manipulation, and metabolic confusion.

Yet the memory of real food persists. It lives in traditional cuisines, regenerative farms, and the biochemical intelligence of our bodies, which still recognize the difference between nourishment and noise. Even after decades of degradation, the human metabolism remains resilient. When given real food—unprocessed meats, wild fish, fresh vegetables, fermented foods, natural fats—it begins to recover astonishingly fast. Inflammation subsides, energy returns, cognition clears. The body remembers.

Reclaiming real food is therefore not nostalgia; it is regeneration. It is how the earth heals through us and how we heal through the earth. When we choose food grown in living soil, we vote for the restoration of the biosphere and the renewal of our own biology. Every meal becomes a small act of reconciliation between humanity and the planet that feeds it.

THE SWEET DECEPTION: HOW AMERICA LOST ITS WAY

"It is difficult to get a man to understand something when his salary depends on his not understanding it."
UPTON SINCLAIR

A Nation Searching for Answers

To understand how America came to fear fat, embrace sugar, and lose its metabolic footing, we must take a long look back—not simply at scientific studies or dietary charts, but at the social climate, scientific culture, and political pressures that shaped modern nutrition. The story does not begin in a laboratory or government office. It begins in a nation confronting a frightening, poorly understood epidemic: the sudden and dramatic rise of heart disease.

At the turn of the twentieth century, heart attacks were uncommon enough that many physicians rarely encountered them. By the 1950s, however, coronary heart disease had become the leading cause of death in the United States. Seemingly healthy men collapsed in their offices, on golf courses, or in their homes. Newspapers carried frequent reports of executives and politicians dying suddenly from "coronary events." Medical journals were filled with confusion, theories, and speculation. Something in modern life, it seemed, was quietly killing Americans.

Then, in September 1955, panic gripped the nation as President Dwight D. Eisenhower, a war hero revered for his leadership of the Allied victory in World War II, suffered a major heart attack while in office. For weeks, the country watched anxiously as updates trickled out of the hospital. Television networks interrupted programming for medical bulletins. If even Eisenhower, a figure of discipline, power, and military resilience could be struck down, what hope was there for the average American?

One detail, known but rarely discussed publicly, hovered in the background: Eisenhower was a heavy smoker, reportedly consuming multiple packs of unfiltered cigarettes a day. By the early 1950s, studies in major medical journals had already linked smoking to coronary artery disease. But in 1955, tobacco companies were powerful advertisers; they funded scientific conferences, influenced medical publications, and shaped public opinion. Their economic weight cast a long shadow. In that climate, few dared to say aloud that the Commander-in-Chief's cigarettes might have contributed to his heart failure. Instead, attention turned toward another potential culprit— something far less politically risky to blame.

Keys, Yudkin, and the Fork in the Road

Into this climate of fear stepped Dr. Ancel Benjamin Keys, a charismatic physiologist with a forceful personality and a talent for persuasive storytelling. Keys was convinced he had identified the primary cause of heart disease: dietary saturated fat. The logic seemed simple: saturated fat raised cholesterol, cholesterol clogged arteries, clogged arteries caused heart attacks. It was a hypothesis that had not yet been rigorously tested, but in a nation desperate for answers it was enough to ignite a movement.

Keys set out to prove his theory through what became known as the Seven Countries Study, a large epidemiologic project comparing diets and heart disease rates across populations. Countries like Japan and Greece, which had lower saturated fat intake, appeared to have lower rates of heart disease. Countries with higher saturated fat intake, such as the United States and Finland, appeared to have higher rates. Keys published his findings with confidence bordering on certainty.

But the story was not so simple. At the time, Keys had access to data from 22 countries, even though only seven were ultimately emphasized in the published analysis. Critics later argued that when broader datasets were considered, the neat correlation between saturated fat and heart disease weakened. Nations like France and Switzerland consumed high levels of saturated fat yet had comparatively low rates of heart disease. Greece—featured in Keys' study as a low-fat exemplar—had undergone economic hardship during the data-collection period and was eating far less food overall.

Years later, investigative journalist Gary Taubes and others would point out that Keys had selectively chosen countries that supported his hypothesis, minimizing those that complicated it. Yet in the 1950s and 1960s, his narrative prevailed. Simple answers often triumph over complex realities. Keys offered an enemy—fat—and in a world desperate for clarity, villainizing fat felt reassuring.

By the early 1960s, Keys' influence was so strong that he convinced major institutions, including the American Heart Association (AHA), to endorse dietary recommendations targeting fat as the primary cause of heart disease. His star was rising. And the nation was about to follow him down a path that would reshape its food system and metabolic health for generations.

While Keys dominated American nutrition debates, British physiologist John Yudkin was studying the same problem—but from a different angle. Yudkin suspected that sugar, not saturated fat, was a major driver of heart disease and metabolic dysfunction. His research suggested that rising sugar consumption was linked to metabolic disruption and worsening cardiovascular markers.

In 1972, Yudkin published *Pure, White, and Deadly*, a book that presented decades of research and issued an early warning: sugar was becoming ubiquitous, habit-forming, and metabolically destructive. But his message threatened powerful interests. Keys mocked him publicly and dismissed his work as "science fiction." Funding bodies withdrew support. His department was reorganized in ways that pushed him to the margins. And while Yudkin's warnings were being buried, the sugar industry was orchestrating a calculated counterattack.

In 1967, the Sugar Research Foundation secretly paid Harvard researchers $6,500—equivalent to roughly $55,000–$60,000 today—to publish two influential review papers that downplayed the risks of sugar and redirected blame toward saturated fat. These papers were presented as objective scholarship from one of the world's premier institutions, shaping national dietary opinion for decades and helping cement the low-fat, high-sugar paradigm that would later devastate public health.

By the time Yudkin retired, he believed he had failed. His life's work had been dismissed, his credibility undermined, and his conclusions buried beneath the rising tide of nutritional dogma. It would take decades, and a growing epidemic of metabolic disease, before the broader scientific conversation began circling back toward his warnings.

Policy Turns into a Nationwide Experiment

The AHA officially adopted Keys' view in 1961. For the first time, Americans were told that saturated fat was dangerous—that butter, eggs, red meat, and full-fat dairy should be limited or avoided. The recommendations rested on a chain of assumptions:

- saturated fat raised total cholesterol
- cholesterol caused heart disease
- lowering saturated fat would reduce heart attacks

These claims were plausible, but they were not proven at the level required to justify a nationwide dietary shift. Several cardiologists warned that the evidence was too weak. But Keys, with his commanding presence and relentless advocacy, overshadowed dissenting voices.

Food manufacturers, ever sensitive to trends and to their own bottom line seized the opportunity. Replacing natural fats with sugar, refined carbohydrates, and industrial seed oils was not only convenient, it was far cheaper. These ingredients cost pennies compared to butter, tallow, or cream, allowing companies to cut production costs while marketing their products as healthier. And so a new era of "low-fat" processed foods was born, an era that improved corporate margins even as it quietly undermined public health.

But when fat is removed from food, something crucial is lost—not only flavor and texture, but satiety and metabolic stability. To compensate, companies added sugar. And with that shift, America's metabolic crisis didn't arrive as a headline. It arrived as a new daily pattern.

The next major turning point came in 1977, when the Senate Select Committee on Nutrition, led by Senator George McGovern, released the Dietary Goals for the United States. These guidelines, written quickly and with limited metabolic expertise, declared that Americans should:

- reduce overall fat consumption
- especially reduce saturated fat
- increase carbohydrate intake
- eat more grains
- limit cholesterol

Scientific debates were ongoing, but political momentum was unstoppable. Policymakers wanted clarity, not nuance. The guidelines became the foundation of national nutrition policy. In 1992, these ideas were distilled into the now-famous USDA Food Pyramid, with grains at the base and fats at the top—the exact opposite of what metabolic science would later show supports stable energy and satiety. America followed these guidelines faithfully. And its health unraveled.

Low-Fat Foods, High-Profit Calories

The low-fat era ushered in an unprecedented expansion of industrialized foods. Grocery store aisles became filled with products engineered to meet low-fat criteria while maximizing consumer appeal, including yogurt sweetened into dessert, margarine made from chemically altered seed oils, fat-free cookies brimming with refined starch and syrup, and cereals coated in sugar and labeled "heart healthy." These products were marketed as scientific breakthroughs. In reality, they were the beginning of a metabolic catastrophe.

By stripping away natural fats and replacing them with sugar and refined carbohydrates, the food industry created products that spiked insulin, destabilized blood sugar, encouraged overeating, and fueled chronic inflammation. For the first time in history, children developed fatty liver disease. Type 2 diabetes appeared in teenagers. Obesity soared. Chronic illnesses multiplied. **Americans were eating less fat than ever before. Yet they were sicker than ever before.** And the next part of the story reveals why these foods became so addictive and so dominant.

The Tobacco Takeover of the American Food Supply

As America embraced low-fat guidelines, an even more consequential shift was happening behind the scenes—one that would permanently alter the nation's relationship with food.

By the late 1960s, tobacco companies faced mounting lawsuits, growing scientific condemnation, and increasing public skepticism. Cigarette executives realized that the era of unchecked profits was ending. They needed a new industry—one where they could apply decades of expertise in addiction science, behavioral conditioning, and sensory engineering. They found it in America's grocery stores.

Philip Morris, maker of Marlboro, launched one of the most aggressive acquisition campaigns in the history of American food. Their timeline reads like the corporate equivalent of a military conquest:

- 1970: Philip Morris acquires Miller Brewing Company.
- 1985: Philip Morris buys General Foods for $5.6 billion (Jell-O, Kool-Aid, Maxwell House, Post cereals).
- 1988: Philip Morris purchases Kraft for $12.9 billion— the largest non-oil corporate takeover in American history at the time.
- 1989: General Foods and Kraft merge to form Kraft General Foods, instantly becoming the largest food company in the United States.

By the early 1990s, one of the world's biggest cigarette manufacturers controlled a massive percentage of the American food supply. And they didn't just provide capital. They provided strategy. Philip Morris installed tobacco-trained executives across Kraft General Foods. Men like Geoffrey Bible, who rose from the cigarette side of the company, eventually became CEO of the entire corporation. Tobacco research scientists were rotated into food product development. They brought with them expertise in creating sensory experiences designed to stimulate cravings—skills honed in the laboratories that refined nicotine delivery.

R.J. Reynolds (RJR) took a similar path. In 1982, RJR acquired Nabisco Brands, forming RJR Nabisco, which controlled Oreos, Chips Ahoy, Ritz, Wheat Thins, and dozens of other snack foods. Many Americans never realized that their cookies and crackers were being engineered by the same corporate scientists who optimized menthol cigarettes.

Tobacco executives directed product strategies. Marketing methods perfected in the cigarette industry: such as emotional branding, flavor testing, and habit reinforcement were applied to food.

The government's war on fat created the perfect opportunity for tobacco-owned food corporations. When fat was removed from processed foods, flavor disappeared. The solution was sugar—cheap, unlimited, legal, and profoundly stimulating to dopamine pathways. Sugar became the replacement for nicotine. Tobacco engineers applied the same behavioral science used to keep smokers hooked:

- "bliss point" formulas to maximize craving
- vanishing caloric density, ensuring foods melt quickly on the tongue
- flavor layering to override satiety signals
- branding strategies designed to create emotional attachment

Entire categories of children's foods including cereals, snacks, and lunchbox products were reformulated under the influence of tobacco-trained sensory engineers. The companies that once optimized nicotine delivery now optimized sugar delivery. This was not a dietary shift. It was a business model. And it reshaped the American palate, metabolism, and disease landscape.

Why the Sweet Deception Lasted

By the time evidence began contradicting the diet–heart hypothesis, the low-fat paradigm was deeply entrenched. It persisted not because it was scientifically secure, but because a powerful ecosystem protected it.

Once the AHA and government guidelines warned against fat, manufacturers saw opportunity. Removing fat from food was inexpensive, and replacing it with sugar, starch, and industrial oils was even cheaper. These ingredients improved shelf life, reduced cost, and increased profit. Sugar was not only cheap—it stimulated repeat buying and reinforced cravings. With tobacco companies owning major food corporations, addiction engineering became normal in food formulation.

Scientific inertia also played a role. Entire careers had been built on the belief that fat caused heart disease. To admit error would mean acknowledging misdirected funding, flawed recommendations, and decades of public messaging built on weak foundations. Nutrition science, influenced by industry and politics, resisted reversal.

Government pride kept the message stable even as evidence evolved. Once national policy declared fat harmful, reversing course would have implied incompetence or negligence. Bureaucracies protect their recommendations, even when the science changes.

Public misunderstanding sealed the narrative. Fat is easy to visualize. It looks like body fat. Sugar does not. It dissolves invisibly, tastes pleasant, and creates no immediate sensory warning. The myth that "fat makes you fat" was simple and intuitive—even if it was metabolically incorrect. And once an idea becomes intuitive, it stops feeling like a hypothesis and starts feeling like common sense. By the time better evidence emerged, the culture wasn't just misinformed—it was trained to defend the story it already believed.

And the media amplified the low-fat message because it was easy to package. Nuance was lost. Complex physiology was reduced to a simple villain and a simple rule. Dissenting scientists were portrayed as fringe, even when they were simply questioning the evidence. Advertising dollars reinforced the story. This feedback loop kept the low-fat narrative alive long after the scientific foundations began to crumble, and long after the health outcomes failed to improve.

When Medical Education Followed the Narrative

The most lasting damage wasn't only what the public believed—it was what doctors were trained to repeat. Once the low-fat framework entered policy, it didn't remain a dietary suggestion. It became a professional baseline. Medical students learned to view saturated fat as a primary driver of cardiovascular risk. Diet history was filtered through that lens. Patients were told to switch to "heart-healthy" grains, avoid red meat, choose skim milk, and treat fat as a hazard requiring constant discipline.

And when a message becomes institutional, it gains authority that science alone can't compete with. Clinical guidelines shape medical board exams. Board exams shape medical training. Medical training shapes the reflexive advice given in every primary care office in the country. In that environment, the low-fat model wasn't just a hypothesis—it was the default script.

Meanwhile, the population did not get healthier. Obesity rates rose. Type 2 diabetes became common. Metabolic syndrome moved from rare to routine. The public was doing what it had been taught to do: avoiding fat, trusting "healthy carbs," and choosing packaged foods engineered to remain palatable even after fat was removed. But the lived results didn't match the promise. The guidelines stayed. The slogans stayed. The labels stayed. And a generation of physicians watched their waiting rooms fill with patients growing sicker, despite following the very advice that was supposed to prevent it.

This is the cruelty of a locked-in narrative: once it is installed into education and policy, it takes decades to unwind—because reversing it requires more than new data. It requires admitting that an entire system taught the wrong fear.

What It Cost and How We Reclaim the Narrative

By the early 2000s, cracks in the diet–heart hypothesis had grown too large to ignore. Major population studies found weaker links between saturated fat intake and heart disease than the public had been led to believe. Randomized trials showed low-fat diets often performed worse in weight loss and metabolic outcomes than expected.

New research highlighted the real drivers of metabolic disease:
- insulin resistance
- chronic inflammation
- excessive sugar consumption

- refined carbohydrate intake
- metabolic inflexibility

Looking back, the public wasn't just misled, it was given the wrong target. Saturated fat became the villain while the modern industrial diet was treated as background noise. Studies with weak design and messy confounding were framed as proof, even when the "evidence" was built from meals like a cheeseburger combo—where the bun, fries, ketchup, and soda likely did most of the metabolic damage. The message was simple: blame the fat. But reality was not. Traditional societies consuming high levels of natural saturated fats—such as the Maasai or Inuit—did not show the same chronic disease patterns seen in populations eating modern processed diets. And as the evidence matured, meta-analyses in major medical journals repeatedly failed to confirm saturated fat as the singular threat it was claimed to be.

The scientific tide shifted. But the damage was already done. Decades after his warnings were ignored, Yudkin's conclusions resurfaced in the scientific literature. Nearly all of his predictions about sugar, insulin, and metabolic illness, were ultimately echoed by later research. His story now serves as a moral compass: science does not always reward truth in real time. Sometimes, truth must wait for the evidence to overwhelm the institutions that once rejected it.

The rise and persistence of the low-fat dogma is not simply a scientific error—it is a lesson in human fallibility. It reveals how deeply flawed incentives, political pressures, industry influence, and media simplification can distort scientific understanding. When dissent is mocked, when evidence is ignored, and when institutions entrench themselves instead of self-correcting, science loses its integrity. And the public pays the price.

Today, nearly 93% of American adults exhibit some degree of metabolic dysfunction. That number is not the result of individual weakness or a collapse in discipline. It is the predictable outcome of a system that reshaped the food landscape, redefined health messaging, and replaced biological wisdom with industrial convenience.

When Americans were told to fear fat, they weren't merely nudged toward different foods—they were pushed into the arms of sugar, refined grains, industrial seed oils, and hyper-processed products. These foods did not emerge innocently. They were engineered by industries whose survival depended on cheap, shelf-stable inputs replacing calories once provided by nutrient-dense fats.

This shift was so profound that entire generations grew up believing that "healthy eating" meant avoiding butter but embracing fat-free flavored yogurt, margarine, breakfast cereals, granola bars, and low-fat cookies. These foods were marketed as virtuous, but they were metabolically destabilizing: loaded with sugar, stripped of fiber, reinforced with industrial oils, and devoid of the satiety that real food provides. To understand how dramatically the landscape shifted, consider the everyday foods that filled American kitchens from the 1970s onward:

- Fat-Free SnackWell's Cookies
- Low-Fat Yoplait Yogurt
- Breakfast cereals promoted as "healthy" foundations
- Margarine and seed-oil spreads replacing butter
- Fat-free salad dressings sweetened with syrups
- Granola bars and energy bars marketed as wholesome
- Fruit-flavored snacks and juices sold as "real fruit"

Through these products, Americans unknowingly shifted from a biologically compatible diet to one that hijacked metabolic signaling. Meals that once contained natural fats, proteins, and fiber were replaced with engineered foods designed for bliss-point appeal—foods that elevate insulin, intensify hunger, and drive repeat consumption rather than satisfaction.

The result was inevitable. Without dietary fat to stabilize hunger and slow digestion, blood sugar surged and crashed throughout the day. Hunger intensified. Snacking became normalized. Insulin levels—never meant to remain constantly elevated—rose chronically. The body shifted into fat-storage mode. Over time, insulin resistance deepened, and with it came the cascade of metabolic disorders that now define modern chronic disease. This was not evolution failing, it was the food environment collapsing.

The modern human physiology is not built for a diet dominated by refined sugar and refined grains. Nor is it designed to handle industrial seed oils as a daily default. Combined, these changes created a metabolic environment the body struggles to resolve: the system becomes overwhelmed, cells become resistant to insulin's signals, energy regulation falters, fat accumulates in the liver, and appetite control becomes unstable.

It is no coincidence that metabolic dysfunction skyrocketed in parallel with the rise of low-fat dietary guidelines and the explosion of processed foods designed to replace fat with sugar.

The Sweet Deception was not a minor scientific misstep, it was a generational detour, a moment when institutions designed to protect public health inadvertently harmed it. What began as a contested hypothesis was soon absorbed into academic research, public policy, medical education, and professional guidelines—until it became less like advice and more like doctrine. For decades, Americans followed recommendations grounded not in metabolic science, but in a rushed consensus built on incomplete evidence, simplified headlines, and the false comfort of a single dietary villain. Now, in 2026, we are still living inside the consequences of that story. And when fat became the enemy, sugar, refined grain, and industrial oils took its place—quietly, profitably, and destructively.

Chronic disease was not the result of individuals failing their diets. It was the natural consequence of diets failing the individual.

Today, we have the opportunity to correct course. Understanding this history empowers us to question inherited nutritional beliefs, reevaluate public health dogma, recognize the metabolic damage caused by processed foods, and restore trust in traditional, whole-food dietary patterns.

The truth John Yudkin tried to warn the world about still echoes today: Sugar—not fat—was the quiet architect of America's health crisis. And the first step toward healing is seeing the deception clearly. Once we see it, we cannot unsee it. And once we understand the metabolic cost of this deception, we are finally free to reclaim our health, our choices, and our future.

CHAPTER EIGHTEEN

THE FACTORY OILS

"If man made it, don't eat it."
JACK LALANNE

A Quest for Truth in Nutrition

The story of how vegetable oils came to dominate our diet is one of industrial influence, flawed science, and public health missteps. In recent years, investigative journalism has played a crucial role in uncovering the truth about these oils, revealing how we were misled into believing they were healthier than traditional fats like butter and lard. This research has challenged long-standing beliefs and exposed the unintended consequences of replacing real, natural fats with factory-processed vegetable oils.

This narrative takes us back through history to explore how vegetable oils, once a byproduct of the agricultural industry, became a dietary staple. The journey uncovers the aggressive marketing, lobbying, and scientific controversies that led to widespread changes in our diet with serious long-term implications. What began as an industrial solution eventually became a public health recommendation—and in many homes, a default ingredient.

The story is important not because fat is dangerous, but because the kind of fat we consume matters. And seed oils did not rise because they were superior foods. They rose because they were profitable, scalable, and easy to embed into almost everything we eat.

From Byproduct to Staple: The Rise of Factory Oils

The origins of vegetable oils can be traced to the early 20th century, a period when American agriculture was rapidly industrializing. As food production became more efficient, companies sought new ways to utilize byproducts, one of which was cottonseed. Initially considered waste, cottonseed found new life when it was discovered that oil could be extracted from it.

In the late 19th century, an enterprising company saw an opportunity to market this cheap oil as a food product. By hydrogenating the oil, they were able to turn it into a solid fat that mimicked the properties of butter and lard. This innovation led to the creation of Crisco, the first vegetable shortening, which was heavily marketed as a healthier, more modern alternative to traditional animal fats.

The success of Crisco paved the way for other seed oils: soybean, corn, safflower, and sunflower to enter the food market. These oils were inexpensive to produce and had a longer shelf life than animal fats, making them attractive to both manufacturers and consumers. Thus began the vegetable oil revolution, which would profoundly alter the American diet.

What makes this shift so consequential is not simply that a new cooking fat appeared. It's that an industrial input began replacing foods humans had relied on for generations. Traditional fats were not inventions. They were part of the food chain. Seed oils, on the other hand, were manufactured products extracted, refined, and stabilized to suit industrial needs.

Over time, the food supply didn't just change its fat source. It changed its entire purpose. Food became something designed to ship, to store, to sell, and to scale—rather than something designed to nourish.

Economics and Scale: Why Seed Oils Won

The dramatic rise of vegetable oils in the 20th century was driven not only by marketing and emerging nutritional theories, but by powerful economic incentives. Compared to traditional fats like butter, lard, tallow, and coconut oil, seed oils were extraordinarily cheap to produce. Animal fats required land, livestock, feed, labor, and refrigeration. By contrast, seed oils allowed manufacturers to turn agricultural byproducts: cottonseed, soybeans, corn germ into an inexpensive, shelf-stable commodity.

Industrial extraction, solvent processing, and chemical refining meant companies could generate enormous quantities of oil at a fraction of the cost of natural fats. Once these processes became standardized, seed oils transformed from industrial waste streams into some of the most profitable ingredients in modern food manufacturing.

This cost advantage reshaped the food economy. Manufacturers, restaurants, bakeries, and processed-food companies rapidly adopted these oils to cut expenses, extend shelf life, and improve margins. Long before the public was told these oils were "heart healthy," industry had already chosen them for financial reasons.

The shift was never driven by nature or nutrition—it was driven by economics. And once the infrastructure was built, it created a feedback loop. The more seed oils were produced, the more products were formulated around them. The more products depended on them, the more the industry had to defend them. This is how an ingredient becomes an institution.

The consumer didn't ask for this transition. It arrived through price incentives and convenience, slowly becoming unavoidable. What began in processed foods eventually made its way into nearly everything.

Hidden in Plain Sight: How Seed Oils Took Over

As production scaled up, seed oils began infiltrating nearly every corner of the American food system. Today, they appear in tens of thousands of products—even items consumers assume contain no added fats. Chips, crackers, breads, cereals, granola bars, cookies, pastries, frozen dinners, sauces, condiments, salad dressings, mayonnaise, coffee creamers, plant-based milks, and even baby foods commonly rely on seed oils.

Restaurants, especially fast-food chains fry nearly everything in soybean, canola, corn, or cottonseed oil because these fats are far more cost-effective than traditional animal fats. They perform predictably in deep fryers and can be purchased in bulk at low cost.

Even foods marketed as "healthy" or "natural" frequently contain these oils. Many consumers don't realize that roasted nuts, hummus, canned fish, and organic dressings often include seed oils as their primary fat source. The industry also obscures their presence under a wide range of labels: vegetable oil, soybean oil, canola oil, sunflower oil, safflower oil, cottonseed oil, corn oil, rice bran oil, grapeseed oil, high-oleic blends, shortening, margarine, hydrogenated oils, or simply "expeller-pressed vegetable oil."

The result is that many people consume significant quantities of seed oils every day—often unknowingly. And because these oils are embedded inside processed foods, they rarely appear as a conscious choice. They become a background exposure repeated, daily, and cumulative.

What Foods Contain Seed Oils

Seed oils are often thought of as something used for frying or found in obvious junk food, but many of the biggest exposures come from foods people assume are neutral, healthy, or barely processed. In many cases, the oil is not added for nutrition, but for texture, shelf stability, flavor consistency, and manufacturing efficiency.

One of the most common hidden sources is restaurant-prepared food, even when it doesn't taste greasy. Grilled meats and vegetables are often cooked on surfaces seasoned with seed oils. Rice, stir-fry dishes, fajita vegetables, sautéed sides, and "light" menu items frequently rely on soybean or canola oil behind the scenes, not because the meal demands it, but because commercial kitchens default to the cheapest fat.

Prepared deli and convenience items are another major category. Many consumers don't realize that rotisserie chicken, pre-marinated meats, seasoned shredded chicken, deli salads (such as potato salad, macaroni salad, coleslaw, and chicken salad), and "grab-and-go" meal kits often contain seed oils as emulsifiers, carriers for seasoning blends, or part of the dressing base, even when they don't look like processed food.

Seed oils also show up heavily in spreadable and "easy" foods that replace traditional cooking. Butter substitutes, whipped spreads, flavored cream cheese blends, cooking sprays, and squeeze-bottle "butter" or "garlic" toppings are frequently built on canola, soybean, sunflower, or blended oils designed to pour smoothly and stay stable at room temperature.

Many people are also surprised to learn how often seed oils appear in packaged flavor systems—foods where the oil isn't the main ingredient, but the delivery mechanism. Spice packets, taco seasonings, powdered sauces, "instant" noodle flavor packs, dry rubs, and shelf-stable marinades often contain vegetable oils or oil-based anti-caking and flavor-binding agents. The oil isn't there to "add fat"—it's there to make the product behave predictably.

Another overlooked category is "health" products sold as lifestyle upgrades. Many keto-labeled snacks, protein chips, collagen snack bars, meal shakes, plant-based protein powders with "creaminess," and functional wellness foods use sunflower, safflower, or canola oils to improve mouthfeel and shelf stability. The label reads clean, but the formulation still relies on industrial fats.

Finally, seed oils increasingly appear in restaurant and packaged desserts that don't seem oily: coffeehouse pastries, flavored whipped toppings, non-dairy foamers, "iced coffee" drinks, and ready-to-drink beverages that use stabilizers and fat blends for texture. In these products, seed oils are often present not as cooking oils, but as engineered ingredients designed to deliver consistency.

This matters because seed oils are not merely "added fat." They often arrive bundled inside foods that are already highly manufactured, where the oil is part of a broader engineered system designed for craving, convenience, and repeat consumption rather than satisfaction.

A World Without Seed Oils: Before 1900

One of the most striking historical facts is that before the 20th century, humans consumed virtually no industrial seed oils. For thousands of years, cultures around the world relied on natural fats: tallow, lard, butter, ghee, coconut oil, palm oil, olive oil, and the fats naturally present in meat and dairy. These fats had nourished human physiology across centuries.

In earlier eras, many chronic metabolic and degenerative disease patterns were far less common—and rarely seen at the scale we accept today. The introduction of seed oils marks one of the most drastic dietary shifts in human history. Never before had people consumed chemically refined oils rich in polyunsaturated linoleic acid as a routine part of daily intake.

The modern rise of chronic disease did not happen in a vacuum. It rose alongside industrial food production, refined carbohydrate intake, and the mass adoption of factory-made oils. The pattern has repeated often enough to warrant serious attention.

Global Uptake and the Parallel
Rise of Chronic Disease

As Dr. Chris Knobbe and other researchers have shown, a global pattern appears to emerge: wherever seed oils entered the food supply and traditional diets were replaced by modern processed foods, chronic illness tended to follow within one to two generations. The United States experienced a sharp increase in metabolic dysfunction after seed oil consumption rose dramatically between 1900 and 2000. India saw similar trends in the 1970s–1990s as traditional ghee was increasingly replaced with cheap industrial oils. Japan and China followed a similar trajectory after World War II as modernization accelerated and processed foods spread.

Conversely, populations that maintained traditional diets—such as the Maasai, Inuit, and Tokelauans tended to experience fewer modern chronic illnesses until industrial oils and processed foods entered their environment. This does not prove a single ingredient caused every outcome, but it reinforces an important point: the human body does not thrive when factory-made oils become unavoidable background exposure.

This doesn't mean every traditional society ate the same foods or followed the same macronutrient ratios. But it does suggest that the modern experiment—the combination of refined starch, sugar, industrial oils, and ultra-processed products—has consequences that appear across geography, culture, and genetics.

Omega-6 Linoleic Acid: A Shift of Unprecedented Scale

The central component of seed oils is linoleic acid (LA), an omega-6 polyunsaturated fat that the human body requires only in small amounts. Ancestral diets typically supplied around 1–3% of calories from LA. Today, the average American may consume 15–20% of calories from LA, representing a 5 to 10 fold increase. Few nutritional changes in history approach this magnitude.

Linoleic acid can accumulate in adipose tissue and persist for long periods, meaning its metabolic impact may be chronic and slow-moving rather than immediate. As LA levels rise, they can alter cellular membranes, disrupt hormone signaling, impair satiety, and increase vulnerability to oxidative stress.

This is not merely about calories. It is about signaling. It is about structure. It is about how the body interprets the environment and how it allocates energy and repair.

Long before linoleic acid became part of the nutrition vocabulary, the broader pattern was already being observed. In the 1930s, dentist and researcher Dr. Weston A. Price traveled to isolated traditional populations around the world and documented what he saw when diets remained rooted in whole, locally sourced foods. He reported low rates of tooth decay, strong physical development, and wide dental arches—signs that seemed to reflect deep nutritional sufficiency.

What stood out even more was what happened when modernization arrived. Price observed that when refined commercial foods entered these communities, particularly white flour, sugar, and industrially produced fats—dental decay rose rapidly, and markers of physical degeneration became more common in the next generation. Price could not measure fatty acid composition the way modern researchers can, and his work cannot isolate one ingredient as the sole cause. But the direction of change was consistent: when traditional diets were displaced, health outcomes declined.

Dr. Knobbe's work can be seen as an extension of this same warning, one now measured through metabolic pathways and biochemical data. It suggests that the rise of seed oils is not a small dietary adjustment. It represents a foundational change in the fats that build our cells and fuel our physiology.

Mechanism: Oxidation, Mitochondria, and Metabolic Damage

The problem isn't fat, it's instability. Seed oils are rich in polyunsaturated fats, which are more prone to oxidation than traditional cooking fats, especially under heat and repeated exposure.

When linoleic acid is heated, metabolized, or oxidized, it breaks down into reactive byproducts, OXLAMs (oxidized linoleic acid metabolites), 4-HNE, and MDA, among others. These compounds can damage cell membranes, proteins, enzymes, and DNA. They also attack cardiolipin, a mitochondrial fat essential for the electron transport chain.

Once cardiolipin becomes enriched with omega-6 fatty acids, mitochondria can become less stable, energy production can falter, and oxidative stress can increase. Over time, this may contribute to insulin resistance, impaired fat metabolism, chronic inflammation, and the metabolic breakdown that defines so much of modern disease.

This is why Knobbe and others argue that seed oils are not merely empty calories. They are metabolically active compounds that can shape cellular signaling, oxidative stress, and long-term metabolic resilience.

Diseases Linked to Excess Seed Oil Exposure

A growing body of evidence associates high seed-oil intake particularly in the context of processed diets with a wide range of chronic conditions, including heart disease, obesity and metabolic syndrome, type 2 diabetes, fatty liver disease, hypertension, autoimmune conditions, neurodegenerative disorders, and age-related macular degeneration (AMD).

Notably, AMD was rarely documented before the widespread adoption of industrial fats. Its rise tracks closely with dietary modernization, offering one of the clearest examples of how diet can influence degenerative disease patterns over time.

The point is not that seed oils alone explain every illness. The point is that the modern food environment changed multiple variables at once, refined sugar, refined flour, industrial fats, hyper-palatable processing, and constant availability. Seed oils may represent one of the most significant chemical shifts inside that larger system.

Industry, Guidelines, and the
Selling of "Heart Healthy"

The rise of vegetable oils was not merely a consequence of technological innovation. It was also the result of strategic marketing and lobbying by the food industry. Companies behind these oils invested heavily in advertising campaigns to convince the public that vegetable oils were healthier than traditional fats. They funded scientific research, supported health organizations, and used their influence to shape public opinion.

A significant turning point came in the mid-20th century when major health organizations began to endorse vegetable oils as part of a heart-healthy diet. These endorsements were often grounded in studies that suggested a link between saturated fat and heart disease—studies that did not always distinguish between different types of fats or control adequately for broader dietary patterns.

The result was a broad cultural message: saturated fat is harmful, cholesterol is dangerous, and vegetable oils are the solution. Traditional fats were vilified. Processed replacements were elevated. And the most industrially refined fats in the food supply were recast as nutritional upgrades.

Shifts in dietary guidelines soon followed. The U.S. government, influenced by major health organizations and industry-backed narratives, issued recommendations encouraging reduced saturated fat intake and substitution with polyunsaturated fats, primarily from vegetable oils.

This shift in dietary advice had profound implications. Traditional fats like butter, lard, and tallow were pushed out of kitchens, while vegetable oils became the foundation of cooking and food manufacturing. Processed foods, increasingly built on these oils, became ubiquitous. Yet despite widespread adoption, metabolic health continued declining, forcing many researchers to revisit the assumptions that shaped these policies.

What "Processing" Actually Means: How Vegetable Oils Are Made

The production of vegetable oil in modern factories is a complex process involving several stages, each designed to extract and refine the oil from seeds or plant sources. The process typically begins with cleaning and preparation of raw materials such as soybeans, sunflower seeds, or rapeseeds. Seeds are cleaned to remove dirt and impurities before being dehulled to separate the outer shell from the inner seed.

Once prepared, seeds undergo pressing or expelling, where mechanical pressure extracts crude oil. In large-scale production, pressing is often enhanced by heating, which increases yield but can also lead to the formation of undesirable compounds that must later be removed.

Crude oil contains impurities—free fatty acids, phospholipids, pigments, and other compounds. To remove these, oils undergo industrial refining stages such as degumming, neutralization, bleaching, and deodorization. Degumming uses water or acid to remove phospholipids that affect stability. Neutralization uses alkaline solutions to reduce free fatty acids. Bleaching uses activated clay to strip pigments. Deodorization applies high heat under vacuum to remove off-flavors, while also reducing natural antioxidants like tocopherols.

Some oils are also hydrogenated—chemically altered to make them more solid and shelf-stable. This process historically produced trans fats, which were later associated with increased cardiovascular risk and eventually removed from many products.

These industrial steps make seed oils usable for mass production. They create uniformity, shelf stability, and a consistent product. But they also reveal the truth: these oils are not simple foods. They are manufactured substances, optimized for storage and industrial utility.

Re-Examining Traditional Fats and the Role of Clinical Research

Recent research has played a critical role in challenging the conventional wisdom that dominated nutritional science for decades. By re-evaluating the evidence, this work has helped restore the reputation of traditional fats like butter, lard, and coconut oil, which were demonized during the rise of vegetable oil promotion.

Traditional fats have been part of human diets for millennia. They are stable, less prone to oxidation, and naturally paired with nutrient-dense foods. In contrast, the long-term effects of consuming large amounts of industrial seed oils remain debated, especially when they are consumed daily as part of a processed dietary pattern.

Over the past several decades, numerous clinical research studies have examined the health effects of vegetable oils, particularly those high in omega-6 fatty acids. Omega-6 fats play essential roles in human biology, including growth, brain development, and immune regulation. But essential does not mean unlimited. Excessive omega-6 intake—especially when omega-3 intake is low—may promote oxidative stress and inflammatory signaling.

Some researchers have raised concerns that linoleic acid can be converted into arachidonic acid, a precursor to pro-inflammatory eicosanoids. Others focus on oxidation and lipid peroxides created when polyunsaturated fats are exposed to heat, light, or repeated reuse.

These concerns do not mean every exposure to seed oils is instantly harmful. They mean the long-term, high-dose, daily saturation of these oils in modern eating patterns may carry risks that were not acknowledged when they were promoted as "heart healthy."

The debate continues because nutrition is complex. But the historical and biochemical questions remain valid. If these oils were truly superior, we would expect the modern shift toward them to improve human health. Instead, the modern era has become an age of widespread metabolic dysfunction, an outcome that should compel investigation, not dismissal.

A Call to Return to Real Food

The rise of industrially produced vegetable oils, marketed as "heart-healthy" alternatives to traditional animal fats, stands as one of the most influential nutrition shifts of the modern era. These oils were aggressively promoted as safer than whole, unprocessed fats, despite limited long-term evidence and significant industry influence.

For decades, beginning in the mid-20th century, vegetable oils like soybean, corn, and canola oil were pushed into the mainstream as "heart-healthy" alternatives. These oils, rich in polyunsaturated fatty acids, were hailed as modern miracles that could reduce cholesterol and prevent disease. But what was rarely emphasized is that these oils are highly refined industrial products. Their extraction methods involve high heat, chemical solvents, and deodorization, resulting in unstable fats prone to oxidation.

The most alarming part of this story is how easily public institutions accepted and promoted these oils as the default. Corporate influence shaped the narrative. Advertising budgets shaped consumer beliefs. Guidelines shaped behavior. And the foods humans had eaten for millennia were replaced by fats that required factories to exist.

The impact on our food system has been immense. Grocery shelves are filled with products laced with vegetable oils. Restaurant meals rely on them. Even "healthy choices" often contain them. Many people now consume seed oils daily without realizing it—not because they chose them intentionally, but because the modern food supply is built on them.

The lesson is clear: we need to return to real, whole foods—foods that nourished humanity for generations long before the rise of factory-made oils and processed food systems. Traditional fats such as butter, tallow, lard, and coconut oil deserve careful reconsideration as stable, natural sources of nourishment.

As we move forward, the history of vegetable oils should serve as a reminder that health cannot be entrusted to the food industry. We must remain skeptical of nutritional dogmas shaped by profit and reinforced by incomplete science. The best path back to metabolic health lies in the simplicity of traditional foods, free from industrial interference and grounded in human biology.

CHOLESTEROL: ESSENTIAL TO HUMAN BIOLOGY

"Form follows function"
LOUIS SULLIVAN

Cholesterol Exists Because Life Requires It

Before we can understand why certain fats destabilize metabolism, we have to revisit the molecule that was blamed for nearly everything that followed: cholesterol. For decades, it was treated as a dietary threat. In reality, it is one of the body's most essential building blocks.

Cholesterol is not a byproduct of modern diets or a malfunction of human physiology. It is a fundamental biological molecule, conserved across evolution because life depends on it. Every nucleated cell in the human body contains cholesterol, not as an optional component, but as a structural requirement.

At the cellular level, cholesterol is an essential constituent of cell membranes. It regulates membrane fluidity, ensuring that membranes are neither too rigid nor too permeable. This balance allows cells to maintain integrity while still permitting the controlled movement of ions, nutrients, and signaling molecules. Without cholesterol, cell membranes lose stability, impairing communication and basic cellular function.

The importance of cholesterol is even more pronounced in the nervous system. Roughly one quarter of the body's total cholesterol is found in the brain, where it is a critical component of myelin—the insulating sheath that surrounds neurons and enables rapid electrical signaling. Cholesterol is required for synapse formation, neuronal repair, and normal cognitive function. During development, the brain synthesizes its own cholesterol locally, underscoring how essential this molecule is to neurological structure and performance.

Cholesterol also serves as the precursor for all steroid hormones. Testosterone, estrogen, progesterone, cortisol, aldosterone, and related signaling molecules are all synthesized from cholesterol. These hormones regulate reproduction, stress response, immune modulation, blood pressure, and metabolism. Without adequate cholesterol availability, normal hormonal signaling cannot occur.

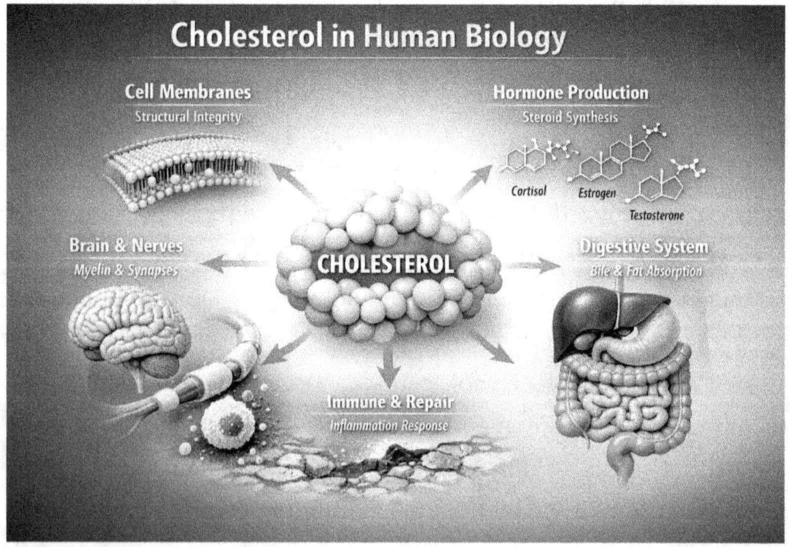

In the digestive system, cholesterol is converted by the liver into bile acids. These bile acids are necessary for the digestion and absorption of dietary fats and fat-soluble vitamins, including vitamins A, D, E, and K. These vitamins play indispensable roles in vision, bone health, immune function, blood clotting, and cellular protection. Impaired bile production compromises nutrient absorption regardless of caloric intake.

Cholesterol also serves as the molecular foundation for vitamin D synthesis. When ultraviolet light interacts with cholesterol-derived compounds in the skin, vitamin D is produced—a hormone-like molecule involved in calcium regulation, immune defense, and gene expression across multiple tissues.

Crucially, the human body tightly regulates cholesterol because it is so essential. The majority of cholesterol present in the bloodstream is synthesized internally, primarily by the liver, rather than absorbed directly from food. This endogenous production reflects biological priority: cholesterol availability is maintained even during periods of dietary scarcity because cellular function depends on it.

Cholesterol as Transport, Repair, and Defense

Cholesterol does not exist in isolation within the bloodstream. It is transported as part of complex particles known as lipoproteins, which move fats, fat-soluble nutrients, and cholesterol between tissues according to need. These particles, often reduced in public discussion to simplistic labels are best understood as delivery and response systems, not toxins.

Low-density lipoproteins (LDL) are a primary means by which cholesterol is delivered from the liver to peripheral tissues. Cells require cholesterol for membrane maintenance, hormone synthesis, and repair, and LDL particles supply this demand. When tissues are growing, stressed, inflamed, or injured, the need for cholesterol increases. Rising LDL levels in the blood often reflect this adaptive redistribution, not an inherent pathological process.

Cholesterol also plays a direct role in tissue repair. Cell membranes are routinely damaged by mechanical stress, inflammation, and oxidative processes. Cholesterol contributes to membrane resilience and repair, helping cells restore structural integrity. In this way, cholesterol behaves less like a passive bystander and more like a response molecule, mobilized when the body is under strain.

Beyond structural repair, cholesterol participates in immune defense. LDL particles have been shown to bind and neutralize bacterial toxins, including lipopolysaccharides released by certain pathogens. These endotoxins can trigger systemic inflammation and severe immune reactions if left unbuffered. By sequestering them, LDL particles help limit inflammatory damage and reduce the severity of immune responses during infection.

This protective role becomes most visible during acute illness. Observational studies have repeatedly shown that individuals with higher cholesterol levels often experience better outcomes in serious infections, including sepsis. In these contexts, cholesterol appears to function as part of the body's innate defense system, mitigating inflammatory overload while supporting cellular repair.

These functions help explain why cholesterol levels frequently rise during periods of illness, injury, or physiological stress. Rather than signaling danger on their own, elevated cholesterol levels may indicate that the body is actively responding to increased demands for repair, transport, and immune protection.

Cholesterol Across the Human Lifespan

Human biology is not static. The demands placed on the body change with growth, aging, stress, and repair, and cholesterol metabolism adapts accordingly. What appears abnormal when viewed through a single numerical lens often makes sense when considered across the lifespan.

In early development, cholesterol requirements are exceptionally high. Rapid cell division, brain growth, and nervous system maturation depend on abundant cholesterol availability. The developing brain synthesizes cholesterol locally because of its critical role in neuronal structure, synapse formation, and myelination. This early reliance underscores a fundamental principle: cholesterol is indispensable wherever growth and organization are taking place.

As adulthood progresses, cholesterol continues to support hormone production, tissue maintenance, and immune readiness. The body regulates cholesterol dynamically, increasing production and transport when repair or stress demands rise. Periods of physical strain, illness, or inflammation often coincide with higher circulating cholesterol levels, reflecting adaptive response rather than dysfunction.

In later life, cholesterol takes on an even more pronounced protective role. Aging tissues experience increased wear, slower repair, and greater vulnerability to infection. At the same time, hormone production declines and immune resilience becomes more fragile. Cholesterol supports all three of these domains: structural repair, hormonal signaling, and immune defense, making its availability increasingly important rather than expendable.

Large population studies examining older adults consistently show that higher LDL cholesterol levels are not associated with increased mortality and in many cases are linked to greater longevity. These findings challenge the assumption that cholesterol exerts uniform risk across all ages. Instead, they suggest that cholesterol may serve as a marker of physiological robustness in later life, supporting resilience rather than undermining it.

This age-related shift helps explain why aggressive cholesterol lowering in elderly populations often fails to deliver the expected benefits and may, in some cases, correlate with poorer outcomes.

Biology prioritizes repair and defense as systems age, and cholesterol appears to be woven into that strategy.

Seen through this lens, cholesterol is not a static risk factor but a context-sensitive molecule, rising and falling in accordance with biological need. Lifespan patterns reveal a consistent theme: when growth, repair, or immune support are required, cholesterol availability increases to meet the demand.

When Cholesterol Signals Dysfunction

Cholesterol becomes associated with disease not because it is inherently harmful, but because certain metabolic conditions alter how it behaves in the body. In these states, cholesterol reflects underlying dysfunction rather than acting as an independent cause. Key markers and conditions that shift cholesterol from adaptive to problematic include:

- **Chronic elevation of blood glucose:** Persistent high blood sugar promotes glycation of LDL particles, altering their structure and impairing normal clearance from circulation.
- **Insulin resistance:** Reduced insulin sensitivity disrupts lipid handling in the liver, increasing production of triglyceride-rich lipoproteins and altering LDL particle characteristics.
- **LDL glycation:** When glucose binds to LDL particles, they become more prone to oxidative damage and inflammatory interaction within arterial walls.
- **Oxidative stress:** Excess reactive oxygen species overwhelm antioxidant defenses, increasing the likelihood that LDL particles will oxidize and provoke immune responses.
- **Elevated triglycerides:** High triglyceride levels signal impaired fat metabolism and excess carbohydrate intake, often accompanying the formation of small, dense LDL particles.
- **Small, dense LDL predominance:** These particles circulate longer, penetrate arterial walls more easily, and are

more susceptible to oxidation compared to larger, buoyant LDL particles.

- **Chronic inflammation:** Inflammatory states increase vascular vulnerability and amplify immune responses to modified lipoproteins.

Viewed together, these markers describe a metabolic environment under strain. In such conditions, cholesterol becomes entangled in pathological processes, not as the initiating factor, but as a participant responding to systemic dysfunction. When glucose regulation is intact, triglycerides are low, inflammation is controlled, and oxidative stress is minimized, cholesterol fulfills its normal roles in transport, repair, and defense without contributing to disease.

Real Food and Cholesterol Function

Cholesterol does not function in isolation. Its role in the body is inseparable from the quality of the foods that accompany it. When food is biologically intact providing fats, proteins, vitamins, and minerals in their natural forms cholesterol metabolism tends to reflect balance, resilience, and repair.

Traditional whole foods supply fats in configurations the human body has long adapted to use. These fats support stable energy delivery, hormone synthesis, and the absorption of fat-soluble vitamins. Vitamins A, D, E, and K rely on dietary fat for absorption, and cholesterol plays a central role in transporting and utilizing these nutrients at the cellular level. Without adequate fat intake, these vitamins cannot perform their regulatory and protective functions.

Nutrient-dense foods also support favorable lipid patterns. When meals are built around whole fats rather than refined carbohydrates, triglyceride levels tend to remain lower, HDL cholesterol higher, and LDL particles larger and more buoyant. In this metabolic context, cholesterol transport remains orderly, responsive, and functional rather than congested or inflammatory.

Real food environments also reduce the conditions that distort cholesterol behavior. Stable blood sugar, lower insulin demand, and reduced oxidative stress limit LDL modification through glycation and oxidation. Cholesterol remains structurally intact and available for its primary roles in membrane maintenance, immune defense, and tissue repair.

Equally important is what real food does not require. It does not depend on chemical stabilization, excessive refinement, or synthetic fortification to appear nourishing. Foods that retain their natural fat content tend to deliver nutrients together, in proportions that align with human digestion and metabolism. Cholesterol functions best within this integrated nutritional matrix.

In this setting, cholesterol is neither suppressed nor exaggerated. It responds appropriately to the body's needs—rising during periods of growth, repair, or stress, and stabilizing when demand subsides. Rather than acting as a threat, cholesterol reflects the state of metabolic alignment between food and biology.

Closing Reflection: Function Over Fear

Cholesterol has been framed for decades as a problem to solve rather than a function to understand. Yet when examined within the broader context of human biology, its role is neither mysterious nor malevolent. Cholesterol is present wherever structure must be maintained, signals must be transmitted, and repair must occur. Its persistence across evolution reflects necessity, not error.

When food aligns with biology, cholesterol behaves predictably and purposefully. It supports membranes, hormones, immunity, and resilience without disruption. When metabolic systems are overwhelmed by excess sugar, chronic inflammation, or nutrient dilution, cholesterol becomes a marker of imbalance rather than its cause. The molecule has not changed; the environment has.

This distinction matters. It shifts attention away from suppression and toward context. It invites a deeper understanding of how food, metabolism, and repair interact over time. Cholesterol does not operate in isolation, and neither does health.

Seen clearly, cholesterol is not something the body struggles against, but something it relies upon. It moves where it is needed, responds when demand increases, and recedes when balance is restored. Its function is quiet, structural, and essential—revealed most clearly when the body is supported rather than stressed.

With this understanding, the narrative of fear gives way to one of design. Cholesterol becomes not a warning sign, but a reminder: health emerges when biology is respected, and when nourishment supports the body's innate capacity to maintain itself.

PART V

Metabolic Disease Through a New Lens

CHAPTER TWENTY

DIABETES: HOW WE GOT THE STORY WRONG

"Type 2 diabetes is not a disease of too much sugar, but too much insulin."

JOSEPH KRAFT

For much of modern medical history, diabetes has been explained, treated, and taught through an increasingly narrow lens. What began as a complex disorder of energy handling gradually became defined almost entirely by blood sugar levels and the tools used to control them.

This shift did not occur because physicians were careless or uninformed, but because each major advance—dietary restriction, insulin therapy, glucose monitoring—solved an immediate crisis while quietly reshaping how the disease itself was understood. Over time, the deeper metabolic story faded from view.

To understand why diabetes came to be managed rather than questioned, we must trace how well-intentioned discoveries slowly redirected attention away from physiology and toward numbers on a chart.

A Century and a Half of Diabetes:
How Our Understanding Evolved

Understanding how diabetes came to be viewed through such a narrow lens requires a brief walk through history. Over the past 150 years, diabetes research progressed from careful clinical observation, to anatomical discovery, to hormonal intervention, and eventually to biochemical insight. Each era revealed an important piece of the puzzle. What follows is not a catalog of names to memorize, but a map of how scientific understanding evolved—and how certain insights were elevated while others quietly faded from view.

In the late eighteenth and early nineteenth centuries, diabetes was approached primarily through observation at the bedside. John Rollo, a Scottish surgeon and physician, was among the first to recognize that diabetes was closely tied to diet and fuel handling rather than being merely a vague wasting disease. In 1797, while serving as Surgeon-General to the Royal Artillery, he published *An Account of Two Cases of the Diabetes Mellitus*. Drawing from patients with what we would now recognize as rapidly progressive, insulin-deficient diabetes, Rollo observed that symptoms worsened dramatically with bread, grains, and other carbohydrate-rich foods. His response—a strict low-carbohydrate, high-fat, animal-based diet—was radical for its time. Rollo's legacy is simple but profound: diabetes was treatable through diet long before it was treatable through drugs. His work reflects an early understanding of diabetes as a disorder of fuel, an insight that would later be obscured rather than expanded.

As anatomy and pathology advanced through the nineteenth century, attention shifted from diet to organs. In 1869, Paul Langerhans identified distinct clusters of cells within the pancreas that would later bear his name. Two decades later, Oskar Minkowski and Joseph von Mering demonstrated that removal of the pancreas produced severe diabetes in animals. These experiments firmly established the pancreas as central to glucose regulation and set the stage for the hormonal era that followed.

The discovery of insulin in 1921 marked one of the most dramatic turning points in medical history. Frederick Banting, Charles Best, John Macleod, and James Collip transformed type 1 diabetes from a fatal diagnosis into a survivable condition almost overnight. Children who once lived only months now lived for decades. This triumph left an indelible imprint on medicine. Insulin became the treatment, and glucose became the target. Around the same time, Elliot P. Joslin emphasized disciplined glucose tracking, patient education, and structured care—an approach that dramatically improved survival while further reinforcing a glucose-centered framework.

Yet even in the early twentieth century, cracks in this simplified model began to appear. In the 1930s, Harold Himsworth demonstrated that not all diabetes was the same, distinguishing insulin-sensitive from insulin-insensitive forms—what we now recognize as type 1 and type 2 diabetes. Around the same time, Wilhelm Falta proposed that insulin resistance, not merely insulin deficiency, played a central role in many patients. These insights hinted at deeper metabolic complexity, but they struggled to gain traction in an era dominated by insulin's undeniable success.

The biochemical era of the mid-twentieth century finally provided tools to examine what had previously been invisible. In the 1950s, Rosalyn Yalow and Solomon Berson developed radioimmunoassay, allowing insulin levels to be measured for the first time. Their discovery was startling: many individuals with type 2 diabetes had *high* insulin levels, not low ones. This overturned the long-standing assumption that diabetes was simply a disease of insulin deficiency. Soon after, Philip Randle introduced the glucose–fatty acid cycle, showing that glucose and fat compete as fuels within the cell. Excess fat oxidation could suppress glucose use, reframing insulin resistance as a problem of fuel competition rather than hormonal failure. Decades later, J.

Denis McGarry expanded this work, arguing that dysregulated fat metabolism, not glucose metabolism, lay at the core of diabetes.

By the late twentieth century, these threads began to converge. In 1988, Gerald Reaven introduced "Syndrome X," later known as metabolic syndrome, identifying insulin resistance as the common link between diabetes, hypertension, and heart disease. Diabetes was no longer an isolated disorder of blood sugar, but part of a broader cardiometabolic disturbance.

At the same time, a parallel movement unfolded outside traditional academic pathways. Beginning in the 1970s, Richard K. Bernstein, an engineer living with type 1 diabetes, demonstrated that tight glucose control was achievable through frequent monitoring and carbohydrate restriction. Long before continuous glucose monitors or large clinical trials, Bernstein showed that reducing carbohydrate intake dramatically lowered insulin requirements and stabilized blood sugar. His advocacy helped bring home glucose monitoring into mainstream care and challenged decades of conventional dietary advice. Though controversial at the time, his work foreshadowed much of what modern metabolic research would later confirm.

Despite these accumulating insights, modern diabetes care has remained largely focused on glucose control rather than metabolic repair. New medications, revised guidelines, and sophisticated monitoring technologies have improved short-term management, yet diabetes prevalence has continued to rise. The upstream drivers—chronic hyperinsulinemia, impaired fat metabolism, and metabolic overload—have often remained secondary considerations. The result is a field rich in tools, but still searching for a unifying framework.

Seen as a whole, the history of diabetes is not a story of ignorance, but of incomplete integration. The necessary pieces were discovered over generations. What has been missing is the willingness to assemble them into a coherent metabolic picture—one that explains diabetes not as a failure of sugar control, but as a disorder of energy handling, fuel congestion, and lost metabolic flexibility.

Before Insulin: Starvation as Treatment

Long before insulin reshaped the landscape of diabetes care, physicians faced a grim and heartbreaking reality. Diabetes, especially what we now recognize as type 1 was almost always fatal. With no way to lower blood sugar and no understanding of insulin's role, the disease progressed relentlessly. Families watched loved ones waste away while physicians worked with little more than observation and intuition. In that world, treatment was not about restoring health; it was about delaying death. It was under these conditions that starvation emerged as the dominant therapeutic strategy.

In the early 1910s, Frederick M. Allen became the leading authority on diabetes treatment. Through experiments on pancreatectomized dogs, animals that reliably developed severe diabetes after removal of the pancreas—Allen observed a consistent pattern. When carbohydrates were introduced, blood sugar rose sharply. When carbohydrates were restricted, blood sugar fell. From these observations, he concluded that carbohydrates were the central driver of the disease.

Allen extended this logic further. If carbohydrates worsened diabetes, then perhaps they should be eliminated entirely. And if the diabetic body struggled to handle any incoming fuel, perhaps all fuel needed to be restricted. From this reasoning emerged starvation therapy, a regimen designed to minimize glucose exposure at almost any cost.

In practice, Allen's approach was severe. Patients were often limited to extremely low caloric intake for prolonged periods, with minimal carbohydrate, fat, or protein. Extended semi-fasting became routine. The immediate goal was simple: keep blood sugar low enough to prolong life. In the short term, this approach sometimes worked.

Glucose levels fell. Urine sugar decreased. But the price was profound physiological fragility.

Children undergoing starvation therapy were often pulled from school because they lacked the strength to sit upright for a full day. Adults who once performed physically demanding work became exhausted by simple movement. Families measured food with clinical precision, knowing that a single additional meal could raise blood sugar dangerously. These were not diets designed to sustain life; they were strategies designed to postpone death.

Allen was not wrong in his core observation. Carbohydrates do raise blood sugar, and restricting them does improve short-term glucose control. But his framework missed a crucial distinction. He did not recognize that fat does not raise blood glucose and does not significantly increase insulin requirements. Fearful that any fuel might overwhelm a diabetic metabolism, Allen restricted fat alongside carbohydrates, overlooking the possibility that fat could have provided a stable source of nourishment when glucose handling was profoundly impaired.

Equally important, Allen applied the same treatment to all forms of diabetes. Children with complete insulin deficiency were treated identically to adults with insulin resistance, two conditions that behave very differently. Without the ability to measure insulin or understand fat metabolism, diabetes appeared to be a single disease with a single solution. The deeper heterogeneity of diabetes remained hidden.

The human cost of this narrow view was immense. Starvation therapy often meant trading the immediate danger of hyperglycemia for the slow deterioration of malnutrition. Patients lost weight rapidly. Muscle mass disappeared. Energy vanished. Even when blood sugar stabilized, the body remained chronically undernourished. Parents faced impossible decisions: follow medical orders and watch their child grow thinner each day, or offer food knowing it might hasten death.

Diaries from the era describe the anguish of children begging for food that parents felt powerless to provide.

Allen's work cemented a belief that would persist for decades—that diabetes was fundamentally and exclusively a carbohydrate problem. Carbohydrates raised blood sugar; therefore carbohydrates were the enemy. This conclusion was understandable in its historical context, but it was incomplete. Diabetes was not merely a failure of carbohydrate handling. It reflected a deeper disturbance in how the body managed energy.

With no insulin and no understanding of fatty acid metabolism, starvation was the only life-prolonging strategy available. In that context, it made sense. But its unintended legacy was profound. It embedded a lasting fear of food, and particularly fat into diabetes care. Even today, echoes of the starvation era remain in the belief that people with diabetes must fight hunger endlessly and accept deprivation as treatment.

The tragedy is not that starvation was tried. It is that it became confused with cure. Starvation was never the solution—it was simply the best option available before medicine understood how energy is meant to move through the human body.

The Fat Thread We Lost

As Frederick Allen's starvation method came to dominate early diabetes treatment, a quieter and more hopeful line of investigation was unfolding in laboratories and clinics. A small group of researchers rejected the idea that people with diabetes were destined to live half-alive under chronic deprivation. Instead, they asked a different question—one that would prove far ahead of its time: what if the problem was not food itself, but the type of fuel the body was being forced to rely on? That question led them toward fat as a stabilizing, insulin-sparing energy source and away from starvation as therapy. Their discoveries challenged starvation itself—and pointed toward a metabolic solution that could have reshaped diabetes care. History, however, moved in another direction.

In the early 1920s at the University of Michigan, Louis Newburgh began noticing clinical patterns that did not fit the prevailing narrative. Many of his patients with diabetes were overweight, at a time when obesity was uncommon, and some showed clear metabolic dysfunction despite measurable insulin. Newburgh began to suspect that the problem in many cases was not a lack of insulin, but a failure to respond to it. This was an early intuition of insulin resistance, articulated decades before the term entered medical language. More importantly, he observed that patients improved not when calories were indiscriminately slashed, but when carbohydrates were reduced and fat intake increased. Energy intake mattered, but energy type mattered more. In an era shaped by starvation logic, this was a radical departure.

Working alongside Newburgh, Phil Marsh helped conduct some of the first structured clinical trials using high-fat, low-carbohydrate diets in people with diabetes. Their 1925 study demonstrated that shifting calories toward fat rather than carbohydrate produced more predictable blood sugar control, improved insulin sensitivity, higher daily energy, and meaningful symptom relief, often without starvation or severe weight loss. Their work challenged a deeply embedded assumption: that people with diabetes could not be trusted with adequate nourishment. Instead, it suggested that the real issue was a mismatch between fuel and metabolic capacity. Carbohydrates demanded insulin many patients could not use effectively; fat did not. In many ways, their findings anticipated modern low-carbohydrate and ketogenic approaches by nearly a century.

Around the same time, physician Rollin Woodyatt was uncovering a discovery that would become foundational across multiple fields of medicine. Woodyatt observed that when carbohydrates were sharply restricted and fat intake increased, the body produced ketone bodies, an alternative fuel capable of sustaining the brain and other organs without high insulin demand. He recognized ketones not as markers of pathology, but as evidence of an evolutionarily conserved energy system—one designed for times when glucose was scarce or poorly tolerated. In the context of diabetes, this insight was revolutionary. It suggested that energy access was still possible, not through deprivation, but through fuel substitution.

Building on Woodyatt's work, Russell Wilder at the Mayo Clinic and Karl Petren in Sweden developed structured, high-fat ketogenic diets for diabetic patients. These were not improvised regimens, but carefully designed therapeutic protocols. Patients experienced reduced glucose fluctuations, lower insulin requirements, improved physical strength, and restoration of daily functioning. Wilder coined the term *ketogenic diet*, recognizing ketone production as a therapeutic feature rather than a side effect, while Petren refined macronutrient ratios to maximize metabolic stability. This was metabolic therapy in its infancy—fuel substitution instead of starvation. Fat did not destabilize these patients; it steadied them.

The discovery of insulin in 1921 changed everything. For the first time, children who would have died within weeks could live for years. The medical community rightly celebrated insulin as one of the greatest breakthroughs in history. But this miracle carried an unintended consequence. The emerging insight that many people with diabetes tolerated fat better than carbohydrates, and that fat-based diets could stabilize metabolism, was quietly sidelined. Insulin made carbohydrates tolerable again, and with that shift, carbohydrate caution faded. Fat began to be viewed with suspicion, and the nuanced understanding of fuel selection was eclipsed by the power of a single hormone. The "fat thread" woven by Newburgh, Marsh, Woodyatt, Wilder, and Petren was overshadowed by insulin's success, even though the metabolic truths they uncovered never stopped being true.

In retrospect, these early researchers illuminated principles that align closely with modern metabolic science. Diabetes was not simply high blood sugar, but impaired access to fuel. Carbohydrates flooded systems already struggling with insulin signaling, while fat provided steady energy with minimal insulin demand. Ketones offered an alternative fuel when glucose handling was compromised. Energy itself was not the enemy, the mismatch between fuel and metabolic capacity was. Their work offered a glimpse of what metabolic medicine might have become. Though overshadowed, the science remained intact, waiting for a later generation to rediscover it.

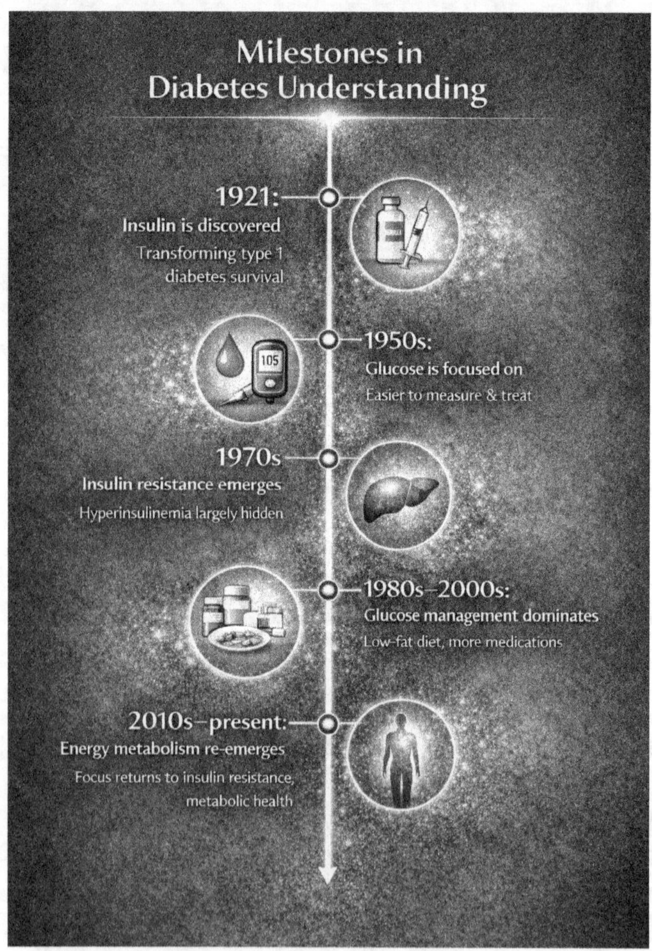

Insulin: Miracle and Turning Point

The discovery of insulin in 1921 marks one of the most dramatic transformations in the history of medicine. From our vantage point today, it is difficult to fully grasp the emotional force of that moment. Before insulin, children diagnosed with what we now recognize as type 1 diabetes often survived only months. Families were instructed to prepare for loss, not for life. Physicians could offer little more than starvation diets—approaches that reduced blood sugar by reducing energy intake to the brink of collapse. Survival was measured in weeks gained, not futures restored.

Then, almost overnight, everything changed. In the winter of 1922, a fourteen-year-old boy named Leonard Thompson lay dying in a Toronto hospital. He was severely malnourished, dehydrated, and slipping away despite the best available care. After receiving the first experimental injections of insulin purified by James Collip, Leonard revived with astonishing speed. His blood sugar fell. Strength returned. Color came back to his face. Energy followed. Nurses reportedly wept. Physicians were stunned. Word spread rapidly across North America and Europe: a fatal disease could now be controlled.

Insulin did more than prolong life—it restored it. Children returned to school. Adults returned to work. Families reclaimed futures they had been told were impossible. The celebration was not only understandable; it was deserved. Insulin remains one of the greatest therapeutic breakthroughs in medical history. Yet like all powerful discoveries, it reshaped not only treatment, but thinking itself—introducing consequences that were not immediately visible.

The scientific achievement behind insulin was extraordinary. Frederick Banting, a young surgeon with an unrelenting idea, partnered with Charles Best, an eager medical student, under the guidance of John Macleod at the University of Toronto. Through months of painstaking experimentation, isolating pancreatic extracts, testing repeatedly in diabetic dogs, refining techniques—they succeeded where others had failed. Insulin could lower blood glucose with remarkable precision. For clinicians, it felt almost magical. A single injection could reverse life-threatening hyperglycemia. A child drifting toward coma could be revived within hours. Patients who had wasted away on starvation diets regained weight and vitality.

But insulin did more than save lives. It altered the therapeutic logic of diabetes care. Before insulin, dietary strategies centered on limiting carbohydrate exposure and, in some cases, relying more heavily on fat—approaches that unintentionally leaned toward metabolic flexibility. After insulin, the emphasis shifted toward matching carbohydrate intake with insulin dosing. Food no longer needed to be avoided; it could be covered. Insulin transformed diabetes from a disease of fuel restriction into a disease of fuel tolerance. In doing so, it quieted much of the curiosity that had begun to emerge about *why* carbohydrates were so problematic in the first place.

Not everyone was convinced this shift was without risk. Frederick M. Allen, whose starvation method had defined the pre-insulin era, welcomed insulin as a breakthrough but warned against misinterpreting its power. In his later writings, Allen emphasized that insulin did not abolish the disease itself. He cautioned that while insulin permitted some carbohydrate intake, it was "not a license for indulgence." Allen feared that insulin's effectiveness would tempt physicians to rely on it as a nutritional hall pass—masking metabolic dysfunction rather than addressing it. Although his own therapeutic approach had been deeply flawed, his warning proved prescient.

That caution, however, was soon eclipsed by optimism. Elliot P. Joslin, the most influential diabetologist of the early 20th century, embraced insulin as a means of restoring normalcy to patients' lives. Having witnessed the physical and emotional toll of starvation diets, Joslin wanted diabetes care to feel humane again. He promoted a "balanced diet," encouraging the inclusion of carbohydrates so long as insulin dosing was adjusted appropriately. He championed glucose monitoring, patient education, and self-management, foundations that still shape diabetes care today.

Yet this push toward dietary normalcy carried an unintended cost. Insulin worked so well that it obscured deeper metabolic questions. Why were patients so intolerant of carbohydrates to begin with? What role did chronically elevated insulin, fatty acid dysregulation, and early insulin resistance play—particularly in adults who were not insulin-deficient? These questions faded into the background as the field focused increasingly on glucose control rather than metabolic context.

As insulin made carbohydrates safer to consume, they also became easier to justify. The stage was set for the gradual liberalization of carbohydrate intake—a shift that would profoundly influence dietary guidelines, clinical practice, and the future trajectory of both diabetes and heart disease.

Geyelin and the Liberalization of Carbohydrates

Rawle Geyelin, an American physician who witnessed the earliest insulin successes, became one of insulin's most enthusiastic champions. He saw something few doctors in history had ever witnessed: children revived from the brink of death, parents clinging to hope instead of preparing for loss, and patients who had endured years of starvation finally eating again. To Geyelin, insulin was not merely a treatment—it was a passport back to life.

Before insulin, even modest amounts of food carried fear. Bread was forbidden. Fruit was dangerous. Desserts were unimaginable. Meals were rationed with the gravity of last rites. Food was not nourishment; it was risk. Insulin shattered that world almost overnight. Patients reintroduced foods they had long been denied—bread, potatoes, fruit, even sweets on occasion. Families described this transition as an emotional resurrection: the return not only of calories, but of pleasure, normalcy, and shared meals. Geyelin embraced this transformation wholeheartedly. In his eyes, insulin had liberated people with diabetes from the tyranny of restriction.

From this optimism emerged a new philosophy—one that would shape diabetes care for the next century. Though never codified as a formal guideline, it became an unspoken rule in clinics and hospitals: eat freely, and adjust insulin accordingly. The message spread through nursing instruction, dietetic education, and patient counseling. It promised simplicity and relief after decades of deprivation. For patients desperate to feel normal again, it was irresistible.

But beneath this promise lay a metabolic cost that was not yet understood. Reintroducing carbohydrates required higher insulin doses, and rising insulin levels carried predictable consequences. Weight gain became common. Hunger intensified. Fat storage accelerated. Insulin resistance quietly deepened. Over time, many patients required escalating doses of insulin to manage the same amount of carbohydrate—a vicious cycle in which the very hormone that saved lives began reshaping metabolism in destabilizing ways. Blood sugar appeared controlled on the surface, while the underlying physiology deteriorated beneath it.

Geyelin never intended harm. He was offering emotional relief, not metabolic restoration. Yet his enthusiasm helped eclipse earlier insights showing that many people with diabetes functioned better on lower-carbohydrate, higher-fat diets that required far less insulin and delivered greater metabolic stability. A generational belief took hold: if insulin could lower glucose, then carbohydrates were no longer the problem. It was a comforting idea and it would dominate diabetes management for decades. But comfort is not the same as healing.

Not everyone shared this unqualified optimism. Charles Best, Frederick Banting's collaborator, understood that insulin was more than a glucose-lowering agent. He recognized its broader effects on hunger, fat storage, liver metabolism, and vascular health. Best cautioned against using insulin as a shortcut around metabolic reality.

He worried that excessive reliance on insulin could promote weight gain, worsen glucose volatility, and contribute to long-term metabolic instability. His call was for balance: insulin used in partnership with thoughtful nutrition, not as a license to ignore physiology.

Best's concerns proved prescient. Insulin was undeniably a miracle—one of the greatest in medical history—but it also ushered in a new therapeutic mindset. Over time, the belief that carbohydrates were harmless so long as insulin was available shaped medical education, hospital nutrition protocols, professional guidelines, and public messaging. Carbohydrates returned to the center of the diabetic diet. Insulin doses rose. Insulin resistance deepened quietly in the background. In many cases, the treatment itself began fueling disease progression.

The tragedy was never insulin. Insulin saved millions of lives. The tragedy was the belief that insulin alone could normalize a metabolism that had lost its ability to process fuel efficiently. Insulin enabled survival, but it did not restore metabolic flexibility. It did not correct fatty acid dysregulation or relieve cellular fuel congestion. And as the medical community grew confident in its ability to control glucose, it gradually stopped asking the most important question of all: why the body was struggling to use its energy in the first place.

That question would not re-emerge until researchers began to examine insulin resistance, fatty acid metabolism, and the metabolic traffic jam at the core of diabetes—opening the door to a very different understanding of the disease.

How Diabetes Became a Sugar Disease

By the time the twentieth century drew to a close, diabetes had come to be understood almost entirely through the lens of blood sugar. High glucose was the defining feature, the diagnostic threshold, and the primary treatment target. This focus did not arise from neglect or error, but from history itself. Glucose was visible, measurable, and actionable. Insulin saved lives by lowering it. Dietary guidance evolved around managing it. Over time, the success of glucose control hardened into certainty: diabetes was a disease of sugar, and the solution was to keep sugar down.

Yet this narrow framing concealed a deeper reality. Long before glucose rises into the diagnostic range, the body is already struggling. Energy crashes after meals. Hunger returns too quickly. Weight accumulates despite effort. Fatigue becomes a daily companion. These early signals are not simply about sugar spilling into the blood. They reflect a system losing its ability to access and move energy efficiently. The bloodstream fills not because fuel is excessive, but because it cannot be delivered where it is needed. Sugar becomes elevated not as the cause of disease, but as its most visible symptom.

This is how diabetes became misnamed. A disorder of energy handling was reduced to a disorder of glucose measurement. Treatment focused on controlling what could be seen, while the forces driving the disease—rising insulin, impaired fat metabolism, and lost metabolic flexibility—remained largely invisible. The history traced in this chapter shows how that happened: how lifesaving discoveries, cultural relief from deprivation, and technological limits gradually narrowed the question from *Why is this body struggling to use energy?* to *How do we lower blood sugar?*

To understand diabetes fully—and to understand why it so often progresses into heart disease and other chronic conditions—we must now step beyond glucose and into metabolism itself. That deeper story begins where history leaves off.

CHAPTER TWENTY-ONE

DIABETES AS AN ENERGY DISORDER

Starving in the Midst of Plenty

*"Cells can be surrounded by nutrients
and still be starving."*
METABOLIC PRINCIPLE

What we call diabetes, heart disease, fatty liver, and many other chronic conditions are not separate problems with separate causes. They are different surface expressions of a shared breakdown: energy handling that no longer works smoothly across tissues. The liver becomes overfilled and begins exporting surplus fuel into the bloodstream. Muscle becomes less responsive to the signals meant to deliver nutrients. Blood lipids drift into riskier patterns. The vascular system absorbs constant metabolic strain. Long before glucose rises into the diagnostic range, the underlying architecture is already under pressure.

This is why diabetes so often appears late in the story. Elevated blood sugar is not the opening act—it is the downstream signal of a system that has been compensating quietly for years. Long before a diagnosis, the body sends subtler warnings: rising insulin, persistent hunger, post-meal fatigue, creeping triglycerides, weight gain that resists effort, and a growing dependence on frequent meals. These are not random symptoms. They reflect a metabolism that is losing flexibility—less able to alternate between fuels, regulate appetite, and sustain stable energy between meals.

At its core, diabetes is not a disease of energy shortage. It is a disease of misplaced energy. Fuel accumulates in places where it disrupts function, inside liver and muscle cells and within expanding fat stores, while high-demand tissues struggle to produce steady output. Insulin rises to compensate. Hunger intensifies. Fatigue deepens. Over time, the system becomes less adaptive. Glucose rises only when the body can no longer maintain control at the edges.

In this chapter, we step beneath blood sugar and into metabolism itself. We examine how competing fuels interact inside cells, why insulin resistance can emerge as a protective response to overload, and how the same upstream disturbance can ripple outward to the heart, liver, vasculature, and brain. The mechanisms that follow—first described in the mid-twentieth century and refined over decades—provide the missing map. They explain why the liver becomes central, why fasting becomes difficult, and why restoring fat access changes the entire trajectory. Once the pattern is visible, diabetes shifts from an "inevitable decline" narrative into a biologically addressable problem—one that improves when the metabolic environment changes.

The Science Catches Up: Fuel Competition and Metabolic Congestion

For decades, clinicians could see diabetes and cardiometabolic disease unfolding long before blood glucose rose. They observed muscles that responded poorly to insulin, livers that continued releasing glucose even when insulin was high, fat tissue that stored energy aggressively while patients remained hungry, and weight gain that occurred without obvious overconsumption. The pattern was clear: energy was present, yet the body struggled to deploy it efficiently.

What medicine lacked was not observation, but explanation. The biochemical tools needed to understand how different fuels interact inside the cell were still developing. That clarity began to arrive in the mid-twentieth century, when researchers shifted their focus away from hormones alone and toward the dynamics of fuel handling inside tissues.

In 1963, British biochemist Philip Randle introduced one of the most important organizing principles in metabolic science: the glucose–fatty acid cycle. Randle demonstrated that glucose and fat do not operate as interchangeable fuels inside the cell. They compete. When circulating fatty acids remain chronically elevated, mitochondria preferentially oxidize fat, and glucose oxidation becomes suppressed—even when insulin is present.

This insight offered a coherent explanation for early insulin resistance. Muscle cells were not "broken," and insulin signaling was not initially defective. The cellular environment was already energy-saturated. Glucose uptake slowed because the intracellular machinery was occupied processing fat-derived fuel. Insulin resistance, in this context, functioned as a protective brake, a way to limit further fuel entry into tissues already operating under metabolic load.

Randle's model reframed insulin resistance as fuel competition rather than hormonal failure. It helped explain why insulin levels often rise early, and why pushing more glucose into tissues through higher insulin exposure can worsen instability rather than restore function. The issue was not a missing signal, but limited capacity for additional incoming fuel.

Nearly three decades later, this framework was extended and sharpened by biochemist J. Denis McGarry. In a landmark 1992 paper published in *Science*, McGarry argued that diabetes had been interpreted backward. Diabetes, he proposed, is not fundamentally a disorder of carbohydrate metabolism, it is a disorder of fat metabolism.

McGarry showed that chronically elevated fatty acids accumulate inside cells, where they interfere with glucose oxidation and insulin signaling. As intracellular fat rises, insulin becomes less effective, prompting the pancreas to secrete more of it. This hyperinsulinemia further reinforces fuel storage and deepens the cycle. Glucose intolerance, McGarry emphasized, is not the initiating injury, it is the late-stage signal of a system that has lost metabolic clearance.

This reframing moved the origin of diabetes from the bloodstream into the cell. The central issue was not excess glucose circulating outside tissues, but excess energy accumulating inside them. The body became metabolically full, yet functionally underpowered—starving in the midst of plenty.

Modern research has repeatedly confirmed this blueprint. Intramuscular fat strongly correlates with insulin resistance. Ectopic fat stored in the liver and pancreas disrupts regulation more than total body fat alone. Mitochondrial overload reduces flexibility in oxidizing both fats and glucose. Importantly, improving fat oxidation and clearing ectopic fat often restores insulin sensitivity even without dramatic weight loss.

Taken together, these findings clarify the sequence of dysfunction. Insulin rises early. Fat accumulates within tissues. Metabolic flexibility narrows. Only later does glucose spill into the bloodstream. Elevated blood sugar is not the beginning of diabetes, it is the visible sign that the buffering system has reached its limit.

This framework also explains the lived experience: hunger that returns quickly after meals, fatigue despite adequate calories, weight that resists effort, difficulty fasting, and fatty liver even in people who are not visibly obese. These are not failures of discipline. They are predictable consequences of metabolic congestion and impaired fuel access.

Once this architecture became visible, the next question was unavoidable: if insulin resistance reflects protective overload, how is the healthy body designed to access stored energy without triggering breakdown? Answering that required turning to a different line of research, one focused not on disease, but on normal human adaptation.

George Cahill and the Physiology of Fuel Switching

While Randle and McGarry clarified how excess intracellular fuel can suppress glucose use, another line of research answered a different but equally important question: how is the human body designed to run when insulin is low?

That question was explored most clearly by physiologist George Cahill during the 1960s and 1970s through meticulous studies of prolonged fasting and metabolic adaptation. Cahill's work restored an essential principle that modern medicine largely forgot: metabolic health depends not on constant glucose intake, but on the body's ability to transition smoothly between fuel sources as insulin rises and falls.

Under normal conditions, the fed state is brief. Insulin rises after a meal, glucose is used or stored, and fat storage is temporarily prioritized. As insulin declines between meals, the body shifts naturally toward fat-derived fuels. Fatty acids are released from adipose tissue, the liver converts a portion of them into ketone bodies, and multiple organs, especially the brain begin using these fuels to maintain stable energy availability even as dietary glucose fades.

Cahill demonstrated that this transition is not an emergency response. It is a conserved, well-regulated physiological state. *Ketones are not a sign of failure, they are evidence that the system is operating as designed:* insulin is low, stored fuel is available, and energy can be delivered without continuous feeding. In this state, glucose demand falls, fat oxidation increases, muscle protein is preserved, hunger quiets, and the body becomes metabolically efficient on its internal reserves.

Cahill's work becomes especially relevant when viewed through the lens of insulin resistance. In these states, insulin often remains elevated even when energy is abundant. Fat release becomes restricted. Ketone production is blunted. The body stays locked in a glucose-dependent mode despite carrying massive stored reserves—leaving hunger active and energy unstable not because fuel is absent, but because access is impaired.

Seen this way, insulin resistance disables the very adaptive physiology Cahill described. The problem is not that fat metabolism is dangerous or abnormal—it is that persistent insulin signaling prevents the body from entering the low-insulin state where fat-based fueling can occur. The capacity to switch fuels becomes constrained, and metabolic flexibility narrows over time.

Cahill's research completes the picture. Randle and McGarry explained how excess energy inside tissues can suppress glucose handling. Cahill showed how recovery becomes possible when insulin falls and stored fuel becomes usable again. Together, they reveal diabetes not as a failure of energy supply, but as a failure of fuel access—a system struggling to shift gears despite being fully fueled.

Why Carbohydrate Reduction Works: Restoring the Body's Natural Fuel Flow

By the time someone develops prediabetes or type 2 diabetes, metabolism is already operating under chronic insulin pressure. Fat oxidation is impaired, liver fat is accumulating, hunger signals are unstable, and day-to-day energy no longer matches the fuel already stored in the body. The problem is not an absence of calories, it is a system that has lost its ability to access and distribute energy efficiently.

Type 2 Diabetes:
Trapped Energy vs Restored Fuel Flow

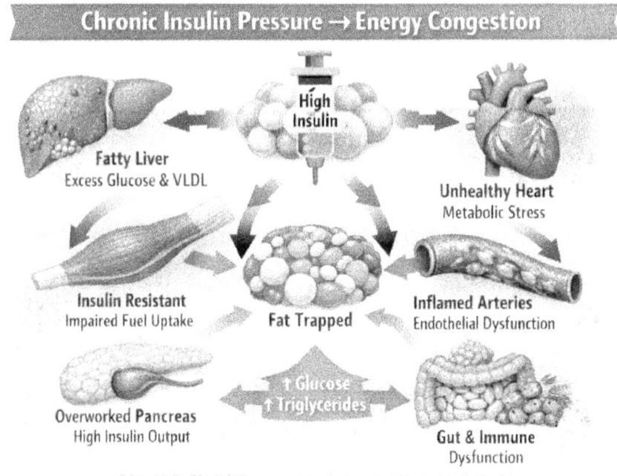

Energy Abundant, Access Impaired

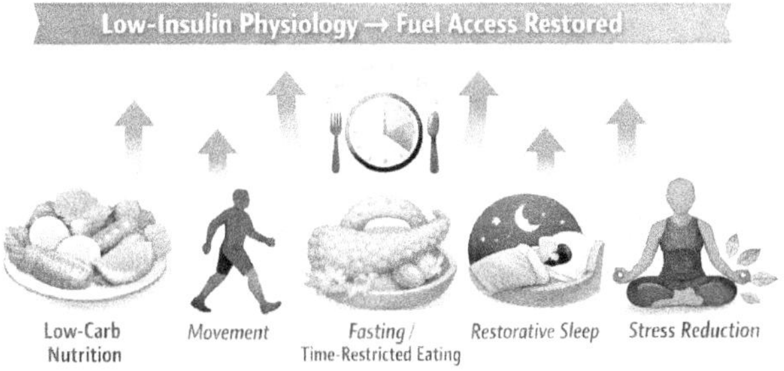

Energy Moves Again

Carbohydrate reduction works because it lowers the primary hormonal driver of this state. Carbohydrates stimulate insulin secretion more than any other macronutrient. In insulin resistance, even moderate carbohydrate intake can produce exaggerated and prolonged insulin responses. When insulin remains elevated, fat release is restricted, the liver continues converting incoming fuel into stored fat, and skeletal muscle becomes less responsive to nutrient delivery signals. Lowering carbohydrate intake reduces insulin demand, and once insulin declines, the bottleneck begins to clear.

One of the earliest and most consequential improvements occurs in the liver. Excess carbohydrate exposure forces the liver to convert glucose into fat, leading to triglyceride accumulation that disrupts glucose regulation, lipid handling, and appetite signaling. When carbohydrate intake is reduced, liver fat often falls rapidly—sometimes within weeks—followed by improvements in fasting glucose and insulin sensitivity. As hepatic congestion resolves, the liver shifts back from being a driver of dysfunction to a regulator of metabolic stability.

Skeletal muscle, the body's largest glucose sink, often recovers in parallel. Muscle insulin resistance is strongly associated with intracellular fat accumulation that interferes with normal fuel handling. As insulin pressure declines and fatty acid oxidation improves, muscle regains flexibility—becoming more responsive to incoming fuel and less dependent on constant glucose supply. Many people experience this change as steadier energy throughout the day and fewer post-meal crashes.

Hunger regulation stabilizes as well. Chronic hyperinsulinemia disrupts satiety signaling and creates hunger that is disconnected from actual energy need. With carbohydrate reduction, glucose swings diminish, reactive lows become less common, and appetite signaling regains coherence. Cravings often soften, fasting becomes tolerable, and food intake frequently decreases without force—because the body can finally draw from internal reserves.

Blood lipid patterns tend to shift in the same direction. Triglycerides typically fall—often quickly—while HDL rises. These changes reflect improved hepatic fat handling and reduced VLDL overproduction rather than dietary fat avoidance. As liver traffic normalizes, the bloodstream begins to reflect metabolic order rather than overflow.

As these processes unfold, medication requirements often decline. Insulin doses are reduced, insulin-stimulating drugs can be withdrawn, and in many cases additional glucose-lowering agents become unnecessary under appropriate supervision. This is what distinguishes metabolic repair from glucose management: as the system improves, support can often be removed rather than escalated.

At its foundation, carbohydrate reduction restores the body's native fuel architecture. Human physiology was designed to alternate between glucose and fat depending on availability and demand. Diabetes and cardiometabolic disease arise when this flexibility is lost and insulin remains persistently elevated. By lowering insulin and reactivating fat oxidation, carbohydrate reduction allows mitochondria to resume their intended role and restores metabolic adaptability.

Every benefit described—improved glucose control, reduced liver fat, normalized hunger, improved lipid profiles, and steadier energy—traces back to a single mechanism: lower insulin reopens access to stored fuel. That is why carbohydrate reduction succeeds where many other approaches fail. It addresses the upstream disruption rather than the downstream symptoms. And once energy can move again, the system regains its capacity to heal.

Clinical Evidence for Remission and Reversal

For much of modern medicine, type 2 diabetes has been framed as a progressive, irreversible condition—something to be managed indefinitely rather than meaningfully improved. Yet over the past two decades, a growing body of clinical evidence has steadily dismantled that assumption. Across diverse interventions—carbohydrate restriction, ketogenic diets, structured fasting protocols, and even bariatric surgery—a consistent pattern appears: when insulin demand falls and metabolic flexibility returns, diabetes often improves and frequently enters remission. This is not fringe theory or isolated anecdote. It is reproducible, well-documented, and increasingly recognized in mainstream research. Type 2 diabetes is not a one-way trajectory; it is a metabolic state, and metabolic states can change.

The most robust and consistent remission outcomes come from carbohydrate-reduction interventions. When carbohydrate intake is lowered, insulin demand drops and hepatic fat handling improves—often accompanied by substantial reductions in HbA1c (A1c) and major decreases in medication requirements. In many trials, participants reduce or discontinue insulin and other glucose-lowering therapies under supervision. Beyond lab markers, patients frequently report steadier energy, calmer appetite, improved cognition, and fewer post-meal crashes—changes that reflect underlying metabolic repair, not merely better glucose control on paper. Long-term follow-up confirms that when carbohydrate restriction is sustained, these improvements can persist.

Support for the same physiology also comes from strategies that are not explicitly low-carbohydrate. The DiRECT trial, which used a low-calorie liquid diet, revealed a key insight: rapid reductions in liver fat can lead to rapid normalization of glucose and insulin sensitivity—sometimes within weeks and before significant weight loss occurs. This underscores the liver as a central driver of type 2 diabetes. Relieve hepatic congestion, and downstream metabolic control often stabilizes.

Real-world clinical data reinforce these findings at scale. Large ketogenic interventions show that even individuals with long-standing, insulin-dependent diabetes can experience sustained improvement when insulin demand is markedly reduced. In these settings, A1c frequently moves out of the diabetic range, insulin requirements fall dramatically or disappear, and outcomes remain stable over time—highlighting the importance of hormonal environment over calorie counting alone.

Even bariatric surgery, though not a strategy advocated here offers unintentional proof of principle. In many cases, glucose improves within days of surgery, far too quickly for weight loss to explain the effect. The earliest change is metabolic: carbohydrate exposure drops, insulin levels fall, and glucose regulation improves almost immediately.

One of the clearest indicators of true metabolic recovery is reduced dependence on medications. Across remission-focused trials, insulin doses often decline sharply, sulfonylureas are discontinued to reduce hypoglycemia risk, and additional glucose-lowering drugs become unnecessary as endogenous regulation improves. This distinction matters: medications can lower glucose, but remission requires a deeper shift in fuel handling and insulin dynamics.

Taken together, these approaches converge on a single conclusion. The methods differ, but the recovery pattern is consistent: reduce insulin pressure, relieve hepatic overload, restore fuel switching, and glucose control frequently follows without force. Type 2 diabetes, then, is not the collapse of a system, it is the congestion of a system. The liver, muscle, pancreas, and adipose tissue are not broken; they are overwhelmed. And when that pressure is removed, these tissues often recover.

Why Diabetes Became a Disease to Manage, Not a Disease to Reverse

Given the growing body of evidence showing that type 2 diabetes can improve, and often enter remission, it is reasonable to ask how lifelong "management" became the dominant medical story for nearly a century. The answer is not a single mistake, but a convergence of scientific assumptions, cultural shifts, technological limits, economic incentives, and institutional inertia. Once these forces aligned, chronic treatment became the default framework, even as some patients quietly achieved metabolic restoration outside the prevailing narrative.

The discovery of insulin in 1921 marked one of the most dramatic breakthroughs in medical history. Type 1 diabetes, once uniformly fatal, became survivable almost overnight. Physicians watched children on the brink of death regain consciousness and strength within hours, a transformation so powerful that it shaped an entire generation of medical thinking. Insulin became synonymous with treatment, and understandably so. But the success of insulin in type 1 diabetes influenced type 2 diabetes care as well, where the early problem is not insulin deficiency but chronic insulin excess and rising resistance. High glucose was interpreted as a signal for more insulin long before insulin resistance was fully understood, reinforcing a glucose-centered paradigm before the deeper metabolic story was integrated.

At the same time, public health messaging was undergoing its own transformation. In the mid-20th century, the fat–cholesterol hypothesis redirected cardiovascular risk toward dietary fat, and low-fat recommendations became widespread. As industry adapted, fats were replaced with refined carbohydrates and sugars—often marketed as healthier alternatives. Because heart disease and diabetes frequently travel together, people with diabetes were placed on the same low-fat, high-carbohydrate dietary framework intended for cardiovascular prevention. The result was a profound mismatch: individuals with insulin resistance were encouraged to rely more heavily on the macronutrient requiring the greatest insulin response, pushing physiology further out of alignment.

Technological limitations reinforced this trajectory. Until the 1970s, clinicians could measure glucose easily, but insulin was largely invisible in routine care. High blood sugar could be tracked, but hyperinsulinemia—the upstream driver in many cases—could not. Treatment naturally focused on what could be seen and adjusted, while the hormonal and metabolic pressure beneath the numbers remained obscured. Diagnostic tools shaped understanding, and understanding shaped treatment.

As diabetes prevalence increased, a vast healthcare infrastructure developed around long-term management. Oral medications, insulin therapies, glucose-monitoring technologies, and later incretin-based drugs improved outcomes and reduced complications. But they also reinforced the cultural expectation that diabetes was permanent and required lifelong pharmacologic control. Economic and regulatory systems evolved to support this chronic care model. Nutrition-based strategies that reduced medication needs often lacked institutional support—not because they lacked physiological merit, but because they did not fit easily into systems designed around maintenance rather than reversal.

Medical training reflected the same priorities. Physicians are highly skilled, yet most receive limited education in nutrition and metabolic physiology. Diabetes care is often taught through protocols: medication algorithms, glucose targets, carbohydrate counting, and standardized dietary guidance. Less attention is devoted to how insulin resistance develops, how hepatic fat initiates dysfunction, or how chronic carbohydrate exposure sustains insulin demand. The result is a system highly effective at controlling glucose, yet often under-equipped to restore metabolic flexibility.

The rise of newer therapies has further strengthened the management narrative. Medications that suppress appetite, reduce glucose, and promote weight loss can be transformative for many patients in the short term. Yet their benefits also reveal a limitation: symptom improvement is not always the same as metabolic repair. When therapy is discontinued, appetite frequently returns and glucose often rises again, highlighting that underlying fuel handling may remain impaired.

The greatest barrier to reversal, however, has not been a lack of evidence—it has been resistance to paradigm change. Reversal requires challenging deeply held assumptions: that glucose is the primary problem, that type 2 diabetes is inevitably progressive, and that nutrition is merely supportive rather than foundational. Moving from symptom suppression to metabolic repair requires new frameworks, new training, and the willingness to update long-standing beliefs. Paradigm shifts are slow, but they are underway.

By the end of the 20th century, nearly every pillar of diabetes care pointed toward lifelong management: low-fat guidelines, glucose-centered diagnostics, medication-first algorithms, and powerful structural incentives. Reversal did not disappear because it was impossible. It disappeared because the system was not designed to look for it. Today that landscape is changing. Patients are achieving remission in clinical practice. Researchers are clarifying the physiology of recovery. Clinicians are beginning to recognize that type 2 diabetes is not a one-way path, but a reversible metabolic state—and the story of management is gradually giving way to the story of restoration.

The Physiology of Reversal: How the Body Heals Once Fuel Flow Is Restored

Reversal is not mysterious, nor is it a rare exception reserved for a fortunate few. It is a predictable physiological response that unfolds when chronic insulin pressure is relieved and the body is once again allowed to access the energy it already possesses. What looks miraculous on the surface: normalized glucose, calmer hunger, declining medication needs—is simply the body returning to its native metabolic design. When fuel flow is restored, the system reorganizes itself in a coordinated way.

Recovery begins when insulin levels stop running high all day. Chronically elevated insulin traps energy inside fat cells, limits access to stored fuel, and forces the body into constant dependence on incoming food. When insulin demand falls through carbohydrate reduction, fasting intervals, or improved sensitivity—energy becomes available again. Hunger quiets, cravings weaken, and steady fuel returns between meals.

The liver is typically the first organ to show measurable change. In insulin resistance and type 2 diabetes, excess carbohydrate exposure drives the conversion of glucose into fat, leading to hepatic congestion and loss of metabolic control. As insulin demand decreases, the liver begins to offload this accumulated fat. With declining hepatic fat, fasting glucose often improves, not because the liver is being suppressed, but because it resumes its intended regulatory role. As the metabolic gateway, liver recovery initiates repair across the system.

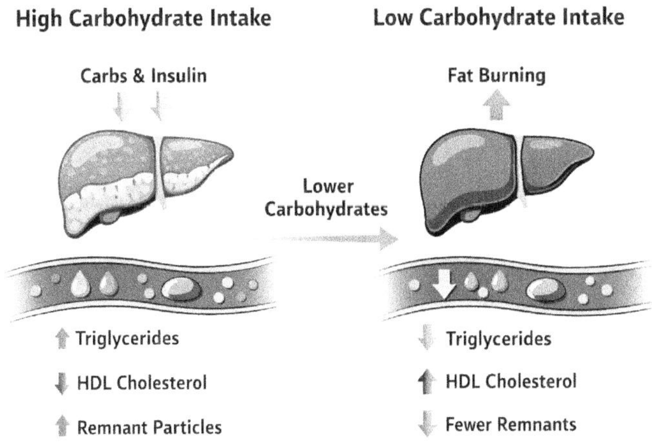

Skeletal muscle often follows. When insulin remains chronically elevated and fat accumulates inside muscle cells, fuel switching becomes impaired and energy output loses flexibility. As intracellular fat declines and mitochondrial function improves, muscle regains the ability to alternate between glucose and fat depending on demand. Many people experience this shift as steadier energy, reduced fatigue, and improved tolerance for movement and exercise.

The pancreas also benefits from this lowered metabolic strain. In type 2 diabetes, insulin secretion is often excessive for years before beta-cell function begins to decline. Hyperinsulinemia reflects overwork, not early failure. As insulin sensitivity improves, the pancreas no longer needs to overshoot. Insulin output becomes more precise and less constant, one reason medication requirements often decrease as recovery progresses.

Appetite regulation stabilizes as well. Chronic hyperinsulinemia disrupts satiety signaling and can produce hunger that is disconnected from true energy need. As glucose swings diminish and fuel access improves, signals such as leptin and ghrelin regain coherence. Fasting becomes tolerable, and eating shifts from urgency back to demand.

One of the most consistent downstream signs of recovery is improved blood lipid patterns, reflecting restored hepatic fat handling. Triglycerides typically fall, HDL cholesterol rises, and remnant particles decline. These changes occur independently of dietary fat avoidance and often indicate that liver energy traffic has begun returning to order.

Inflammation may quiet alongside these changes. Glucose variability, fatty liver, and metabolic overload place chronic inflammatory stress on tissues. As stability returns, many individuals notice clearer thinking, reduced joint pain, improved mood, and faster recovery from activity.

Perhaps the clearest indicator of reversal is the ability to reduce or discontinue medications. As insulin demand drops and tissue responsiveness improves, pharmacologic support becomes less necessary to maintain control. Under appropriate medical supervision, many individuals taper insulin and other glucose-lowering therapies. This is not management, it is repair.

The core mechanism is simple: when insulin pressure lifts, fuel flow returns. The body is not defective, it has been overwhelmed by chronic metabolic load. Remove that load, and the system begins functioning as designed again. Type 2 diabetes is not a fixed identity. It is a reversible metabolic state.

Yet modern medicine increasingly relies on therapies that improve visible markers while leaving the underlying metabolic environment unchanged, creating the next dilemma in chronic disease care.

Surface Control and the Illusion of Resolution

Modern diabetes care has become remarkably effective at improving visible markers of disease. Blood glucose can be lowered, A1C normalized, appetite suppressed, and weight reduced, often quickly. But those outcomes do not automatically signal healing. They show that glucose can be managed, even while deeper physiology remains unchanged.

The body does not experience disease through laboratory values alone. It experiences energy availability. When insulin remains chronically elevated and fat oxidation stays suppressed, cellular access to stored fuel remains limited, even as numbers improve on paper. Liver fat may persist, muscle insulin resistance may continue, and metabolic flexibility may remain constrained beneath a reassuring dashboard.

This distinction matters because it explains a central paradox of modern care: why symptoms can improve while long-term risk quietly remains. Surface markers respond to intervention, but unless the metabolic environment shifts, the forces that created diabetes remain in place.

True reversal requires more than numerical improvement. It requires restoring the conditions under which fuel can be stored and released appropriately, signaling can normalize, and tissues can regain flexibility. Without that shift, management replaces repair, and the same upstream pressure continues to affect other organs.

Seen through this lens, diabetes is not the beginning of the story, but one expression of a broader breakdown in energy distribution. The liver, muscle, pancreas, vasculature, and heart are not separate victims of separate failures, they are participants in the same metabolic traffic jam. When fuel cannot be routed correctly, function declines. When the bottleneck clears, repair begins.

This is why diabetes so often travels with heart disease, fatty liver, hypertension, and vascular dysfunction. And it is why focusing on glucose alone is rarely enough. The core disturbance lies upstream—in how energy is handled, stored, and accessed across tissues. Once that architecture is understood, the pattern becomes difficult to ignore: restore metabolic flexibility, and multiple systems improve together; ignore it, and dysfunction spreads.

Diabetes is not an isolated malfunction. It is a signal, a warning flare that energy has become trapped in the wrong places. Listen to that signal, and the path forward comes into focus: not only for blood sugar, but for the heart, the vessels, and the entire metabolic system that depends on energy moving where it is needed most.

THE CELL DANGER RESPONSE

Why the Body Gets Stuck in Disease

"Healing does not occur until the danger has passed."
ROBERT NAVIAUX

A New Lens for Chronic Illness

What if many chronic illnesses are not failures of the body—but expressions of protection?

Across conditions that appear unrelated on the surface fatigue, brain fog, autoimmune flares, metabolic dysfunction, post-viral syndromes—a common pattern quietly emerges. The body behaves as though it senses ongoing threat. Energy production shifts. Repair slows. Communication between cells grows guarded. Systems meant for growth and renewal move into conservation mode. This is not chaos. It is strategy.

At the most fundamental level of biology, life depends on energy, not just how much is available, but how safely it can be used. When cells perceive danger whether from infection, injury, inflammation, toxin exposure, or metabolic overload—they do not simply push harder. They pull back. Energy is redirected away from growth, reproduction, and long-term repair toward short-term survival. In this state, the body is not broken. It is cautious.

Modern medicine often interprets this protective shift as dysfunction: systems that are "stuck," metabolism that is "slowed," immunity that is "overactive" or "confused." But from an energy-first perspective, these patterns make sense. They reflect a biological decision: *it is not safe to return to full energy flow yet*. This chapter explores that decision.

Known as the Cell Danger Response, this conserved cellular state represents a temporary suspension of normal energy-driven cooperation between cells. It is a metabolic posture designed to preserve life under threat. When properly resolved, it gives way to healing, regeneration, and restored energy. When unresolved, it can quietly shape the long arc of chronic illness.

Understanding this response reframes disease itself, not as an enemy to suppress, but as a signal to interpret. And it brings us closer to the central question of this book: **How does the body decide when it is safe to return to energy?**

Mitochondria as the Body's Danger Sensors

This question, how the body decides when it is safe to return to energy has been explored in depth by physician-scientist Dr. Robert K. Naviaux, professor of medicine, pathology, and genetics at the University of California San Diego. Building on decades of research across mitochondrial biology, immunology, and metabolomics, Naviaux helped formalize a conserved biological response—an ancient, evolutionarily preserved survival program—that links energy metabolism, immune signaling, and healing: the Cell Danger Response.

Importantly, this expanded view of mitochondria does not stand alone. Over the past several decades, a growing body of research has reshaped how these organelles are understood. Rather than functioning solely as ATP-producing structures, mitochondria are now recognized as central regulators of cellular signaling, stress sensing, immune activity, and metabolic coordination across tissues.

Within this broader scientific context, Naviaux's work offers a coherent framework for understanding how mitochondria translate environmental and metabolic information into a unified biological response.

Far from passive energy factories, mitochondria operate as active environmental sentinels. They continuously monitor both internal and external conditions: signals related to infection, oxidative stress, inflammation, nutrient availability, toxins, and physical or metabolic injury. When these signals indicate threat, mitochondria initiate a coordinated defensive program designed to protect the cell and, by extension, the organism.

At the cellular level, this response involves a shift in energy metabolism. Oxidative phosphorylation—the efficient, oxygen-dependent process that supports growth, repair, and long-term function—is downregulated. In its place, cells rely more heavily on glycolysis, a faster but less efficient pathway better suited for short-term survival. This metabolic shift does not simply reduce energy production. It changes how energy is used and communicated.

As intracellular ATP production declines, mitochondria release ATP and related purinergic molecules into the extracellular space. Outside the cell, ATP no longer functions as fuel, it becomes a signal. These molecules act as chemical alarms, activating purinergic receptors on neighboring cells and immune pathways, broadcasting the message that danger is present.

As long as these danger signals persist, the priorities of the organism change. Cellular cooperation gives way to isolation. Growth, reproduction, and long-term repair are placed on hold. Resources are conserved. Survival takes precedence. In this context, reduced energy is not a failure of biology—it is a deliberate protective state, maintained until conditions are perceived as safe again.

Energy Translation—the Cell Danger Response

For a long time, low energy has been treated as a sign that something is broken. This research suggests the opposite is often true. When the body senses danger—whether from illness, inflammation, toxins, injury, or metabolic stress—it does not fail. It adapts. Cells shift into a protective mode, and mitochondria stop prioritizing long-term output in favor of short-term survival.

In that state, energy isn't gone. It is being withheld. ATP is used less as fuel and more as a warning signal, alerting nearby cells and activating immune pathways that something isn't safe yet. Growth and repair pause, not because the body has given up, but because it is waiting.

As physician-scientist Robert K. Naviaux has shown, healing begins not when energy is forced, but when the environment finally feels safe enough to move forward again. In that light, fatigue can be understood differently. Energy does not return simply by pushing harder. It returns when the danger signals quiet and the system is free to resume repair.

When Protection Becomes Pathology

This defensive physiology is what Naviaux terms the Cell Danger Response (CDR)—an ancient, evolutionarily conserved program designed to limit viral replication, contain injury, and buy time for repair. Under normal circumstances, the CDR is temporary. Once danger resolves, cells restore oxidative metabolism, reopen communication networks, and reintegrate into healthy tissue function.

Problems arise when this protective response does not complete its cycle. When the CDR remains active, a chronic loop can emerge. Mitochondrial metabolism stays constrained. Cellular communication remains muted. The cell becomes locked in a low-energy, defensive posture, no longer responding to immediate threat, yet unable to resume normal growth and repair.

At the level of the whole organism, this state is experienced not as acute illness, but as persistent dysfunction. Energy remains low. Inflammation simmers. Recovery stalls. Healing never fully arrives. Naviaux associates this prolonged, unresolved defensive state with the physiology underlying many forms of chronic illness.

The Brain's Version of the Cell Danger Response

This same survival physiology plays out throughout the body, but it becomes especially visible in the brain—because the brain is both energy-hungry and communication-dependent. It cannot "coast" when fuel delivery becomes unstable.

When neurons face an energy crisis, whether from insulin resistance, inflammation, trauma, oxidative stress, or mitochondrial dysfunction—glucose metabolism declines. That energy shortfall acts as a local danger signal. The Cell Danger Response activates, reducing communication between neurons, slowing synaptic activity, and impairing the networks required for memory, focus, and emotional stability.

> A destructive loop can form: energy deficit → perceived threat → CDR activation → deeper energy deficit

What many people experience as brain fog may be one of the earliest and most recognizable signs of this protective shutdown—not because the brain is failing, but because it is shifting into conservation mode.

Conditions That Reflect Persistent CDR Activation

Although the mechanism of the Cell Danger Response may sound abstract, its physiology appears across many conditions familiar to readers. When this protective response fails to resolve, the resulting state does not express itself as a single disease. Instead, it shows up as recognizable patterns of dysfunction that modern medicine often categorizes separately.

Variations of this unresolved defensive state have been described in association with a wide range of conditions, including:

- Autism spectrum disorder
- Myalgic encephalomyelitis / chronic fatigue syndrome (ME/CFS), often marked by post-exertional malaise
- Post-viral syndromes, including Long COVID
- Post-traumatic stress disorder (PTSD) and trauma-related states
- Traumatic brain injury
- Metabolic syndrome and insulin resistance
- Polycystic ovary syndrome (PCOS) and infertility linked to metabolic stress
- Autoimmune conditions characterized by relapsing inflammatory flares
- Early cognitive decline and persistent brain fog
- Chronic pain syndromes, including fibromyalgia

These conditions differ widely in their triggers, timelines, and clinical presentations. What unites them is not a single cause or diagnosis, but a shared physiological posture: inflammation that fails to fully resolve, mitochondrial stress that constrains energy production, disrupted cellular communication, and persistent danger signaling.

From this perspective, chronic illness is not best understood as continuous damage or degeneration. Rather, it reflects a biological state in which the body remains guarded—unable to complete the transition from defense back to repair. In this light, the question is no longer why the body is malfunctioning, but why it has not yet received the signals required to safely stand down and return to energy.

To understand chronic illness through the lens of energy, it helps to first understand the body's natural healing cycle. The diagram below illustrates **salugenesis**—the process by which cells move from danger, through defense and repair, and back into healthy reintegration. Under normal circumstances, this cycle completes. Chronic illness emerges when the process stalls, leaving the body trapped in a prolonged defensive state.

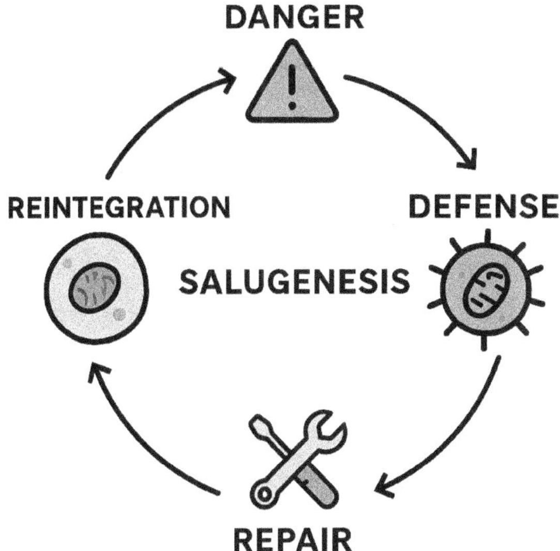

ME/CFS: A Modern Echo of a Survival Program

When danger signals do not resolve, the body may shift into a prolonged state of energy conservation. Metabolism slows, repair is postponed, and activity becomes limited, not because the system is failing, but because it is protecting itself.

One of the clearest examples of this state in humans is Myalgic Encephalomyelitis/Chronic Fatigue Syndrome (ME/CFS). People with ME/CFS are often told they are simply tired, deconditioned, or depressed. But the biology tells a very different story.

In a landmark 2016 study, researchers analyzed hundreds of chemical markers in the blood of people with ME/CFS. What they found was striking. Instead of a scattered pattern of deficiency or breakdown, almost every major metabolic pathway was turned down at the same time.

Not just one fuel system. Not one nutrient class. *Across the board, the body's chemistry was quieter.* Lipids. Amino acids. Energy-related metabolites. Antioxidant signals. All were reduced together. This was not the pattern of a system failing. It was the pattern of a system intentionally slowing itself down.

To make sense of this pattern, researchers looked to biology for comparison. In nature, some organisms enter a temporary survival state when conditions become dangerous. One well-studied example is the *dauer* stage in the tiny worm *C. elegans*. When food is scarce or the environment becomes hostile, the worm does not die. Instead, it slows its metabolism, stops growing, and conserves energy until conditions improve. When safety returns, normal life resumes.

A similar process may occur in humans. When cells remain locked in the Cell Danger Response, the body can enter a hypometabolic state—a low-energy mode designed to protect against further harm. Energy is conserved. Activity is limited. Repair is postponed. Seen through this lens, chronic fatigue is not simple exhaustion. It is a biological pause. The problem arises when the pause does not end.

Long after the original trigger, an infection, toxin, injury, or severe stress has passed, the body continues to behave as though danger remains. Energy stays restricted. Communication between cells remains muted. The system never fully restarts.

Patients are left trapped in low energy, unable to tolerate exertion, unable to recover, and unable to return to normal function, not because their bodies are broken, but because their biology has not yet received the signal that it is safe to move forward.

Mitochondria Coordinate and Sustain the Danger Response

Sensing danger is only the first step. For the Cell Danger Response to persist, that signal must be communicated and reinforced across cells and tissues. This coordinating role is one of the less appreciated but most important functions of mitochondria. Beyond producing energy, mitochondria act as signaling hubs. They integrate multiple stress signals, including:

- Redox imbalance
- Viral or microbial fragments
- Inflammatory cytokines
- Environmental toxins
- Nutrient insufficiency
- Psychological and emotional stress

When these signals accumulate, mitochondria respond by releasing ATP into the space outside the cell. Inside the cell, ATP is energy. Outside the cell, ATP becomes a danger signal.

This extracellular ATP activates purinergic receptors—such as P2X and P2Y—on nearby cells. In effect, it broadcasts a message: *danger is still present.* As long as these receptors continue to detect elevated ATP, cells remain in defensive mode, even if the original trigger has passed. In this state, the body is not simply inflamed or fatigued. It is biologically prevented from moving forward. Growth, regeneration, and long-term repair are postponed because the system has not received a clear signal that it is safe to stand down. Healing requires more than the absence of harm. It requires the silencing of danger signals.

Energy Translation—How the Danger Signal Persists

Detecting danger is only the beginning. For the Cell Danger Response to continue, that message must be shared and repeated across cells. Mitochondria coordinate this by releasing ATP outside the cell. Inside, ATP powers work. Outside, it acts as a warning signal. As long as neighboring cells continue to detect extracellular ATP, they remain in defensive mode, even after the original threat is gone.

A simple analogy helps: this is like a fire alarm that keeps sounding. The fire may be out, but until the alarm is silenced, the building cannot return to normal operation. This is why healing requires more than removing the initial insult. The danger signals themselves must quiet before growth, repair, and long-term recovery can resume.

The Suramin Studies: A Glimpse into Restored Communication

Once the Cell Danger Response is understood as a problem of persistent danger signaling, an obvious question follows: *What happens if that signaling is quieted, even briefly?*

To explore this question, Naviaux investigated an approach known as antipurinergic therapy. In simple terms, this strategy asks whether blocking specific cellular danger receptors could turn down the alarm system and allow normal communication to resume. Suramin, a century-old drug originally developed to treat African sleeping sickness, became a research tool for this investigation. Suramin is known to block purinergic receptors—the same receptors that respond to extracellular ATP, the chemical "danger signal" released by stressed or threatened cells.

In a small 2017 pilot study published in *Annals of Clinical and Translational Neurology*, a single low dose of suramin led to temporary improvements in language, social engagement, and metabolic markers in children with autism spectrum disorder. Naviaux was careful in how he interpreted these findings. He did not present suramin as a cure, nor as a treatment ready for widespread use. Instead, the study served a more important scientific purpose: proof of concept.

It demonstrated that when danger signaling is reduced, even briefly, cells can begin to communicate again. Metabolism can shift away from defense. The healing cycle can restart. In other words, the Cell Danger Response is not irreversible. It is not permanent, only persistent.

Importantly, interventions that quiet danger signaling are not meant to replace the body's own healing processes. Their role is temporary. By reducing the intensity of the alarm, they may create a brief window in which normal communication can resume. What follows: repair, rebuilding, and reintegration must still be carried out by the body itself. In this sense, such interventions do not *cause* healing. They allow healing to begin. Long-term recovery depends on whether the organism can adapt, restore energy flow, and re-establish the conditions of safety required to complete the healing cycle.

Salugenesis: Completing the Body's Natural Healing Cycle

The true goal of healing is not simply the suppression of inflammation or the management of symptoms. It is the completion of salugenesis, the body's natural process of generating health. Salugenesis describes the sequence by which the body moves through and out of a defensive state:

1. Defense – energy is conserved and danger is prioritized
2. Repair – damaged structures are stabilized and restored
3. Rebuilding – tissues regain strength and function
4. Reintegration – cells reconnect and resume cooperation

As this process unfolds, inflammation subsides and gives way to tissue reconstruction. Mitochondrial networks reconnect. Metabolic flexibility returns. Cells begin working together again rather than operating in isolation.

This is the completion of the Cell Danger Response, the moment the organism is no longer bracing for threat. Energy can safely flow again. Communication is restored. Equilibrium returns.

Why Modern Medicine Often Struggles with the Second Half of Healing

Modern medicine is highly effective at addressing pathogenesis—identifying pathogens, suppressing inflammation, and managing acute symptoms. These approaches save lives and are essential in many settings. Yet they do not always address the biological processes required for full recovery.

From the perspective of the Cell Danger Response, healing involves more than removing what causes harm. It also requires the restoration of communication between cells, the coordinated signaling that allows tissues to repair, reorganize, and reintegrate after a threat has passed.

When danger signals persist, the body remains locked in a protective state. Symptoms may be controlled, but the deeper biological transition from defense back to growth never fully occurs. As a result, recovery can stall—not because the initial injury or illness remains, but because the system has not received a clear signal that it is safe to move forward.

When this communication remains disrupted, the body's ability to self-correct can be limited, even in the presence of advanced medical interventions. Symptoms may improve, but the underlying defensive state can persist. Viewed this way, some chronic conditions do not reflect ongoing injury alone, but an incomplete transition from defense to repair. Treatment addresses the immediate problem, while the deeper recovery process remains unfinished.

Chronic Illness as a Breakdown in Biological Communication

When chronic illness persists, the problem is not always ongoing damage or degeneration. From a systems perspective, it often reflects a breakdown in biological communication—the coordinated signaling that allows cells, tissues, and organs to function as an integrated whole.

In this state:

- Cells exchange fewer signals with one another
- Mitochondrial activity becomes rigid rather than responsive
- Energy use shifts from cooperation toward isolation
- Tissues lose coordinated organization

- Immune signaling becomes exaggerated, blunted, or misdirected
- Hormonal rhythms grow irregular

These changes are not random. They reflect a body operating under conditions of perceived threat, where communication is restricted and flexibility is sacrificed for protection. When this cellular coordination breaks down, the organism may survive, but it cannot fully thrive.

Notably, this biological state often parallels the lived experience of patients. Many describe withdrawal, overwhelm, persistent fatigue, cognitive clouding, and a sense of disconnection from themselves and their environment. Internal signaling mirrors external experience. Across levels: cellular, tissue, organ, and whole person the pattern is consistent: The body behaves as though it does not yet trust that it is safe.

Energy Translation—When Symptoms are Misread

When my wife's energy collapsed, the explanation offered was simple: depression. Medication was suggested. When I described my own fatigue, loss of motivation, and mental fog, the answer was the same. On paper, it made sense. We were tired. Slowed. Disconnected. But it didn't feel like sadness. It felt like being stalled. We were still showing up. Still trying. But time slipped away. Motivation wouldn't engage. Rest didn't restore. Effort didn't translate into energy. Something deeper wasn't working.

Looking back, we weren't failing to cope. Our bodies were operating in a protective state, conserving, limiting output, waiting. There were other life stresses, yes. But beneath them was a shared sensation we both recognized: our systems hadn't shut down. They had paused. We didn't need to be pushed forward. We needed to change the conditions so the safety signals could return.

Energy as Information: How Cells Coordinate Healing

Energy is not only a biochemical resource, it is also a means of biological communication. When energy production and use remain flexible, information flows efficiently between cellular systems. When energy becomes constrained by toxins, trauma, nutrient deficiency, chronic infection, or sustained metabolic stress—communication is altered.

The Cell Danger Response represents this constrained state. Energy is not absent, but held in reserve. Metabolism shifts toward defense. Signaling becomes cautious and limited. Cells reduce cooperation until conditions are perceived as safe again. Salugenesis describes the restoration of this flow—the gradual reopening of communication pathways that allow coordinated healing to resume:

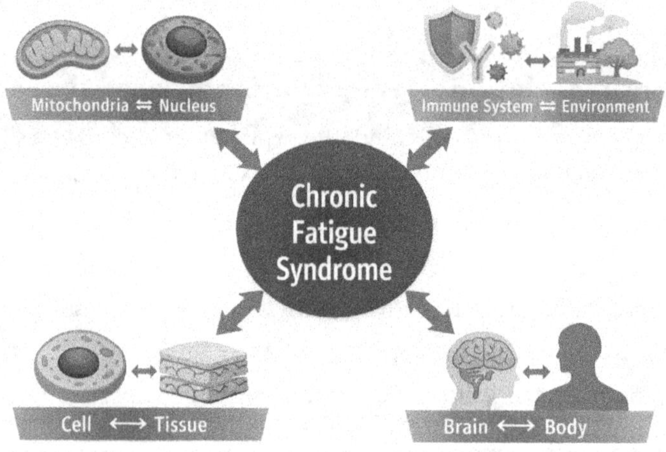

From this perspective, healing is not simply the elimination of pathology. It is the reorganization of energy and information across biological systems.

From Molecular Biology to Whole-Body Recovery

Naviaux's work helps connect molecular biology with a systems-level understanding of chronic illness. It offers a way to understand why conditions such as chronic fatigue syndromes, autism spectrum disorders, trauma-related states, infertility associated with metabolic stress, PCOS, metabolic syndrome, post-viral illness, and cognitive decline often resist simple explanations or single-target treatments.

These conditions are not random failures of the body. They reflect variations of a shared physiological state: The body remains in protection. The body has not yet received sufficient signals of safety. From this viewpoint, recovery requires more than removing harmful inputs. It requires restoring the signals—metabolic, immunologic, and environmental—that allow the organism to move out of defense and back into coordinated function.

The Return to Energy: Preparing
the Ground for Repair

Seen through this lens, chronic fatigue is not merely low stamina, but sustained energy conservation. Brain fog reflects early danger signaling in neural energy networks. Metabolic and reproductive disruptions reflect the prioritization of survival over long-term investment.

The solution is not to force performance or override symptoms. It is to quiet the alarms, restore communication, and allow energy to move safely again. This is the essence of *salugenesis*—Naviaux's term for the body's ability to complete the healing cycle once conditions of safety are restored.

What remains is a practical and biological question: How does the body reopen these pathways, safely and without triggering further stress? The answer lies not in suppressing defense, but in gently reshaping mitochondrial function itself. In the next chapter, we turn to one of the body's most important built-in mechanisms for doing exactly that: mitochondrial uncoupling.

CHAPTER TWENTY-THREE

MITOCHONDRIAL UNCOUPLING—A PORTAL BACK TO DESIGNED RENEWAL

"Efficiency is not always optimal for survival."
EVOLUTIONARY BIOENERGETICS

Introducing Mitochondrial Uncoupling

Mitochondria do far more than generate energy. They help decide whether a cell stays in defense—or returns to repair, renewal, and normal function. After the Cell Danger Response has served its purpose, the body needs a safe way to exit protection mode. One of the most important mechanisms supporting that transition is mitochondrial uncoupling.

In simple terms, uncoupling is a built-in pressure-relief system inside the energy engine of the cell. Instead of forcing mitochondria to extract the maximum ATP from every unit of fuel, uncoupling allows them to trade a small amount of efficiency for stability. Energy throughput rises, pressure falls, and fewer damaging byproducts accumulate. This is not a failure of metabolism. It is a protective adjustment—one that can restore the internal conditions that make healing possible.

The diagram that follows illustrates this shift. On the left is the coupled state, efficient, but vulnerable to overload when the system is under chronic metabolic stress. Pressure builds, reactive oxygen species rise, and the cell remains "on edge." On the right is the uncoupled state where some energy is intentionally released as heat, pressure is reduced, and stress chemistry quiets. In the center is the book's guiding theme: Return to Energy. Radiating outward are five major benefits that tend to appear together when uncoupling is gently activated: lower oxidative pressure, better fuel handling, mitochondrial renewal, cellular cleanup with stress adaptation, and protection of vulnerable systems like the brain and immune network.

This process is not rare, extreme, or experimental. It is part of normal human physiology. Mild ketosis from fasting or reducing dietary sugar, temperature variation such as brief cold exposure, and other low-level metabolic challenges can all signal mitochondria to enter this restorative mode. These inputs do not overwhelm the system—they redirect it.

When uncoupling turns on, the effects are not isolated. They cascade. Cellular stress drops. Repair pathways reengage. Cleanup systems activate. The body begins moving from "survival mode" back toward resilience and function. What follows is a closer look at what each spoke represents—and why these changes matter in everyday health.

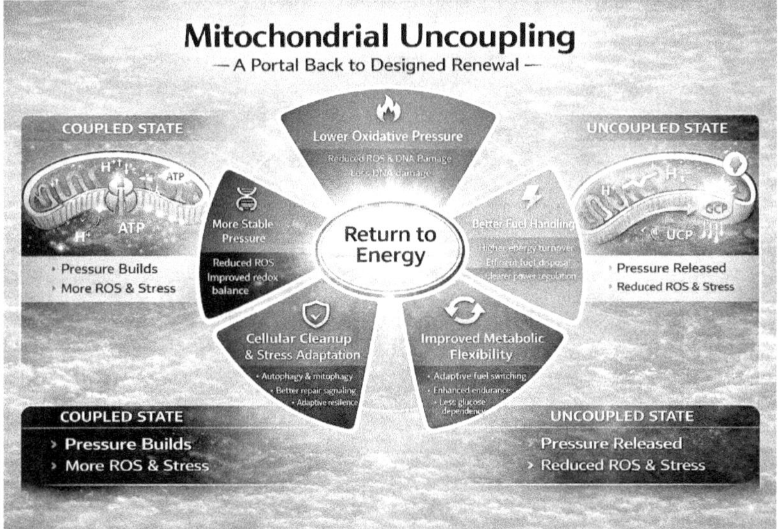

Mitochondrial Uncoupling Benefits—The Portal Effect

Mitochondrial uncoupling is not one benefit, it is a shift in internal conditions. When cellular pressure drops and energy signaling stabilizes, multiple systems improve at the same time. That is why uncoupling behaves less like a single "switch" and more like a portal: once the cell is no longer operating under threat, it becomes safe enough to repair.

The image above shows this visually. In the coupled state, mitochondria squeeze every drop of usable energy from fuel. That efficiency is valuable, but under chronic overload it can come with a hidden cost: pressure builds inside the respiratory chain, electrons leak, and oxidative stress rises. Over time, this contributes to the very pattern people recognize as modern chronic illness, fatigue that doesn't lift, inflammation that won't resolve, cravings that cycle, stubborn weight gain, brain fog, poor sleep, and symptoms that feel disconnected but share a common root.

In the uncoupled state, mitochondria "exhale." Some of that pressure is released. The cell becomes calmer and cleaner. And once the internal environment is safer, repair becomes possible again. Below are the five major spokes shown in the diagram—not as separate promises, but as connected outcomes of one central improvement: stable energy flow.

Lower Oxidative Pressure

When mitochondria are forced to run at maximum efficiency under high metabolic load, pressure rises in the electron transport chain and electron leakage increases. That leakage forms reactive oxygen species (ROS), which can damage proteins, membranes, and DNA. Uncoupling lowers that pressure by letting some energy dissipate safely as heat. The result is less unnecessary ROS and a better redox environment—more balance between oxidants and antioxidants, and less biochemical "friction" inside the cell.

> **Why this matters:** Lower oxidative pressure means less accumulated wear. Less internal damage. Less strain that forces the body to stay defensive. Over time, this supports healthier aging, recovery capacity, and cellular stability.

Better Fuel Handling

When uncoupling increases energy throughput, cells burn fuel more effectively and become less dependent on constant glucose availability. Fuel is handled, cleared, and regulated with less congestion. As that happens, insulin signaling often becomes more responsive and blood sugar tends to stabilize. This is where many "everyday wins" appear: fewer crashes after meals, fewer cravings, less appetite volatility, and more consistent energy through the day. A metabolism that can smoothly switch between glucose, fat, and ketones is no longer trapped in sugar dependency.

> **Why this matters:** Better fuel handling is what makes weight loss finally feel possible, hunger feel manageable, and

energy feel steady. It is a foundational reversal of the pattern behind insulin resistance, fatty liver, and metabolic syndrome.

Mitochondrial Renewal

Uncoupling doesn't just change how mitochondria perform—it can influence how mitochondria are maintained. When energy signaling improves, the cell becomes more willing to clear damaged mitochondria through mitophagy, and more capable of building new ones through mitochondrial biogenesis. This is quality control at the cellular level: fewer broken engines, more functional engines, and a stronger energy network across tissues.

> **Why this matters:** Many chronic symptoms behave like an infrastructure problem. When the cellular "power grid" becomes unreliable, everything downstream begins compensating. Renewal restores reliability—and reliability is what allows the body to rebuild.

Cellular Cleanup + Stress Adaptation

Cells recover not only by producing energy, but by clearing what is damaged. Uncoupling supports autophagy, the recycling system that removes worn-out proteins and dysfunctional components before they accumulate into dysfunction. At the same time, uncoupling aligns with a powerful biological pattern: hormesis. Small, controlled challenges can train the system to respond with greater stability rather than collapse. Instead of being overwhelmed by stress, the body learns to adapt, repair, and strengthen.

> **Why this matters:** This is what people experience as "I bounce back faster." Recovery improves. Inflammation resolves more cleanly. The body becomes less fragile—less reactive—and more resilient.

Protection of Vulnerable Systems

Certain tissues are highly sensitive to energy instability and oxidative stress. The brain is one of them. Neurons require steady power and stable signaling. When energy drops or oxidative burden rises, cognition can degrade quickly. By lowering oxidative strain and improving mitochondrial stability, uncoupling supports neuroprotection—helping preserve clarity, memory, focus, and emotional regulation under metabolic stress. The immune system also benefits, because immune cells rely on mitochondrial signaling to regulate activation, resolution, and rebuilding.

> **Why this matters:** This is where "brain fog," mood instability, poor sleep, frequent illness, and inflammatory flares begin to make sense. When cellular energy becomes more stable, vulnerable systems stop operating on the edge.

A Single Pattern Behind Many Benefits

These spokes are not separate claims stacked on top of each other. They are connected outcomes of one central shift: when mitochondrial pressure decreases and signaling stabilizes, the cell becomes safer, and safe systems can repair. This is why metabolic healing often feels like a cascade. Energy steadies. Cravings fade. Inflammation quiets. Sleep improves. Movement returns. The body begins doing what it was designed to do: protect when necessary, then rebuild when safe.

The Scientific Evidence—Four Studies That Illuminate the Portal

When we speak of mitochondrial uncoupling as a built-in safeguard, this is not metaphor alone. It is grounded in decades of careful biochemical research. A small number of key studies help clarify this mechanism, each illuminating a different aspect of how uncoupling protects cells from overload and restores balance.

A comprehensive review by Sébastien Demine and colleagues, published in the early 2020s, lays much of the modern foundation for understanding uncoupling. Drawing on a wide range of experimental data, the authors describe uncoupling as a central regulator of cellular health—coordinating autophagy, mitophagy, lipid metabolism, redox balance, and programmed cell death. Rather than representing wasted energy, uncoupling emerges as a protective response: a way for mitochondria to dissipate excess energetic pressure and prevent the buildup of damaging reactive oxygen species. In this view, uncoupling functions as a reset mechanism—one that preserves cellular integrity when metabolic demand runs high.

The work of Ma and colleagues, extends this concept into adipose tissue. Their studies show that ketone bodies—particularly β-hydroxybutyrate—actively promote mitochondrial uncoupling in fat cells. Ketones are not merely alternative fuels; they act as signaling molecules that increase mitochondrial respiration, upregulate uncoupling proteins, and allow excess energy to be released as heat. These findings help explain why ketosis shifts metabolism away from storage and toward flexibility, and why fasting and low-carbohydrate states so reliably activate protective metabolic pathways.

The relevance of uncoupling becomes even clearer in the nervous system. In a widely cited review, Maalouf and colleagues examined the neuroprotective effects of ketogenic metabolism. Their analysis shows that ketones enhance mitochondrial performance in neurons, reduce oxidative stress, stabilize redox balance, and activate survival signaling pathways. These effects help protect the brain against excitotoxic injury and neurodegenerative processes, preserving cognitive function under conditions of metabolic stress.

Direct experimental evidence from the brain itself comes from the work of Sullivan and colleagues, whose studies focused on the hippocampus. They demonstrated that ketogenic diets increase the expression of uncoupling proteins, particularly UCP2, UCP4, and

UCP5 in neurons. This increase was associated with lower reactive oxygen species and improved mitochondrial efficiency in brain tissue. What appears outwardly as a dietary shift, reducing sugar and increasing fat produced measurable molecular changes that shielded some of the brain's most metabolically vulnerable regions.

Taken together, these studies converge on a consistent conclusion. Mitochondrial uncoupling is not an accident of evolution or a flaw in efficiency. It is a conserved safety mechanism—one that appears across tissues and organ systems, activated by conditions such as fasting, ketosis, cold exposure, and physical exertion. Through uncoupling, cells reduce oxidative stress, renew damaged mitochondria, regulate lipid metabolism, preserve DNA integrity, and protect neural tissue. Science describes this process in the language of physiology; faith may recognize it as thoughtful design—a body equipped with built-in responses that protect, restore, and guide it back toward balance. In both cases, the message is the same: life is safeguarded by mechanisms that transform stress into renewal.

Living the Portal—Practical Pathways into Mitochondrial Uncoupling

Science shows us that mitochondrial uncoupling is not an abstract concept or a laboratory curiosity. It is a normal physiological process that can be engaged through simple, repeatable actions woven into daily life. These practices do not overwhelm the body; they gently signal mitochondria to shift out of constant defense and back toward repair and balance.

Long before we understood the biology, many of these actions—fasting, movement, rest, and restraint—were already embedded in human tradition. Modern science now helps explain why they work. The figure that follows summarizes the most practical entry points into uncoupling: small daily choices that promote steadier energy flow, lower internal pressure, and create the conditions for renewal.

Fasting and Time-Restricted Eating

Fasting is one of the most reliable ways to produce ketones, and ketones are more than fuel, they act as metabolic signals that support uncoupling. When the body shifts from sugar dependency toward fat-based energy, mitochondria respond by opening a controlled safety valve that lowers stress and steadies energy production.

- **Begin gently:** Start with a 12-hour overnight fast (for example, finishing dinner at 7:00 p.m. and not eating again until 7:00 a.m.).
- **Progress gradually:** Extend to 14 or 16 hours as comfortable. Many people find an 8-hour eating window (16:8 fasting) sustainable.
- **Occasional deeper fasts:** A 24-hour fast once per week or once per month can further elevate ketones and strengthen renewal pathways.
- **Spiritual connection:** Fasting has always been more than metabolic. It clears the mind, focuses prayer, and awakens the soul. Now we can see that it also awakens the mitochondria.

Low-Carbohydrate, Low-Sugar Nutrition

The modern diet is saturated with sugar and refined starches. These constant glucose spikes keep mitochondria tightly coupled and under pressure. Reducing these inputs allows the body to enter mild ketosis more often, nudging the system toward repair rather than constant defense.

- **Prioritize whole foods:** Proteins, healthy fats, non-starchy vegetables, nuts, and seeds.
- **Limit refined sugars:** Avoid sodas, candies, pastries, and white breads that push glucose into repeated highs and lows.
- **Embrace healthy fats:** Olive oil, avocado, fatty fish, and other stable fats provide consistent fuel that supports mitochondrial calm.

Cold Exposure and Temperature Variation

Cold awakens brown adipose tissue (BAT), which is rich in uncoupling proteins. The cold is not punishment; it is instruction. It tells the body to burn energy safely as heat, turning stored fuel into warmth rather than stress.

- **Cold showers:** End a warm shower with 20–30 seconds of cool or cold water. Over time, extend as tolerated.
- **Outdoor air:** Spend time outside in cooler seasons with lighter clothing, letting the body experience natural temperature variation.
- **Cold immersion:** For the adventurous, cold plunges or ice baths provide stronger activation, but they are not required to gain benefit.
- **Balance with warmth:** The cycle of cold and rewarming trains resilience, both in body and in spirit.

Exercise as Hormetic Stress

Movement challenges the mitochondria, not to break them, but to strengthen them. Exercise raises oxidative demand briefly, which paradoxically stimulates renewal pathways such as biogenesis, autophagy, and uncoupling. This is hormesis: stress transformed into strength.

- **Endurance training:** Walking, jogging, cycling, or swimming creates steady energy demand that supports mitochondrial efficiency.
- **Interval training:** Short bursts of intense effort followed by rest push mitochondria to adapt and build capacity.
- **Strength training:** Lifting weights or using bodyweight builds muscle and improves glucose regulation and metabolic flexibility.
- **Start where you are:** Even a brisk 15-minute walk is enough to begin awakening the portal.

Rest, Sleep, and Rhythm

Uncoupling is not only about stress, it is also about renewal. Sleep and circadian rhythm align mitochondrial turnover with hormonal cycles, lowering oxidative burden and supporting repair.

- Prioritize 7–9 hours of quality sleep each night.

- Keep a steady rhythm, rising and resting at similar times each day.
- Limit late-night eating, which disrupts both metabolic signaling and circadian cues.

Daily Integration—Small Steps with Eternal Meaning

The portal does not require extremes. It requires faithfulness in small, daily steps.

- Skip a late-night snack and extend your overnight fast.
- Choose water over soda.
- End a shower with a cool rinse.
- Take a walk after dinner.
- Rest with intention and protect your sleep.

Each of these steps is a way of unlocking the mercy written into your biology. Each is a prayer in action, a practical way of aligning body and spirit with God's design. Over time, these small practices become a new baseline quietly shifting the body from chronic stress toward stable energy, repair, and resilience.

The Deeper Message

Science names the mechanisms—ketones, uncoupling proteins, proton gradients, redox balance. Faith names the meaning—mercy, renewal, resilience, grace. Both are describing the same underlying truth: life has been safeguarded with a capacity to transform pressure into protection. Fasting, eating wisely, cold exposure, movement, and rest are not merely lifestyle strategies. They are ways of cooperating with the body's design—signals that tell the system it is safe to release defense and return to repair.

When these practices are lived, the body is not forced or overridden. It is allowed to do what it was designed to do. Energy begins to move more freely again. Stress chemistry quiets. Balance re-emerges. What follows is not just more energy, but a return to coherence, to renewal, resilience, and the steady equilibrium that makes life feel whole.

At the cellular level, mitochondria constantly balance efficiency with protection. Under heavy metabolic strain, the system can become too tightly "compressed" trying to conserve every unit of energy while operating under pressure. Over time, that strain increases oxidative stress and keeps the body behaving as though danger is still present.

Uncoupling offers a different path. It is controlled release—a natural way for mitochondria to lower pressure, reduce damaging byproducts, and stabilize energy signaling with aboveng. Not because the body is failing, but because it is protecting itself. In that safer internal environment, repair can restart. Cleanup pathways reengage. Renewal becomes possible. Uncoupling is the cellular equivalent of an exhale—an intentional release that makes sustained health possible.

Mitochondrial Uncoupling Research

The four studies discussed earlier reveal how mitochondrial uncoupling operates at the molecular and tissue level. To understand why this mechanism exists at all and why it has persisted across species and decades we need to step back and follow the arc of uncoupling research itself.

In 2000, biochemist Michael D. Brand published a pivotal review in *Experimental Gerontology* titled "Uncoupling to survive? The role of mitochondrial inefficiency in aging." The paper challenged a long-standing assumption in bioenergetics: that mitochondria are optimized to maximize efficiency at all times. Brand proposed a radically different view. Mitochondria, he argued, are intentionally somewhat inefficient. A portion of the fuel they process is not converted into ATP but is instead released as heat through mitochondrial uncoupling. What appeared to be waste, Brand suggested, might actually serve a protective function.

Central to this argument was the recognition that proton leak is a normal and substantial component of mitochondrial respiration. In idealized models, protons pumped across the inner mitochondrial membrane by the electron transport chain, return exclusively through ATP synthase to generate ATP. In real biological systems, however, a significant fraction of these protons re-enter the mitochondrial matrix without producing ATP. Measurements in rat liver mitochondria indicated that roughly 20–25% of respiration was devoted to this leak, while in skeletal muscle the figure could reach 35–50%. These findings implied that a surprisingly large share of resting metabolism was committed to controlled inefficiency.

Brand proposed that this apparent inefficiency functioned as a defense mechanism. By lowering the electrical potential across the inner mitochondrial membrane, uncoupling reduced the likelihood that electrons would stall within the respiratory chain and prematurely react with oxygen to form reactive oxygen species (ROS). Because ROS can damage proteins, lipids, and DNA, their accumulation accelerates aging and increases disease risk. In this framing, uncoupling acted as a pressure-release valve—sacrificing some efficiency in order to prevent oxidative overload. At the time, Brand acknowledged that much of the evidence was indirect, and that definitive proof linking uncoupling to improved health or longevity in living organisms was still emerging. Later research would suggest that this defensive response may also be one of the key ways repair and recovery reengage once threat subsides.

Over the following years, researchers tested these ideas through multiple lines of investigation. Genetically engineered mice lacking specific uncoupling proteins (UCPs) provided some of the first direct evidence. Animals deficient in UCP2 or UCP3 often exhibited higher levels of ROS in certain tissues, supporting the idea that uncoupling proteins play an active role in cellular defense. These effects varied by tissue and physiological context, highlighting that uncoupling is regulated—not uniformly beneficial in every setting.

Exercise physiology added further insight. Endurance training was shown to stimulate mitochondrial biogenesis, increasing the total number of mitochondria in muscle. This expansion reduced the workload on individual mitochondria while simultaneously promoting a state of mild uncoupling. Together, these adaptations lowered oxidative stress while preserving or improving performance—a dual strategy in which the body both builds more engines and allows each one to operate under less strain.

The rediscovery of metabolically active brown adipose tissue (BAT) in adults during the early 2000s expanded the relevance of uncoupling beyond background proton leak. BAT is rich in UCP1, a protein specialized for uncoupling respiration to generate heat. Cold exposure studies demonstrated that activating BAT improved glucose handling and lipid metabolism, confirming that uncoupling plays a central role in human thermoregulation and broader metabolic health.

Nutritional research also converged on uncoupling. Calorie restriction, intermittent fasting, and ketogenic diets were consistently associated with shifts toward mild uncoupling. Plant-derived compounds—including resveratrol, green tea catechins, and capsaicin—were found to influence mitochondrial membrane potential and thermogenesis. These findings suggested that mitochondrial efficiency is not fixed, but responsive to dietary and environmental signals—echoing Brand's early insight that uncoupling is a tunable feature of physiology.

Hormonal regulation further integrated uncoupling into whole-body control systems. Thyroid hormone was shown to increase proton leak and upregulate uncoupling proteins, while adrenaline activated brown fat during cold exposure or acute stress. Together, these pathways connected mitochondrial uncoupling to endocrine responses that coordinate energy use with environmental demand.

By the 2010s, a broad consensus had emerged: mild uncoupling plays a protective role by limiting oxidative stress, but its benefits depend on balance. Too little uncoupling allows damaging pressure to build; too much undermines ATP availability and cellular function. The optimal range varies by tissue and context, adding nuance to Brand's original proposal.

In the past decade, attention has shifted from establishing the existence of uncoupling to understanding how it can be safely influenced in daily life. Human studies now show that endurance exercise, controlled cold exposure, fasting, and specific dietary patterns can nudge mitochondria toward healthier uncoupling states. This work has reinforced the concept of metabolic flexibility, the ability to switch efficiently between fuel sources, as a cornerstone of resilience and healthy aging.

Takeaway

Mitochondrial uncoupling is now best understood as a built-in safety and adaptation system. What once appeared to be wasted energy is increasingly recognized as a strategic trade-off: relinquishing maximal efficiency in order to reduce oxidative stress, preserve cellular integrity, and support long-term resilience. Over the past two decades, evidence has accumulated that mild, regulated uncoupling contributes to healthy aging, recovery, and endurance. As Brand suggested at the turn of the century, biology often values balance over perfection—and what appears inefficient on the surface may be one of the mechanisms that helps sustain life over time. In other words, uncoupling is not a detour from energy—it is one of the most reliable pathways back to it.

CHAPTER TWENTY-FOUR

THE BRAIN'S ENERGY CRISIS

How Metabolic Dysfunction Unifies Mental and Neurodegenerative Disease

*"The brain is an energy-intensive organ,
and when energy fails, function fails."*
NEUROENERGETICS PRINCIPLE

Most people don't recognize an energy problem at first. It often begins quietly: attention slips, emotional bandwidth shrinks, and motivation fades. Then brain fog arrives—followed by the unsettling sense that memory is no longer reliable.

Brain-related disorders are increasing in ways that no longer fit traditional medical categories. Early challenges with attention and emotional regulation can give way to brain fog and memory decline in midlife, and in some cases progress to neurodegenerative disease later on. The pattern suggests continuity rather than a collection of isolated conditions.

For decades, psychiatry and neurology have treated these problems as separate—one focused on mood and behavior, the other on memory and cognition. Yet a growing body of research suggests these disorders are not isolated at all, but interconnected expressions of systemic dysfunction within the brain.

This chapter draws on work across nutrition, neurology, psychiatry, and metabolism to explore how these seemingly distinct conditions intersect—and what becomes visible when brain energy metabolism is placed at the center of the discussion.

Alzheimer's Disease as a Metabolic Breakdown of the Brain

Alzheimer's disease is increasingly understood as a breakdown in the brain's ability to generate and regulate energy, rather than a condition driven solely by genetics or protein accumulation. This metabolic perspective challenges the long-standing view that Alzheimer's is primarily unavoidable or purely protein-driven. Instead, it reframes the disease as one rooted in impaired energy metabolism, with amyloid plaques and tau tangles emerging as visible markers rather than primary causes. Among the clearest and most accessible syntheses of this growing body of research is Amy Berger's book The Alzheimer's Antidote (2017).

One way this metabolic breakdown has been described is through the concept sometimes referred to as "type 3 diabetes," a framework used to highlight brain-specific insulin resistance and disrupted glucose handling. In this model, neurons struggle to transport, respond to, and efficiently metabolize glucose even while glucose remains abundant in circulation. Fuel is present but inaccessible. This mismatch creates a chronic, progressive state of energetic stress that weakens synaptic connections, disrupts communication between neural networks, and contributes to the cognitive decline characteristic of Alzheimer's disease.

Importantly, this impairment does not occur in isolation. Metabolic dysfunction in the brain unfolds alongside chronic inflammation, oxidative stress, disrupted hormonal signaling, and declining efficiency of cellular energy systems. These processes reinforce one another. Impaired insulin signaling constrains glucose utilization. Oxidative stress damages cellular structures, inflammatory pathways stiffen metabolic responsiveness, and declining energy production undermines neuronal repair and communication. The result is a brain struggling not only with the amount of available fuel, but with the coordination and reliability of its metabolic processes.

From this perspective, Alzheimer's diverges from the dominant neurological narrative. Conventional models often frame metabolic changes as secondary, downstream consequences of structural damage driven by protein pathology. The metabolic model inverts this sequence, proposing that prolonged energetic instability may precede and contribute to protein accumulation rather than simply result from it. This shift in causality reframes how prevention, progression, and treatment are understood.

A central concept within this framework is metabolic flexibility—the brain's ability to adapt to different energy sources depending on availability and demand. In a healthy state, neurons can draw on glucose, ketones, and lactate as conditions change. Alzheimer's disease may represent a loss of this flexibility: the brain becomes increasingly dependent on glucose while simultaneously losing the ability to use it effectively. Long before structural damage is visible, vulnerability to energy shortfall increases.

Within this context, supplying alternative fuels—particularly ketone bodies—has emerged as a potential strategy to support compromised brain metabolism. Ketones, produced naturally during fasting, carbohydrate restriction, or ketogenic diets, can provide an efficient fuel when glucose utilization falters. Research summarized by Berger and others suggests that increasing ketone availability through dietary approaches, medium-chain triglycerides (MCTs), or exogenous ketones may support cognition, attention, and daily function in some individuals with early or moderate Alzheimer's disease.

Crucially, ketones are not presented as a replacement for glucose, but as a metabolic bridge an auxiliary fuel that can stabilize energy availability when glucose metabolism is impaired. This approach contrasts with the historical emphasis in neurology on symptom-targeted pharmacology, drawing attention instead to the metabolic environment in which neurons attempt to function.

Seen through this lens, Alzheimer's disease aligns with broader metabolic epidemics, including insulin resistance, obesity, type 2 diabetes, and chronic low-grade inflammation. Rather than an isolated neurological condition, it may represent the long-term convergence of systemic metabolic imbalance on the brain. From this perspective, nutritional and lifestyle interventions move from the margins to the foundation of both prevention and therapeutic strategy.

Born Ketogenic: Ketones, Development, and the Aging Brain

A central insight in brain energetics is that ketone metabolism is not an emergency workaround, but a normal and deeply conserved feature of human physiology—especially during periods of rapid brain growth or metabolic stress. This perspective becomes particularly clear when viewed through developmental biology, where ketones play a foundational role in early neurological development. One of the most accessible clinicians to bring this insight into the Alzheimer's conversation is Mary T. Newport, whose work bridges neonatal physiology, metabolic flexibility, and lived clinical experience.

In neonatology, ketone utilization is not theoretical. During late pregnancy and early infancy, the human brain relies heavily on ketones to support membrane formation, myelination, synaptogenesis, and rapid cellular growth. Infants are, in effect, born in a mildly ketogenic state. This is not pathological; it is normal human physiology. Breast milk reflects this design, containing substantial amounts of medium-chain triglycerides (MCTs), which the infant liver rapidly converts into ketones that cross the blood–brain barrier efficiently and serve as a clean, high-yield fuel during a critical window of neurological development.

This developmental framework raises an important question: if ketones are essential during early life, might they also serve a compensatory role later in life when glucose metabolism becomes impaired? That question became personal for Newport through her husband, Steve. Steve Newport was in his mid-50s when he began showing signs of mild cognitive impairment—difficulty with tasks, memory lapses, disorientation, and declining executive function. Clinical evaluation confirmed early-onset Alzheimer's disease. Despite medications, structured routines, and lifestyle changes, his condition continued to worsen. Standard pharmaceutical treatments offered only modest and temporary benefit.

As Newport reviewed the literature, she encountered evidence suggesting that while the Alzheimer's brain often struggles to metabolize glucose, its capacity to use ketones remains relatively preserved. This observation reframed the problem: if neurons are energy-deprived despite abundant circulating glucose, could ketones provide an alternative fuel capable of supporting function?

Her search led to research on MCT oil and the development of AC-1202 (later commercialized as Axona), a medical food designed to raise ketone levels without requiring prolonged fasting or strict carbohydrate restriction. MCTs—particularly C8 and C10 fatty acids—are rapidly transported to the liver and converted into ketones. Coconut oil, while less potent than purified MCT formulations, provided a practical and accessible starting point.

Newport began supplementing Steve's diet with coconut oil and concentrated MCT oil, carefully documenting his response. The changes were noticeable. Within hours of administration, she observed improvements in alertness, verbal expression, emotional engagement, and task initiation. Over subsequent weeks and months, she reported gains in reading, drawing, gait, facial expression, and orientation—functional changes that drew attention from clinicians and researchers around the world.

Her detailed observations formed the foundation of What If There Was a Cure? The Story of Ketones, which combined Steve's case with emerging research on brain glucose hypometabolism and ketone-based metabolic support. Families resonated with the book not because it promised a cure, but because it offered a coherent energy-based framework. Some individuals reported meaningful improvements; others experienced subtler stabilization. The common thread was not reversal, but improved energetic support in a compromised system.

Steve's course was not linear. Participation in an experimental drug trial led to a severe seizure, an event from which he never fully recovered. Following this neurological insult, the earlier gains associated with ketone support were eclipsed, and his condition declined until his death. Newport has been candid about this outcome, emphasizing both the promise and the limits of metabolic therapy in a vulnerable brain.

Her continued advocacy reflects a balanced conclusion: supporting brain energy metabolism does not eliminate all risk or halt all progression, but it may meaningfully influence function, quality of life, and resilience. Through lectures, research collaboration, and clinical guidance, Newport has remained a prominent voice in advancing ketone-based strategies for cognitive decline. What distinguishes her contribution is not a single intervention, but a unifying insight drawn from physiology, caregiving, and careful observation: ketones are not exotic fuels. They are a built-in metabolic reserve—used at birth, called upon during scarcity, and potentially lifesaving when glucose metabolism fails. In this sense, her work aligns with a broader movement in metabolic neurology: Alzheimer's disease reflects, at least in part, a failure of brain energy supply, and restoring metabolic balance may offer a meaningful avenue for support even late in the disease process.

Rethinking Mental Illness Through a Metabolic Lens

A growing body of research suggests that many psychiatric conditions may be better understood through the lens of brain energy metabolism rather than as isolated disorders of mood, thought, or behavior. This perspective is articulated clearly by Christopher M. Palmer in Brain Energy, where he proposes a unifying metabolic framework for mental illness. While earlier sections focused on Alzheimer's disease, this model extends the same energy-based logic across a much broader spectrum of psychiatric and neurological conditions.

From this viewpoint, conditions such as depression, anxiety, bipolar disorder, schizophrenia, OCD, PTSD, and related syndromes are not fundamentally separate disease entities. Instead, they can be seen as different expressions of a shared brain energy crisis. When cellular energy systems are compromised, through factors such as poor diet, chronic stress, sleep disruption, toxins, infections, medications, or genetic vulnerability—neuronal signaling becomes inefficient. Communication between networks degrades, regulation falters, and symptoms emerge.

Modern psychiatric diagnosis, however, is largely organized around symptom clusters rather than underlying biology. Disorders are defined by patterns of mood, thought, and behavior, and treatment is typically aimed at modifying neurotransmitter signaling to reduce distress. For many patients, this approach offers meaningful relief. Yet responses are often incomplete or transient, and individuals frequently cycle through multiple diagnoses and medications over time. From a metabolic perspective, these limitations do not reflect a failure of care, but a deeper mismatch: symptom-directed treatments cannot reliably restore function when the brain's core energy systems remain impaired.

Across different stages of life, this energy failure may appear in different forms. In children and young adults, it can present as emotional instability, impulsivity, or learning difficulties. In adults, it may emerge as burnout, treatment-resistant depression, or psychosis. In later life, it can converge with cognitive decline and dementia. Rather than representing unrelated disorders, these patterns suggest a continuum of vulnerability shaped by the brain's ability, or inability, to meet its energy demands. Energy failure does not act in isolation. The brain also depends on intact immune and clearance systems to manage inflammation and misfolded proteins. When these house-keeping processes falter, inflammatory debris accumulates, further impairing energy production and reinforcing a self-perpetuating cycle:

- impaired energy production reduces cellular clearance
- impaired clearance amplifies inflammation
- inflammation further weakens energy production

This framework carries important implications. Rising rates of ADHD, anxiety, depression, and early-onset bipolar disorder may not represent random epidemics of "chemical imbalance," but early warning signs of metabolic strain. Younger brains are not immune to the vulnerabilities that later manifest as dementia; they are simply expressing them at a different developmental stage.

Viewed this way, treatment strategies must address metabolism itself rather than only downstream symptoms. Approaches that support energy regulation—nutritional strategies such as ketogenic or low-carbohydrate diets, physical activity that enhances metabolic resilience, and sleep that supports clearance and repair—begin to make sense as foundational interventions rather than adjuncts. They act on the system that underlies both psychiatric and neurodegenerative disease.

The brain energy crisis model invites a rethinking of psychiatry and neurology. Instead of dozens of discrete disorders, we may be confronting a shared mechanism: impaired brain energy metabolism compounded by disrupted immune clearance. From the child struggling with attention, to the adult battling anxiety or depression, to the elder facing Alzheimer's disease, the common thread is the same. The challenge ahead is to move beyond managing symptoms in isolation and toward restoring the brain's energy systems.

Glucose Hypometabolism and Brain Energy Rescue

At the center of the brain's energy crisis is glucose, the primary fuel in most modern diets. Neurons depend on specialized glucose transporters, primarily GLUT1 at the blood–brain barrier and GLUT3/GLUT4 on neurons and glial cells to import glucose from the circulation. In states of insulin resistance, aging, or neurodegeneration, these transport systems become less efficient and intracellular glucose handling deteriorates. Fuel remains abundant in the blood, yet fails to reach or be effectively used by neurons.

Long-chain fatty acids do not cross the blood–brain barrier in meaningful amounts. This constraint makes ketones uniquely important. Produced during fasting, carbohydrate restriction, or ketogenic diets, ketone bodies bypass many of the limitations imposed by impaired glucose transport and insulin resistance. They enter the brain via monocarboxylate transporters (MCTs) and provide an efficient alternative fuel when glucose metabolism falters.

Much of what is known about this process comes from the work of Stephen Cunnane, whose research has been central to defining brain glucose hypometabolism as a core feature of conditions such as Alzheimer's disease (AD) and mild cognitive impairment (MCI). Using dual-tracer positron emission tomography (PET) imaging, his group demonstrated that glucose uptake in the brain is reduced in these conditions—not merely as a consequence of diminished neuronal activity, but in a pattern consistent with primary metabolic dysfunction.

This observation led to the concept of *brain energy rescue*. If neurons cannot effectively metabolize glucose, energy metabolism may be partially restored by supplying alternative fuels. Cunnane's studies show that ketogenic interventions, including MCT supplementation and ketogenic drinks, can increase brain energy availability and are associated with measurable improvements in certain cognitive outcomes, particularly in individuals with MCI.

The significance of this finding becomes clear when considering the brain's energetic demands. Although it accounts for only about 2% of body weight, the brain consumes roughly 20% of the body's resting energy, almost all of it derived from glucose under standard dietary conditions. In healthy neurons, glucose crosses the blood–brain barrier and is metabolized to generate ATP—the energy required for electrical signaling, synaptic plasticity, and cellular repair. In neurodegenerative and neurodevelopmental conditions such as AD, MCI, epilepsy, and some psychiatric disorders, this finely regulated system begins to fail.

In this state of *glucose hypometabolism*, neurons are unable to import or metabolize glucose efficiently. Proposed mechanisms include reduced expression or function of glucose transporters, brain insulin resistance (sometimes described as "type 3 diabetes"), and impaired cellular energy production. The consequences are widespread. Energy-deprived neurons cannot sustain the ion gradients required for normal electrical activity, increasing vulnerability to excitotoxic injury. Synapses, among the most energy-intensive structures in the body lose the capacity to release and recycle neurotransmitters effectively. Oxidative stress increases, inflammatory pathways are activated, and microglia and astrocytes release cytokines that further impair metabolic function. Over time, these processes may accelerate amyloid-β and tau accumulation, deepening the energy deficit in a self-reinforcing cycle.

The crucial insight from this work is that, even as glucose metabolism declines, the brain's capacity to utilize ketones often remains intact. Ketone bodies such as β-hydroxybutyrate and acetoacetate cross the blood–brain barrier via monocarboxylate transporters (MCT1, MCT2, MCT4), which remain functional in aging and dementia. Once inside neurons, ketones bypass insulin-dependent pathways and provide an efficient source of energy that supports neuronal signaling and maintenance.

This understanding opens new therapeutic possibilities. Rather than focusing exclusively on removing pathological proteins, clinicians may attempt to support brain function by restoring energy availability. Strategies explored in this context include nutritional ketosis through ketogenic diets or intermittent fasting; MCT oil and specialized ketogenic drinks, which raise ketone levels without strict dietary restriction and have demonstrated cognitive benefits in Cunnane's kMCT (ketogenic medium-chain triglyceride) trials; and exogenous ketone supplements as a direct means of increasing ketone availability.

Beyond ketones themselves, this framework emphasizes broader strategies that support metabolic resilience: regular physical activity to enhance energy capacity, sleep optimization to support clearance and repair, targeted nutritional support, and dietary patterns that reduce systemic insulin resistance and inflammation. Emerging approaches—including ketogenic medical foods, intranasal insulin, and multi-target metabolic therapies—are being explored as adjuncts to conventional treatment.

Taken together, this expanding field of metabolic neurology introduces cautious but genuine optimism. By reframing neurodegenerative disease as a problem of impaired energy supply rather than inevitable neuronal loss, this work suggests that supporting brain metabolism may slow decline, preserve function, and particularly in early stages—delay progression. In this sense, brain energy rescue does not promise reversal, but it restores something equally important: therapeutic possibility.

Reversing Cognitive Decline Through Systems-Based Metabolic Repair

A systems-based interpretation of Alzheimer's disease views cognitive decline not as the failure of a single pathway, but as the collapse of multiple interacting networks that support synaptic function, repair, and energy metabolism. This perspective gained traction as repeated single-target drug strategies—particularly those focused narrowly on amyloid removal—failed to meaningfully alter disease progression. The limitations of that approach suggested that Alzheimer's is not a linear problem, but a multifactorial one.

From this vantage point, Alzheimer's can be understood as a form of network failure. Cognitive decline emerges when the balance between synapse formation and synapse loss shifts persistently in the wrong direction. In this model, amyloid and tau are not merely toxic byproducts, but downstream responses to deeper physiological stress. They appear in a brain that has already entered a state of energetic and metabolic compromise.

A central insight of this framework is that the brain continuously integrates multiple signals—metabolic, inflammatory, hormonal, trophic, and toxic—to determine whether synaptic connections are maintained or dismantled. When energy availability is sufficient and the internal environment is supportive, synapses are preserved. When energy falters and stress signals dominate, the brain shifts into a protective, regressive mode that sacrifices synaptic complexity in favor of survival.

Seen this way, Alzheimer's disease is not driven by a single cause, but by multiple converging insults, many of which are metabolic in nature. These include insulin resistance, impaired glucose utilization, chronic inflammation, oxidative stress, hormonal decline, nutrient deficiencies, sleep disruption, and exposure to environmental toxins, often unfolding together over decades. Individually, none may be sufficient to cause disease. Together, they can overwhelm the brain's capacity to sustain energetic and synaptic integrity.

This systems perspective was formalized clinically through the work of Dale E. Bredesen, who developed a multi-factorial approach known as the ReCODE protocol. Rather than targeting a single mechanism, this model evaluates and addresses dozens of interacting variables simultaneously: nutritional, metabolic, inflammatory, hormonal, and lifestyle-related. The unifying principle is not a specific intervention, but restoration of the conditions required for neuronal energy production and repair.

Importantly, energy metabolism occupies a central position within this network. When neurons cannot efficiently generate ATP, synaptic repair falters, plasticity declines, and vulnerability to degeneration increases. In this respect, the systems model converges closely with other lines of evidence presented in this chapter: imaging studies of glucose hypometabolism, psychiatric models of brain energy failure, developmental reliance on ketones, and nutritional strategies that restore metabolic flexibility.

Early case series and observational studies reported by Bredesen and colleagues described stabilization or improvement in cognitive performance among some individuals with mild cognitive impairment or early Alzheimer's disease when multiple metabolic factors were addressed together. While these findings do not constitute definitive proof and require larger controlled trials, they challenged the long-held assumption that cognitive decline is universally irreversible.

Crucially, this approach is not framed as a cure. It is a proof of concept: if the brain is supplied with adequate energy, reduced inflammatory burden, and restored metabolic signaling, functional recovery may be possible—particularly early in the disease process. Even in later stages, supporting metabolic health may slow progression or preserve remaining cognitive function.

What distinguishes this contribution is not a single discovery, but a strategic shift. Instead of asking which molecule to block, the systems model asks what the brain requires in order to function. This reframing moves Alzheimer's disease out of the category of inevitable decline and into the broader domain of chronic, systems-level metabolic dysfunction—complex, yes, but potentially modifiable.

In the context of this chapter, the systems approach reinforces a unifying theme: whether viewed through imaging, psychiatry, developmental physiology, nutritional biochemistry, or systems biology, many disorders of the brain share a common vulnerability. When energy metabolism falters, cognition falters with it. Restoring metabolic balance does not guarantee reversal, but it reopens the door to prevention, stabilization, and meaningful intervention.

One Story, Many Faces

Taken together, the evidence points to a coherent conclusion:

- Alzheimer's disease is not merely a protein problem; it is fundamentally a problem of fuel.
- The brain is designed from early development to use ketones as a major energy source, particularly when glucose metabolism is compromised.
- Psychiatric and neurological conditions—from depression and bipolar disorder to schizophrenia and dementia—may represent different surface expressions of the same underlying brain energy crisis.
- Glucose hypometabolism and impaired energy regulation sit near the center of this crisis, while the brain's capacity to use ketones often remains relatively preserved.

This framework does not dismiss genetics, environment, trauma, or psychosocial factors. Instead, it places them within a larger context: anything that disrupts metabolic signaling or constrains cellular energy pushes the brain toward an energetic tipping point. The resulting symptoms, whether panic, psychosis, cognitive fog, or memory loss— are not separate failures, but different manifestations of the same underlying vulnerability.

The chapters that follow build on this foundation. Understanding brain disease as a disorder of energy metabolism is not the end of the story—it is the beginning of one. Much of what determines brain health operates beneath the surface, unseen and unmeasured, long before symptoms emerge. The critical question becomes not only what we see when function fails, but what lies underneath—and how we restore energy before the system collapses.

THE METABOLIC ICEBERG

A New Lens on Alzheimer's, Parkinson's, ALS, and Huntington's

"The greatest danger in times of turbulence is not the turbulence—it is to act with yesterday's logic."
PETER DRUCKER

If disorders of the brain reflect a failure of energy metabolism, then plaques, tangles, Lewy bodies, and motor neuron loss are not the starting point of disease, they are its visible aftermath. The more pressing question becomes not whether the brain enters an energy crisis, but how early that crisis begins and how long it remains hidden.

Across decades of neuroscience, mitochondrial biology, and metabolic research, a consistent pattern has emerged: neurodegenerative diseases begin long before clinical symptoms appear. Memory loss, movement disorders, and neuronal death represent late-stage events in a process that often unfolds silently for years, or even decades. At its core, this process reflects a progressive breakdown in cellular energy metabolism: an erosion of the brain's ability to generate, distribute, and flexibly use fuel.

One useful way to conceptualize this long preclinical phase is through the idea of a metabolic iceberg, a framework articulated clearly by Matthew C. Phillips and Martin Picard in a recent synthesis

(2024). What clinicians and patients ultimately observe the symptoms that lead to diagnosis, form only the exposed tip. Beneath the surface lies a much larger and largely invisible structure of metabolic stress, including impaired energy handling, redox imbalance, disrupted cellular signaling, and maladaptive stress responses. These submerged processes accumulate quietly over time, setting the stage for disease to finally emerge.

Viewed through this lens, Alzheimer's disease, Parkinson's disease, amyotrophic lateral sclerosis (ALS), and Huntington's disease no longer appear as isolated pathologies. Despite their distinct clinical features, each reflects a shared vulnerability: an energy-intensive organ attempting to function amid progressively constrained or dysregulated metabolic capacity. The defining differences between these disorders lie not in whether energy failure occurs, but in how, where, and when it manifests.

What follows applies the metabolic iceberg framework across four major neurodegenerative diseases. While symptoms differ, the underlying pattern is the same: long before diagnosis, metabolic stress accumulates beneath the surface.

Quick Comparison: The Four Metabolic Icebergs

- **Alzheimer's (AD):** progressive fuel scarcity and impaired glucose utilization *(energy shortage)*
- **Parkinson's (PD):** extreme energetic demand with cumulative oxidative load *(energy overload)*
- **ALS:** energy misdistribution across long motor neuron networks *(delivery failure)*
- **Huntington's (HD):** inherited mitochondrial toxicity and metabolic rigidity *(genetic constraint)*

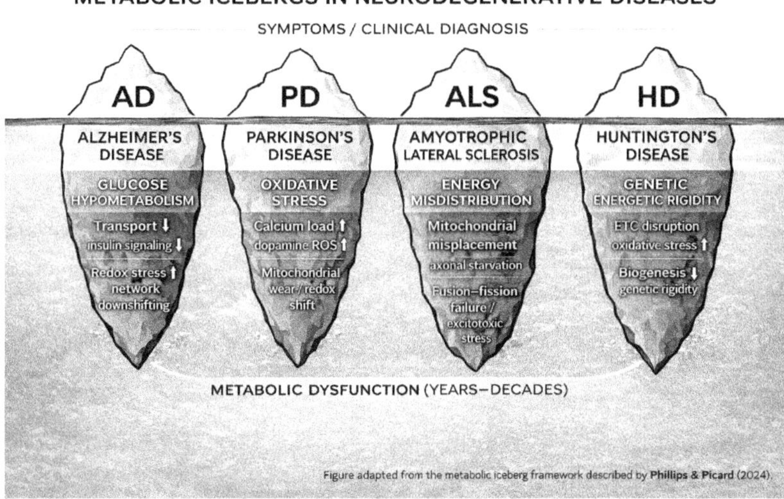

The Alzheimer's Iceberg (AD): Glucose Hypometabolism and Progressive Energy Failure

Alzheimer's disease can be understood as a disorder of impaired brain energy metabolism long before it becomes a disorder of synapse loss, memory failure, or cognitive decline. One of the most consistent and reproducible findings in Alzheimer's research comes from fluorodeoxyglucose positron emission tomography (FDG-PET), which reveals a measurable decline in cerebral glucose metabolism—particularly in the temporal, parietal, and posterior cingulate regions—often 10 to 20 years before clinical symptoms emerge.

These hypometabolic patterns are observed not only in individuals with diagnosed Alzheimer's disease, but also in cognitively normal adults who later progress to mild cognitive impairment (MCI) and Alzheimer's dementia. In many cases, regional glucose hypometabolism precedes detectable amyloid deposition, tau pathology, or structural brain atrophy. This temporal sequence strongly suggests that energetic failure is not simply a downstream consequence of Alzheimer's pathology, but an early and predictive event in the disease trajectory. This early metabolic disruption forms the submerged base of the Alzheimer's iceberg.

The Submerged Layers of the Alzheimer's Iceberg

Viewed through a metabolic lens, Alzheimer's disease begins with a gradual erosion of the brain's ability to acquire, process, and efficiently use energy. These changes accumulate silently over years, shaping neuronal behavior long before outward symptoms appear.

Fuel availability and transport

- Chronic, brain-wide energy insufficiency
- Impaired glucose transport across the blood–brain barrier
- Reduced insulin signaling and diminished neuronal glucose uptake

Cellular energy utilization

- Inefficiencies in oxidative energy production
- Disrupted redox balance and declining NAD+ availability
- Elevated oxidative stress with reduced antioxidant capacity

Network-level compensation

- Synaptic pruning to reduce energetic demand
- Decreased neuronal firing and network complexity
- A progressive shift toward an energy-conserving functional state

Because neurons are among the most energy-demanding cells in the human body, even modest and sustained reductions in energy availability force adaptive tradeoffs. To remain viable, neurons reduce their functional footprint. Synapses are pruned, signaling becomes less flexible, and large-scale neural networks gradually simplify. Importantly, these changes are initially compensatory, allowing neurons to survive under constrained energetic conditions.

Only later do the familiar pathological hallmarks of Alzheimer's disease emerge, amyloid-β accumulation, tau hyperphosphorylation, microglial activation, neuroinflammation, and hippocampal atrophy. In this framework, plaques and tangles are not primary initiators of disease, but downstream manifestations of prolonged metabolic stress.

From Silent Energy Failure to Clinical Symptoms

As Alzheimer's disease advances, mitochondrial quality control deteriorates further, reactive oxygen species accumulate, inflammatory signaling intensifies, and neuronal communication becomes increasingly fragile. Cognitive symptoms—memory impairment, disorientation, slowed processing speed, and executive dysfunction—represent the visible portion of the iceberg finally breaking the surface after years or decades of submerged metabolic decline.

From this perspective, Alzheimer's disease is less a sudden neurodegenerative event and more a slow-burning energy crisis unfolding within the brain's most metabolically vulnerable networks.

Implications of the Metabolic Iceberg Model

The metabolic iceberg framework helps explain why strategies that support brain energy metabolism appear most effective early in the disease course. Interventions that enhance metabolic flexibility, such as increased ketone availability, physical activity, caloric cycling, and other hormetic stressors—may help supply alternative fuels or strengthen cellular energy systems when glucose utilization begins to falter.

Therapeutic ketosis, intermittent fasting, and related metabolic strategies do not target plaques or tangles directly. Instead, they act beneath the surface, addressing the submerged mass of the iceberg by improving cellular energy availability and resilience. While these approaches are unlikely to reverse advanced neurodegeneration, they may help slow progression or preserve function when applied during the long preclinical phase of energetic vulnerability.

In this model, timing matters. Once the iceberg's tip has fully emerged, restoring energy becomes far more difficult. Long before that point, however, Alzheimer's disease may be best understood and most effectively addressed as a disorder of progressive energy failure rather than isolated protein pathology.

Alzheimer's may begin as a long, silent decline in brain fuel use. By the time memory fails, the underlying energy deficit has often been building for years.

The Parkinson's Iceberg (PD): High-Energy Neurons Overwhelmed by Oxidative Load

Parkinson's disease follows a distinct metabolic trajectory, yet it fits squarely within the iceberg framework. Unlike Alzheimer's disease, where early energy insufficiency dominates, Parkinson's disease arises from a paradox of extraordinary energetic demand coupled with cumulative oxidative stress. The submerged structure of the Parkinson's iceberg centers on a small but critically important population of neurons: dopaminergic neurons within the substantia nigra pars compacta.

These neurons are metabolically exceptional. Even under healthy conditions, they operate near the upper limits of cellular energy tolerance. Their vulnerability stems not from inactivity, but from the way they generate and sustain activity over a lifetime. Dopaminergic neurons are uniquely stressed because they:

- fire autonomously and continuously, independent of synaptic input
- rely heavily on voltage-gated calcium channels to maintain pacemaking activity
- require substantial ATP output to preserve ionic gradients

- generate large quantities of reactive oxygen species (ROS) as a byproduct of dopamine metabolism and mitochondrial respiration

Together, these features create a hyper-energetic, high-oxidative intracellular environment. For decades, dopaminergic neurons can maintain function despite this cost. Early in life, total ATP production may remain adequate, but the cumulative burden of oxidative byproducts slowly erodes mitochondrial integrity. Mitochondrial DNA sustains damage, electron transport chain complexes destabilize, calcium buffering capacity declines, and redox balance shifts unfavorably. This is the submerged Parkinson's iceberg: energetic excess gradually transforming into energetic collapse.

As mitochondrial resilience wanes, dopaminergic neurons fail disproportionately compared to neighboring neuronal populations. Their loss is not sudden, but progressive, reflecting years of accumulated oxidative and metabolic strain. Once compensatory capacity is exhausted, the visible tip of the iceberg emerges—manifesting as the hallmark motor features of Parkinson's disease:

- resting tremor
- bradykinesia
- muscular rigidity
- impaired gait and postural instability
- autonomic dysfunction

By the time these symptoms become clinically apparent, a substantial fraction of dopaminergic neurons has already been lost.

Amplifiers of the Submerged Iceberg

Environmental and systemic stressors can accelerate this submerged process. Exposure to pesticides, solvents, heavy metals, and herbicides adds oxidative pressure to neurons already operating near their metabolic limits. Epidemiological associations between Parkinson's disease and agricultural or industrial exposures support this amplification model.

Non-environmental factors further erode resilience. Sleep deprivation, chronic psychological stress, and systemic inflammation increase oxidative burden and impair mitochondrial repair mechanisms. Each factor subtly enlarges the submerged iceberg, shortening the window during which neurons can sustain compensatory function.

Metabolic Resilience and Mitohormesis

A key insight emerging from contemporary mitochondrial research is that vulnerability does not arise from energy demand alone, but from insufficient adaptive capacity. In this context, *mitohormesis*—the strengthening of mitochondrial defenses through controlled, recoverable stress—becomes particularly relevant.

Interventions such as physical exercise, intermittent fasting, heat exposure, and mitochondrial-supportive nutrition introduce manageable metabolic stress that upregulates antioxidant systems, enhances mitochondrial quality control, and improves energy efficiency. Rather than increasing energy output indiscriminately, these strategies reduce oxidative strain per unit of energy produced, easing the metabolic burden on vulnerable neurons.

Within the iceberg framework, mitohormetic stressors may help "melt" portions of the submerged Parkinson's iceberg—not by reversing neuron loss, but by reinforcing mitochondrial resilience before collapse occurs. As with Alzheimer's disease, timing is critical. Once the iceberg's tip has fully emerged, opportunities for meaningful intervention narrow dramatically.

Parkinson's begins where energy demand is highest—neurons that run "hot" for decades. Over time, oxidative strain and mitochondrial wear turn high performance into gradual collapse.

The ALS Iceberg: Energy Misdistribution, Axonal Starvation, and Mitochondrial Disarray

Amyotrophic lateral sclerosis (ALS) does not present with memory loss or tremors, but with the progressive failure of voluntary muscle control driven by motor neuron degeneration. Within the metabolic iceberg framework, ALS represents a distinct form of energy failure, not primarily a deficit of total energy production, but a failure of energy delivery, organization, and maintenance. In ALS, neurons may initially generate sufficient ATP, yet be unable to transport, deploy, or sustain that energy where it is most urgently required. The submerged iceberg is therefore defined by spatial and organizational breakdown rather than simple energetic scarcity.

The Submerged Layers of the ALS Iceberg

The metabolic vulnerability of ALS centers on the extreme architecture of motor neurons and the systems required to sustain them over long distances.

Mitochondrial distribution and transport

- Impaired axonal transport of mitochondria along microtubules
- Failure to deliver energy to distal axons and neuromuscular junctions

Mitochondrial dynamics and quality control

- Disrupted fusion and fission dynamics
- Defective mitophagy with impaired removal of damaged mitochondria
- Accumulation of dysfunctional mitochondrial fragments

Inter-organelle signaling and excitotoxic stress

- Breakdown in endoplasmic reticulum–mitochondrial communication

- Impaired calcium handling and buffering
- Glutamate excitotoxicity and heightened neuronal stress

Inflammatory amplification
- Chronic inflammation at neuromuscular junctions
- Progressive loss of motor unit integrity

Motor neurons possess some of the longest axons in the human body, extending from the spinal cord to distal muscle fibers. Maintaining function across these distances requires exquisitely coordinated mitochondrial trafficking, fusion, repair, and signaling. When this system falters, energy becomes misdistributed, available in one region yet critically absent in another.

The result is a metabolic gradient of failure. Distal axonal segments and neuromuscular junctions become energy-starved first, initiating degeneration that progresses proximally toward the cell body. Importantly, this process can unfold even while proximal neuronal structures appear metabolically intact.

From Axonal Starvation to Clinical Disease

As mitochondrial organization deteriorates, motor neurons lose their ability to sustain synaptic transmission and muscle innervation. What ultimately rises above the surface—muscle fasciculations, progressive weakness, spasticity, dysarthria, and dysphagia—reflects the late-stage consequences of prolonged mitochondrial disarray rather than an abrupt neurodegenerative event.

By the time clinical symptoms are evident, extensive metabolic disruption has already occurred across axonal networks that can no longer be repaired or reinnervated.

Implications of the ALS Iceberg Model

Within this framework, ALS highlights a critical principle: energy availability alone is insufficient if energy cannot be delivered, maintained, and regulated at the cellular periphery. Stabilizing mitochondrial dynamics and preserving axonal energy distribution become central goals, even in the absence of curative therapies.

Lifestyle and metabolic strategies that support mitochondrial quality control—adequate sleep, nutrient-dense nutrition, resistance exercise within tolerance, and metabolic therapies including ketone support—aim to reinforce the submerged iceberg rather than reverse neuron loss. These approaches do not claim to cure ALS, but to preserve mitochondrial organization, reduce secondary stressors, and slow the cascade of axonal starvation.

This architecture-based view of energy failure, emphasized in contemporary mitochondrial research including work by Matthew C. Phillips and Martin Picard, clarifies why ALS cannot be understood through energy quantity alone. When cellular power plants cannot reach their destination, even an energy-rich system can fail.

ALS is not only a problem of making energy—it's a problem of delivering and sustaining energy across long motor neuron networks. When distribution fails, function fails.

The Huntington's Iceberg (HD): Genetic Mitochondrial Toxicity and Energetic Rigidity

Huntington's disease (HD) is genetically determined, yet its clinical course is profoundly shaped by metabolism. Unlike Alzheimer's disease, Parkinson's disease, or ALS—where metabolic failure emerges from accumulated or acquired stressors—HD begins with an inherited molecular disruption that directly impairs cellular energy systems. An expanded CAG repeat in the huntingtin gene produces a mutant protein that interferes with mitochondrial function, gene expression, and energy regulation from early life onward.

Within the metabolic iceberg framework, Huntington's disease represents a form of genetic mitochondrial toxicity, in which energetic stress is imposed continuously and systemically long before overt neurodegeneration becomes apparent.

The Submerged Layers of the Huntington's Iceberg

The huntingtin mutation disrupts energy metabolism across multiple, interconnected levels:

Mitochondrial structure and trafficking

- Impaired mitochondrial transport along axons and dendrites
- Abnormal mitochondrial localization within neurons
- Reduced capacity to meet localized energy demands

Mitochondrial respiration and redox balance

- Altered electron transport chain function
- Reduced ATP synthesis efficiency
- Elevated oxidative stress with impaired antioxidant defenses

Calcium handling and excitability

- Disrupted calcium buffering and signaling
- Increased vulnerability to excitotoxic stress

Transcriptional and metabolic regulation

- Widespread dysregulation of metabolic gene expression
- Impaired mitochondrial biogenesis signaling
- Reduced adaptive metabolic flexibility

Together, these disruptions produce a state of energetic rigidity—a diminished ability to adjust fuel utilization, mitochondrial output, or stress responses as demands change. Unlike the adaptive metabolic slowing seen in other neurodegenerative conditions, Huntington's disease constrains flexibility at a fundamental genetic level.

Early Formation of the Submerged Iceberg

The submerged Huntington's iceberg forms early—often during adolescence or young adulthood—decades before clinical symptoms emerge. Neuroimaging and metabolic studies demonstrate reduced cerebral glucose metabolism and mitochondrial dysfunction well before measurable brain atrophy or motor impairment. During this prolonged preclinical phase, neurons operate under chronic energetic strain, gradually accumulating damage that remains largely invisible. This extended silent period helps explain why structural degeneration can appear abrupt once symptoms manifest. By the time the iceberg's tip emerges, the metabolic foundation has already been compromised for years.

From Energetic Rigidity to Clinical Disease

When compensatory capacity is exhausted, the visible manifestations of Huntington's disease rise above the surface:
- involuntary choreiform movements
- progressive cognitive decline
- emotional instability and psychiatric symptoms
- executive dysfunction
- eventual severe motor and cognitive impairment

These features reflect not only neuronal loss, but the collapse of metabolic systems that once sustained neuronal resilience. The striatum and associated cortical networks, regions with high energetic demand and limited redundancy are particularly vulnerable to this long-standing metabolic stress.

Variability in Onset and Progression

The iceberg model helps explain the striking variability in age of onset and disease progression among individuals with the same genetic mutation. While CAG repeat length establishes a broad framework for risk, metabolic resilience appears to influence how long the iceberg remains submerged. Individuals with stronger mitochondrial quality control, greater baseline metabolic flexibility, or more effective cellular repair mechanisms may delay the transition from silent dysfunction to overt disease.

Implications of the Huntington's Iceberg Model

Although the genetic cause of Huntington's disease cannot currently be altered, the metabolic iceberg framework highlights potential avenues for modifying disease expression. Interventions that support mitochondrial turnover, enhance biogenesis, reduce oxidative burden, or provide alternative energy substrates, such as ketone-based metabolism may help stabilize the submerged iceberg and slow its ascent.

These strategies do not claim to reverse the genetic mutation or halt disease entirely. Rather, they aim to reinforce the energetic systems that determine how rapidly genetic stress translates into clinical decline. In Huntington's disease, perhaps more than in any other neurodegenerative condition, metabolism shapes the tempo of inevitability.

Huntington's starts with a genetic disruption that creates early metabolic rigidity. Symptoms appear late, after years of chronic energetic stress have quietly reduced resilience.

Mitohormesis: A Unifying Strategy Across All Icebergs

A central insight emerging from contemporary mitochondrial research is that mitochondria are adaptive systems. When stress is excessive or chronic, mitochondrial function deteriorates. When stress is brief, intermittent, and followed by recovery, it can strengthen mitochondrial capacity and reinforce the very systems that fail early in neurodegenerative disease. This adaptive response is known as *mitohormesis*, a concept articulated clearly in work by Matthew C. Phillips and Martin Picard, and increasingly supported across metabolic and neurological research.

Mitohormesis occurs when manageable metabolic challenges signal cells to initiate repair, regeneration, and expansion of energy capacity. Rather than accelerating degeneration, these signals activate protective pathways that enhance resilience at the level where neurodegenerative disease quietly begins—long before symptoms surface. Common mitohormetic stimuli include:

- aerobic and resistance exercise
- brief high-intensity efforts
- ketogenic metabolism and nutritional ketosis
- fasting and time-restricted eating
- heat exposure (such as sauna use)
- cold exposure
- mitochondrial-supportive nutrients (including creatine, spermidine, omega-3 fatty acids, and magnesium)
- circadian alignment and restorative sleep

Although these interventions differ in form, they converge on shared cellular responses. Mitohormetic signaling increases metabolic flexibility, activates energy-sensing and repair pathways (including AMPK and PGC-1α), strengthens antioxidant defenses, enhances mitochondrial turnover, and reduces chronic inflammatory signaling. Within the iceberg metaphor, mitohormesis acts beneath the surface—shrinking the hidden mass and delaying the rise of the symptomatic tip. The key insight is simple but consequential: the earlier metabolic reinforcement begins, the more of the iceberg can be reduced before it becomes clinically visible.

A Shared Root Problem with Different Manifestations

At first glance, Alzheimer's disease, Parkinson's disease, ALS, and Huntington's disease appear unrelated. Their symptoms differ. Their timelines vary. Their triggers—genetic, environmental, or age-related—are distinct. Yet when viewed through the metabolic iceberg framework, these conditions reveal a shared foundation. Across all four disorders, the same underlying vulnerabilities recur:

- mitochondrial dysfunction
- impaired energy distribution
- redox imbalance
- metabolic inflexibility
- failure of adaptive stress-response pathways
- disrupted cellular communication
- inflammation driven by energetic collapse

What differs is not whether the brain's energy systems fail, but where and how that failure concentrates most intensely:

- **Alzheimer's disease** reflects progressive energy scarcity and impaired fuel utilization
- **Parkinson's disease** reflects energy overload paired with cumulative oxidative strain
- **ALS** reflects energy misdistribution and breakdown of mitochondrial quality control
- **Huntington's disease** reflects inherited mitochondrial toxicity and rigid metabolic regulation

Once energy instability begins, disease-specific patterns unfold. But the underlying vulnerability—the brain's dependence on finely tuned, resilient energy systems—is universal. This shared foundation helps explain why seemingly distinct neurodegenerative diseases respond to similar metabolic signals, and why strategies that support mitochondrial resilience may influence the trajectory of many conditions at once. The iceberg model does not erase important distinctions—it reveals the common energetic terrain upon which they all develop.

CHAPTER TWENTY-SIX

FUELING MINDS IN AN ANXIOUS AGE

*"In matters of health, experimentation guided by
care is often wiser than certainty guided by dogma."*
CLINICAL PHILOSOPHY

A Crisis Taking Shape in a New Generation

Over the past decade, mental health challenges on college campuses have shifted from episodic concerns to a sustained public-health emergency. National surveys show sharp increases in depression, anxiety, attention difficulties, sleep disorders, and perceived stress among college-age adults. Counseling centers report demand far exceeding capacity, while prescriptions for antidepressants, anxiolytics, and sleep medications among young adults have risen steadily. Many students function academically while describing a parallel internal experience marked by low mood, cognitive fatigue, emotional blunting, and persistent exhaustion. These trends accelerated in the mid-2010s and intensified further after 2020, at rates that exceed population growth and cannot be fully explained by diagnostic awareness, social media exposure, or academic pressure alone.

Prevailing explanations for rising mental-health challenges among students emphasize social media exposure, academic pressure, and the volatility of modern life. These factors matter and are well documented. Yet over the past decade, a growing body of research has pointed toward a deeper biological layer underlying many of these struggles. Many young adults now exhibit markers of metabolic dysregulation—high dietary glycemic load, irregular eating patterns, impaired insulin sensitivity, disrupted sleep–wake cycles, and chronically activated stress responses. These physiological conditions directly affect the brain's energy environment. Beneath psychological symptoms, there is often neurological strain driven by unstable energy availability.

If mental health depends, at its foundation, on the brain's capacity to produce steady and reliable energy, then the widespread deterioration of metabolic health among young people becomes highly relevant. From this perspective, the crisis unfolding on college campuses is not only emotional or cultural in origin, but increasingly metabolic, a mismatch between modern lifestyles and the energetic demands of the human brain.

It is within this context that the KIND Study (Ketogenic Intervention for Neuropsychiatric Disorders), led by Jeff Volek and colleagues, becomes especially important. Its question was deceptively simple: could changing how students fuel their bodies—and therefore their brains—meaningfully improve mood, resilience, and mental clarity? The answer, as the study found, was not only yes—but yes in ways that were rapid, measurable, and clinically significant.

The Brain as an Energy Organ

Understanding why diet can influence mental health requires rethinking the brain not as an abstract center of thoughts and emotions, but as a physical organ with immense metabolic demands. We often describe depression, anxiety, or irritability as psychological experiences—but they are also energetic states.

The human brain accounts for only about two percent of body weight, yet it consumes nearly twenty percent of the body's resting energy. Even modest disruptions in fuel availability can compromise how neurons fire, how networks synchronize, and how neurotransmitters are released and recycled.

Under modern dietary patterns dominated by refined carbohydrates, the brain becomes heavily dependent on glucose—a fuel that fluctuates sharply with diet, stress, and daily routines. When meals are carbohydrate-heavy and irregular, blood sugar rises and falls in pronounced waves. Insulin spikes become frequent. Over time, this volatility destabilizes the brain's energy environment. Neurons pass through periods of relative fuel deprivation despite abundant glucose in circulation. This mismatch—fuel present but functionally inaccessible—creates metabolic stress with far-reaching consequences. Students often describe symptoms such as:

- mental fog or difficulty concentrating
- sudden waves of anxiety
- trouble regulating mood
- reliance on caffeine just to feel baseline
- mid-afternoon crashes
- feeling overwhelmed by relatively small stressors

These experiences are often framed as emotional or psychological problems. The physiology suggests something more fundamental. They are also predictable consequences of an unstable energy supply within the brain.

Ketones: A More Reliable Fuel for a Stressed Brain

When carbohydrate intake is reduced and the body shifts toward fat metabolism, the liver produces ketone bodies—including beta-hydroxybutyrate (BHB)—which readily cross the blood–brain barrier and serve as an alternative fuel for neurons. Ketones are not, as they are often portrayed, an emergency backup. They are an evolutionarily conserved energy source that supported the human brain during fasting, seasonal scarcity, and long intervals between meals.

Why Ketones Support Brain Function

1. Stable fuel supply Ketones provide a steady, non-fluctuating energy source. Unlike glucose, which rises and falls with dietary intake and stress, ketones remain stable for hours once produced, supporting smoother and more consistent neuronal activity.

2. Efficient energy production Ketones generate ATP efficiently. They yield more usable energy per unit of oxygen while producing fewer reactive oxygen species, allowing cellular energy systems to operate with less oxidative strain.

3. Calmer neural signaling Ketones shift neurotransmitter balance toward GABA, the brain's primary inhibitory neurotransmitter. Individuals in nutritional ketosis frequently report reductions in anxiety, mental noise, and sensory overstimulation.

4. Reduced inflammation Ketones reduce inflammation both systemically and within the brain—an effect that matters deeply, as neuroinflammation is increasingly implicated in depression, anxiety, and mood instability.

5. Mitochondrial resilience Ketones promote mild mitochondrial uncoupling, a process now recognized as protective rather than wasteful. By lowering oxidative stress and activating repair pathways, uncoupling enhances cellular resilience over time.

Taken together, ketones offer a powerful means of stabilizing the brain's energy environment, particularly in individuals whose glucose metabolism has become unreliable.

The KIND Study: How It Was Designed

This intervention was designed to test these ideas directly in a real-world college environment. Researchers recruited 24 undergraduate and graduate students reporting clinically meaningful symptoms of depression, anxiety, perceived stress, or emotional dysregulation—conditions that were intentionally treated as the population of interest rather than grounds for exclusion.

Participants entered a single-arm ketogenic diet intervention lasting 10–12 weeks. This timeframe was selected to align with an academic term while allowing sufficient time for metabolic adaptation. Sixteen participants completed the full protocol, a completion rate that compares favorably with other behavioral and dietary interventions conducted in college settings.

Students received structured ketogenic support throughout the intervention, including dietary education, food selection guidance, meal-planning strategies, and ongoing encouragement to navigate the practical constraints of campus life. Blood beta-hydroxybutyrate (BHB) levels were measured regularly to objectively verify adherence. Among study completers, BHB levels met or exceeded 0.5 mmol/L— the threshold for nutritional ketosis—during approximately 73 percent of measured time points, confirming consistent achievement of the intended metabolic state.

Mental-health outcomes were assessed using validated psychological instruments, tracking changes in depression severity, anxiety symptoms, perceived stress, emotional regulation, and cognitive clarity over the intervention period. This design allowed psychological changes to be evaluated alongside objectively measured metabolic shifts, rather than inferred from self-report alone. At its core, the study asked a simple but consequential question: **If metabolic stability improves, will emotional and cognitive stability follow?**

What the Study Found: Mood Follows Metabolism

The findings were both notable and internally consistent. Among the 16 participants who completed the ketogenic intervention, validated mental-health assessments showed statistically significant reductions in depressive symptoms, along with meaningful improvements in anxiety, perceived stress, and emotional regulation over the 10–12-week study period. Participants also reported increases in emotional resilience, calm, and overall psychological well-being.

Beyond mood, the intervention produced measurable physiological and cognitive changes. Participants experienced reductions in body mass index (BMI) despite the absence of calorie restriction or weight-loss goals, suggesting improved metabolic efficiency rather than simple energy deprivation. Cognitive testing demonstrated improvements in attention, executive function, and mental clarity, aligning closely with participants' subjective reports of improved focus and reduced cognitive fatigue.

Equally important was the speed of response. Improvements in mood, emotional stability, and cognitive function emerged within the first several weeks of the intervention—earlier than the typical response timeline associated with first-line antidepressant therapies. Many participants described feeling less reactive, more emotionally grounded, and better able to manage academic and social demands as the study progressed.

Taken together, these outcomes reinforce a central principle of metabolic psychiatry: when the brain's energy environment becomes more stable, improvements in mood, cognition, and emotional regulation often follow.

Why the Ketogenic Diet Improved Mental Health

Several interconnected mechanisms likely explain the KIND Study's findings. Ketones provide the brain with a stable, high-efficiency fuel, smoothing the energy fluctuations that destabilize mood, attention, and emotional regulation. At the cellular level, ketone metabolism supports more efficient ATP production while reducing oxidative stress. This improved energetic environment helps normalize neuronal signaling and network stability.

Ketones also influence neurotransmitter balance, particularly along the GABA–glutamate axis. By favoring inhibitory signaling, they tend to dampen neural overactivation—a process closely tied to anxiety, irritability, and emotional reactivity. In parallel, ketones reduce systemic and neuroinflammation, a factor increasingly recognized as central to depression, fatigue, and cognitive dysfunction.

Finally, ketones promote mild mitochondrial uncoupling. Once viewed as wasteful, uncoupling is now understood as a protective adaptation: it lowers mitochondrial membrane potential, limits reactive oxygen species production, and activates cellular repair pathways. This same resilience-building response is observed with exercise, fasting, and other hormetic stressors. In this sense, ketosis functions as more than a fuel source. It acts as a metabolic signal that promotes cellular resilience and renewal.

Broader Implications: Changing the Future of Mental Health Care

Taken together, the results suggest that many young adults struggling with mental health may also be struggling with energy. Their symptoms need not be interpreted as personal failure, weakness, or lack of resilience, but may instead reflect a state of metabolic overload—an energetic system strained by poor diet quality, chronic stress, disrupted sleep, and irregular routines.

If this perspective holds, then mental health care must expand its scope. Counseling and medication remain important and often necessary tools, but they may be insufficient when the brain's underlying energy environment remains unstable. Nutritional strategies that support steady fuel availability and restore metabolic flexibility deserve a more central role in both prevention and treatment.

Universities are uniquely positioned to lead this shift. By improving food environments, offering nutrition education, incorporating metabolic screening, and supporting dietary interventions as part of student wellness programs, institutions can address mental health upstream rather than only responding after symptoms escalate. The stakes are substantial: the long-term cognitive, emotional, and professional trajectories of an entire generation may depend on restoring a metabolic environment that better supports brain resilience.

Challenges and Realities

Ketogenic diets can be powerful, but they are not effortless. Students may encounter social friction around eating, limited food options on campus, and the transient discomfort that can accompany metabolic adaptation. Responses also vary. Some individuals experience changes quickly, while others improve more gradually. Genetic factors, sleep quality, stress load, and physical activity all influence outcomes.

Concerns about long-term sustainability are common. However, emerging evidence suggests that meaningful benefits do not require strict or permanent carbohydrate elimination. Moderate carbohydrate reduction, intermittent ketosis, or cyclic ketogenic approaches may provide many of the same advantages. The objective is not dietary perfection, but metabolic stability that can be maintained over time.

The Larger Story: Returning to Energy

The KIND Study ultimately reinforces the central theme of this book: modern life has quietly disconnected us from the natural energy rhythms that once shaped human biology. For most of human history, metabolism was dynamic. Our ancestors regularly shifted between glucose and ketones through fasting, physical activity, seasonal scarcity, and changing food availability. These cycles were not incidental—they forged metabolic resilience.

Today, those rhythms are largely absent. Constant access to refined carbohydrates has created near-continuous glucose dependence, with ketosis occurring rarely and often only during illness or extreme deprivation. The result is a fragile metabolic state—one that leaves the brain more vulnerable to stress, emotional volatility, and cognitive fatigue. Reintroducing ketones, then, is not simply a dietary preference. It represents a reconnection to an older and more robust metabolic design. In that sense, it is not about restriction, but restoration—a deliberate return to energy.

Conclusion: A Doorway to a Brighter Future

The KIND Study is more than a research project. It serves as a signal for a broader rethinking of mental health—one that places brain energy at the center of mood, cognition, and emotional regulation. The findings demonstrate that stabilizing brain energy through ketogenic metabolism can produce rapid, meaningful improvements in mood, focus, and emotional stability in young adults. They also suggest that the mental health crisis is not solely psychological, social, or cultural in origin. It is, in important ways, biological.

If the brain is fundamentally an energy organ, then how it is fueled matters. The ketogenic diet is not a universal solution, nor does it replace other forms of care. But it opens a doorway—one that points toward improved resilience, restored metabolic stability, and a generation better equipped not merely to endure the pressures of modern life, but to meet them with clarity and strength.

PART VI

Cancer, Cells, and Energy Failure

CHAPTER TWENTY-SEVEN

CANCER: A BROKEN ENERGY STATE

"Cancer is, above all, a metabolic disease."
OTTO WARBURG

We all know someone who has died from cancer. It may be a parent, a spouse, or a sibling—someone whose illness rearranged the emotional structure of the household. Sometimes it is a friend or coworker we watched deteriorate over months or years. Other times it's someone more distant, yet their name and story linger long after their funeral. The grief has layers. For the patient, there is suffering, uncertainty, and fear. For the family, there is helplessness, confusion, and the unspoken dread that the next scan or lab result might bring devastating news.

Cancer reshapes lives in ways few other conditions do. It pulls people into unexpected roles—caregiver, advocate, medical interpreter—roles they never imagined and certainly never asked for. Days are spent sitting in hospital rooms, waiting for labs, scans, and treatment decisions, measuring time in test results and physician rounds. This is the most advanced cancer care system ever assembled, built on more than a century of accumulated knowledge, research, and clinical practice. And yet, even within that sophistication, patients and families often encounter a profound sense of helplessness. Cancer takes more than physical health. It takes time, peace, independence, and the quiet confidence people once had in their own bodies.

In the United States, the burden of cancer is immense and unrelenting. According to the American Cancer Society's *Cancer Facts & Figures 2025*, an estimated 2,041,910 new cancer cases will be diagnosed this year, and approximately 618,120 Americans are projected to die from the disease. That translates to about 1,700 cancer-related deaths every day—nearly one American every minute. Despite decades of scientific progress and modest declines in age-adjusted mortality rates, cancer remains the second leading cause of death nationwide, accounting for nearly one in five deaths.

And for all the tremendous human cost, the disease itself remains poorly understood by the public. Most people experience cancer not as biology, but as a sequence of tightly scheduled medical events— imaging, biopsies, pathology reports, appointments with oncologists, meetings with surgeons, and discussions of radiation, chemotherapy, and immunotherapy. Care unfolds across a growing cast of specialists, each responsible for a narrow domain, with information moving between teams faster than it can be processed by the patient. The pace is dizzying. The language is foreign. The fear is suffocating. Even within a highly advanced system, patients are often carried forward by momentum rather than understanding. One procedure leads to another. There is little space to ask deeper questions. There is almost no time to understand the nature of the disease itself.

Instead, medicine often treats cancer like an invading enemy— something to be found and destroyed. The approach prioritizes eliminating visible disease, with harm to the body anticipated and often accepted as part of treatment. In this frame, cancer becomes depersonalized—reduced to scans, lab markers, and standard treatment plans. The person behind the diagnosis is treated, but not always understood. Their biology is managed, but rarely explained. And for most patients, the deeper cause of the disease remains unknown. **At some point,**

the question becomes unavoidable: do we truly understand what cancer is? For all the urgency, the action, and the certainty surrounding treatment, there is remarkably little space to ask deeper questions about the nature of the disease itself. Cancer is treated aggressively, studied endlessly, and spoken about with confidence—but understanding is often assumed rather than examined. **This way of thinking extends beyond the clinic and into culture itself.** The familiar slogans "fight cancer," "raise awareness," "find a cure" have become cultural fixtures. Yet the outcomes have barely budged. Since President Richard Nixon declared a national "War on Cancer" in 1971, billions have been poured into research, infrastructure, and therapies. Nearly every country has its own version of this fight, often involving massive financial and institutional commitments. But despite all of it, cancer remains one of the leading causes of death worldwide.

We run the same races. We walk for cancer. We host the same fundraisers. We wear the same ribbons. The loop feels unbroken: **fundraise → fight → funeral.** So where are the breakthroughs? Where is the cure? Where is the progress that justifies the hope? The gap between effort and outcome forces an uncomfortable question: **Do we truly understand what cancer is?**

The Standard Story: Cancer as a Genetic Disease

For more than half a century, the defining story of cancer has been largely accepted as this: cancer is a genetic disease caused by mutations in DNA. This view came to be understood as foundational—not because it was ever definitively proven, but because it was repeated so consistently, so authoritatively, and so publicly that it became almost a cultural reflex. If someone develops cancer, it is widely assumed that something must be wrong in their genes—something inherited, damaged, or corrupted. Over time, this assumption became deeply woven into medical thinking, shaping nearly every treatment pathway, research grant, and clinical decision.

But this story did not arrive as settled science. It took decades to form and solidify. Its roots trace back to the mid-20th century, as molecular biology emerged and DNA moved to the center of biomedical research. After Watson and Crick described the structure of DNA in 1953, the scientific world became captivated by the idea that genes could explain not only inheritance, but disease itself. Genetics began to feel like a master key. The more DNA was decoded, the more it seemed that the deepest causes of illness would finally be revealed.

Cancer research adopted this framework eagerly. In the 1970s and 1980s, scientists identified the first oncogenes—genes that, when mutated or overexpressed, appeared to drive uncontrolled cell growth. Cancer now seemed explainable at the molecular level. If altered genes could push cells toward unchecked division, it followed that repairing, blocking, or inhibiting those genes might stop the disease.

The discovery was widely seen as a breakthrough. It offered clarity in a field long defined by uncertainty. Researchers soon identified tumor suppressor genes—genes that normally restrain growth and, when disrupted, allow cancer to emerge. From this work, a simple and compelling framework took shape:

- **Oncogenes**—"accelerators" stuck in the on position
- **Tumor suppressors**—"brakes" that had failed

This model was elegant. Simple. Mechanistic. And it exploded into a full-blown research paradigm. By the 1990s, gene-centric thinking dominated oncology. Vast sums were poured into sequencing tumors, cataloging mutations, and organizing them into the first **cancer gene atlases**. Institutions built entire research programs around this framework. Pharmaceutical companies designed drugs specifically to target these mutated proteins. Medical students were taught that "cancer is caused by accumulated genetic mutations." Public messaging reinforced it. Advocacy groups repeated it. Even fundraising campaigns wrapped themselves around the idea of finding "the genetic cure."

And then came the **Human Genome Project**—the moment that promised to cement this theory into fact. When the initiative launched in 1990, the world was told that by mapping the full sequence of human DNA, scientists would identify the master genetic switches behind diseases like cancer. The expectation was enormous: once the genome was decoded, cancers would reveal their signature mutations. Each tumor type would have a definable, consistent genetic fingerprint. Once we had those fingerprints, we would design drugs to target them, interrupt them, and cure the disease at its root.

The optimism was overwhelming. Editorials in *Science* and *Nature* declared that medicine was on the cusp of a revolution. Congressional hearings promised breakthroughs. Pharmaceutical companies reoriented entire pipelines toward gene-targeted treatments. The public was told that the "mystery of cancer" would soon be solved. *But when the sequencing data started pouring in, the picture was not what anyone expected.*

Instead of finding clear, reproducible mutation patterns, researchers found chaos. Tumors of the same type had wildly different mutation profiles. The same cancer in two different patients often shared no common genetic drivers. Even more puzzling, many mutations branded as "drivers" were found in completely healthy tissue with no cancer present at all. Large-scale projects like The Cancer Genome Atlas (TCGA) and the International Cancer Genome Consortium (ICGC) kept discovering more complexity, not less.

Instead of the predicted few hundred cancer-causing mutations, genome-wide sequencing revealed thousands to tens of thousands of mutations within individual tumors. Instead of predictable patterns, there was randomness. Instead of clarity, confusion.

Still, the field pressed on, driven by massive institutional momentum. Genetic thinking was so deeply embedded in research funding, training programs, and pharmaceutical strategy that abandoning it was unthinkable. The logic became circular: if we just sequence more tumors, build bigger databases, use more computational tools, eventually the patterns will reveal themselves. But the patterns never consistently came together.

Meanwhile, gene-targeted therapies—heralded as the future of oncology—struggled in clinical practice. Drugs designed to inhibit specific mutations showed early promise in a few select cancers but quickly encountered resistance, relapse, or complete failure. Many extended life by only months. Some worked only in a fraction of patients with the same mutation. Others failed despite apparently perfect genetic matches. Still others worked initially but stopped when cancer cells adapted through alternative metabolic pathways.

By the early 2000s, gene-editing tools like CRISPR were widely celebrated as the next phase of the cancer revolution. The hope returned: if we could cut out or repair the mutations, the disease itself could be stopped. When mutated genes were corrected, cancer cells persisted. When new driver mutations were inserted into healthy cells, they didn't turn cancerous.

Something deeper was at play.

This was not a failure of effort. It was a failure of the underlying model. But because the genetic theory had become foundational—scientifically, institutionally, and financially—the field doubled down. Billions continued to flow into gene-targeted research. Medical conferences still organized sessions around mutation-based subtypes. Precision oncology programs kept profiling tumors and recommending targeted drugs, even though the majority of patients gained little or no benefit. The genetic story persisted not because it was working, but because the system had been built around it. What began as an inspiring theory hardened into an unquestioned doctrine. Yet beneath the surface, the data kept pointing to a different truth—one that would become impossible to ignore. And that leads to the next section.

The Cracks in the Genetic Narrative

For decades, the genetic theory of cancer held such a powerful grip on research that few stopped to question its underlying assumptions. It appeared elegant, scientific, and consistent with a broader cultural belief that DNA governs nearly everything about who we are. Over time, this framework became more than a hypothesis—it became the default lens through which cancer was studied, treated, and taught.

In medicine, this progression is not unusual. When a theory offers a coherent explanation and produces workable interventions, it is often adopted long before it is definitively proven. Scientific theories are not declarations of absolute truth; they are models—best current explanations built from available evidence, always provisional, always subject to revision as new data emerge. The somatic mutation theory of cancer followed this familiar path. It was accepted because it made sense, not because it was ever conclusively settled.

The somatic mutation theory rests on a simple premise: that cancer originates from the gradual accumulation of random DNA mutations in the nucleus of a single body cell. The term *somatic*—from the Greek *sōma*, meaning body—distinguishes these mutations from inherited germline changes. Within this framework, cancer is understood primarily as a genetic accident, with cellular metabolism, tissue organization, and environmental context treated as secondary effects rather than driving forces.

As genome sequencing technologies advanced, the data increasingly strained the genetic model. Tumors were found to carry far more mutations than expected, often with little consistency across patients—or even within different regions of the same tumor. What had been framed as a small set of causal errors began to look more like widespread genetic noise.

A more serious contradiction soon followed. In the 1980s and 1990s, increasingly sensitive sequencing revealed that many mutations thought to cause cancer were also present in normal, healthy tissue.

Researchers identified mutated oncogenes in skin, colon, breast tissue, and even in the blood of individuals with no signs of disease. These cells behaved normally. They divided normally. They did not form tumors. This directly challenged the idea that mutations alone were deterministic triggers of malignancy. If mutations were sufficient to cause cancer, their presence in healthy tissue should not have been so common—or so inconsequential.

As sequencing methods improved, the expectation was that clearer patterns would emerge. Instead, the opposite occurred. Tumors with the same clinical diagnosis—such as breast or colon cancer—continued to display entirely different mutational profiles, even as datasets grew larger and more detailed. Appeals to "complexity" became common, but complexity alone does not erase patterns; it reveals them. What stood out instead was the persistent absence of a consistent genetic signature.

By the late 1990s, as the Human Genome Project neared completion, many believed this problem would finally be resolved. With the full human genome in hand, researchers expected to overlay tumor sequences and identify the decisive genetic switches driving cancer. Instead, as sequencing deepened, the genetic disorder only intensified. More data did not bring clarity—it magnified the chaos.

Researchers attempted to preserve the paradigm by introducing new categories: "driver mutations," "passenger mutations," "subclonal mutations," "founder events," and "evolutionary trajectories." Each term was meant to account for why the genetic theory failed to align with the data. Over time, however, these distinctions functioned more as patches than solutions—a growing vocabulary of exceptions that exposed the model's limitations rather than resolving them. Then came the clinically devastating realization: Even when tumors did have identifiable driver mutations, targeting them often failed.

Precision oncology—the primary clinical expression of SMT—promised to match drugs to mutations. The idea was simple: if a tumor expressed a certain mutated protein, drugs designed to inhibit that pathway would kill the cancer. But the reality was far more disheartening. Drugs that worked in laboratory settings often failed in patients. Others initially reduced tumor size but stopped working after a few months as cancer cells switched metabolic pathways or activated alternative survival mechanisms. For many cancers, the genetic target was simply irrelevant to disease progression.

Studies began to reveal that most targeted therapies extended life by only weeks or months, not years. Many worked only in a small fraction of patients with the "correct" mutation. Others showed early promise but failed after metabolic adaptation. Some required combinations so toxic that the therapy itself became worse than the disease. Instead of validating the genetic model, precision oncology exposed its fragility. These failures were not isolated. They were systemic.

When the Experiment Broke the Theory

Another devastating blow came from nuclear–cytoplasmic transfer experiments—some dating back to the 1960s, others replicated across decades. These experiments directly tested the core assumption of SMT. If the nucleus truly contained the cancer-causing mutations, then transferring the nucleus of a cancer cell into a healthy cell should have produced malignant behavior.

But the results showed the opposite. Hybrid cells with cancer nuclei behaved normally as long as they inherited healthy mitochondria. Conversely, when healthy nuclei were placed into cytoplasm with damaged mitochondria, the resulting hybrid behaved like cancer.

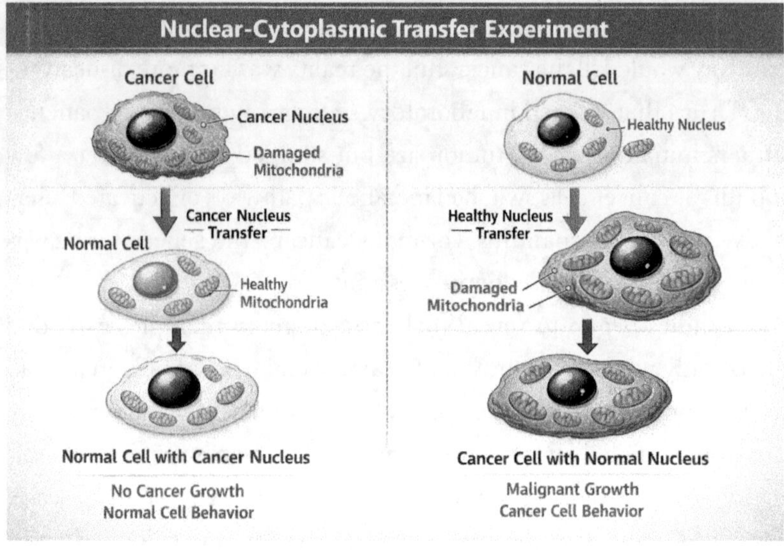

These results were not marginal. They were foundational. They suggested that the nucleus—including its DNA—does not determine malignancy. Instead, the cellular environment, the mitochondria, and the energetic state do.

Yet despite this evidence, the research establishment remained committed to the genetic paradigm. Billions in grants, countless careers, entire pharmaceutical pipelines, and decades of institutional momentum depended on it. The theory had become too big to fail— scientifically, financially, and politically. But the cracks kept widening.

Why do some cancers arise with no identifiable genetic trigger? Why do identical twins not share identical cancer risks? Why do cancers behave differently even when they share the same mutation? Why do tumors behave so inconsistently across patients and tissues? Why do cancer cells require such extraordinary amounts of fuel?

The genetic theory offered no coherent answers to these questions. By the time next-generation sequencing made it possible to analyze tumor genomes in extraordinary depth, the contradictions had become undeniable. Some cancers had so many mutations that no rational pathway could link them. Others had almost none—yet behaved aggressively. Some tumors mutated rapidly. Others hardly mutated at all. If mutations were the engine of cancer, why was the mutational landscape so chaotic, inconsistent, and biologically incoherent?

At this point, the evidence began pointing toward a different interpretation—one that explained the inconsistency rather than resisting it. Mutations looked less like causes and more like side effects. They appeared to be the genetic residue of a deeper cellular dysfunction. That deeper dysfunction was not in the DNA. It was in the mitochondria. The cell's energy system had collapsed. The genetic chaos was merely the debris.

This realization forces a deeper question. If cancer cannot be consistently explained by mutations alone, the search for causality must move beyond genes and DNA. The failure is not merely one of incomplete mapping, but of misplaced emphasis.

Cancer as a Metabolic Disorder: The Energy Origin

If the somatic mutation theory leaves us with contradictions, then we must look deeper, beyond genes and nuclei, beyond the familiar metaphors of "brakes" and "accelerators." We must go to the level that quietly governs everything else: energy. In modern medicine, we often start with the parts we can name and measure, genes, pathways, receptors, without first asking a simpler question: **what does a cell need in order to function at all?**

Before a cell divides, before a gene is expressed, before a protein folds or a membrane forms, the cell must first generate usable energy. Without it, nothing in biology proceeds—no repair, no signaling, no growth, no differentiation. This is not a philosophical claim; it is a constraint written into life itself. Long ago, complex cells emerged because they acquired a new way to generate energy—through the partnership we now call mitochondria. That energy surplus made higher-order life possible, and only then could the layers we often treat as primary—gene regulation, epigenetic control, coordinated tissue behavior—operate at scale. Energy is not a downstream consequence of life; it is the precondition that makes life possible. **This understanding is not new—it predates modern genetics and once sat at the center of how scientists thought about cellular health and disease.**

Long before cancer was studied genetically, scientists understood that the essence of cellular health is energy production. In the 19th century, the physiologist Claude Bernard emphasized that "the fixity of the internal environment" is the foundation of life. Louis Pasteur observed that yeast cells ferment glucose when deprived of oxygen—but stop fermenting and resume respiration when oxygen is reintroduced (later called the Pasteur effect). Early biochemists therefore studied cancer not as a genetic problem, but as a metabolic anomaly—cells behaving as though they were suffocating, even when surrounded by oxygen.

Then came Otto Warburg in the 1920s. Warburg discovered something astonishing: cancer cells ferment glucose into lactic acid even when oxygen is plentiful. This aerobic fermentation was not a random curiosity—it was a defining feature of cancer. Warburg proposed that cancer begins with damage to the cell's respiratory machinery, especially the mitochondria. When respiration fails, the cell falls back to fermentation, an ancient survival mechanism that predates mitochondrial evolution. Warburg's theory was bold, elegant, and rooted in observable, repeatable biology. But as genetics gained cultural dominance in the 1960s and 1970s, metabolism became a secondary consideration. The rise of molecular biology shifted focus away from energy and toward DNA. Genetic reductionism captured the imagination of institutions, and Warburg's insight receded into the background—not because it was disproven, but because a new scientific story had taken over. Yet the evidence remained: **When energy metabolism breaks, everything else breaks.**

Energy Is Structure

A cell's architecture is shaped by its energy supply. The mitochondrial membrane potential is not merely a voltage; it is the structural heartbeat of the cell. It powers the synthesis of ATP, the folding of proteins, the maintenance of ion gradients, and the organization of the cytoskeleton. When mitochondrial respiration falters, the cell's physical structure begins to destabilize. Membranes stiffen. Redox balance collapses. Organelles lose their functional integrity. Even the nuclear membrane becomes distorted when ATP is scarce. *It's no different than a home losing steady power: when energy drops, systems still exist, but they no longer function properly.*

This is why mitochondrial dysfunction is not a minor footnote in cancer biology—it is the first domino. When the energy system wobbles, the cell's architecture adapts to survive, often in ways that later appear malignant.

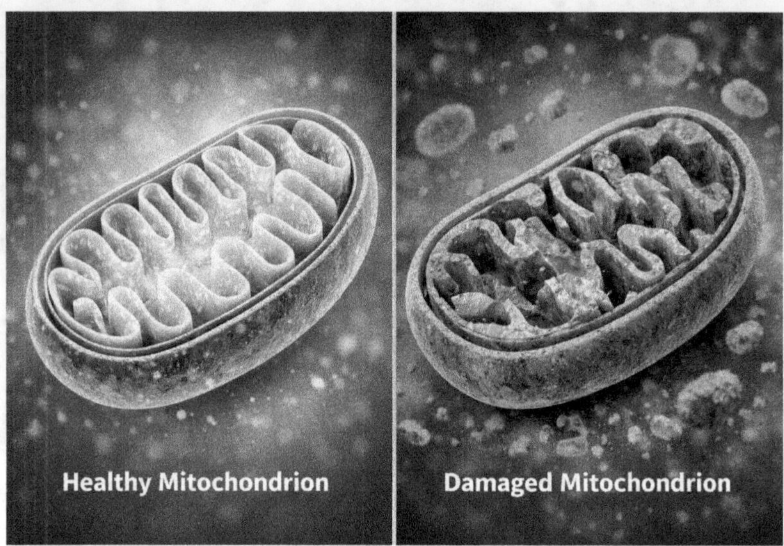

The contrast is visible at the level of the mitochondrion itself: healthy cells maintain dense, orderly cristae that support respiration, while metabolically impaired cells show swollen, fragmented cristae—structural collapse driven by energetic failure.

Energy Is Communication

Within every cell, thousands of signals move constantly between mitochondria and the nucleus. These signals help determine which genes are expressed, which proteins are synthesized, which pathways activate, and which remain silent. Healthy mitochondria are not simply ATP factories, they are metabolic conductors, coordinating the timing and flow of cellular messages.

When respiration—the process by which mitochondria generate usable energy from oxygen and nutrients— collapses, those messages become distorted. The nucleus receives incomplete or misleading information. Pathways normally regulated by ATP levels, redox balance, and mitochondrial metabolites begin to misfire. The cell's internal logic unravels. Genetic instability—long treated as the cause of cancer—appears instead to be the downstream consequence of disrupted cellular communication.

Energy Is Repair

DNA is constantly damaged by normal cellular processes. Thousands of breaks occur each day through oxidation, replication stress, and metabolic byproducts. A healthy cell repairs this damage quickly and efficiently—but only if it has adequate energy. DNA repair is one of the most energy-intensive processes in biology. When ATP levels fall and NAD+ levels collapse due to faulty respiration, repair enzymes slow or stop. Mutations accumulate not because the genome is inherently unstable, but because the energy required to maintain it is no longer available. This explains the chaos seen in tumor genomes. The mutations are not the initiating event. They are the genetic shrapnel of an energy system in decline.

Energy Is Identity

A cell's identity—whether it becomes a liver cell, a muscle cell, or a neuron—depends on metabolic cues. Stem cells rely heavily on glycolysis; differentiated cells rely on mitochondrial respiration. This metabolic choice shapes gene expression, epigenetic patterns, and cellular behavior. When mitochondrial function fails, cells revert toward a more primitive, less differentiated state. They lose their specialized identity. They become more stem-like, more proliferative, and more aggressive. This shift is not driven by new mutations, but by metabolic regression. Cancer cells are, in many ways, metabolically immature cells trapped in a fermentative, survival-first state.

Energy Is Fate

When a cell can no longer produce energy through respiration, it faces a choice: die, or adapt. Cancer represents the most extreme form of adaptation—a desperate pivot toward fermentation. This shift is not random. It is not accidental. It is a survival mechanism encoded deep within evolutionary history—an ancient metabolic fallback that predates oxygen-rich life. From this perspective, cancer is not a genetic mistake. It is a metabolic strategy:

- less efficient
- less regulated
- less stable
- but capable of keeping the cell alive when respiration fails

This strategy carries costs. Fermentation requires enormous amounts of glucose and glutamine. It disrupts redox balance. It produces lactic acid. It destabilizes the genome. It forces cells into rapid division to maintain energy flow. From this vantage point, the hallmarks of cancer—uncontrolled growth, angiogenesis, acidity, and genetic chaos—are no longer mysterious. They are predictable outcomes of metabolic desperation.

Energy Collapse Explains What Genetics Cannot

When viewed through the metabolic lens, cancer's incoherent genetic landscape suddenly makes sense:

- Why do tumors contain thousands of seemingly random mutations? Because DNA repair is energy-dependent.
- Why do cancers rely so heavily on glucose? Because fermentation cannot meet ATP demands without massive substrate intake.
- Why do PET scans illuminate tumors using radioactive glucose? Because cancer cells are metabolically addicted to sugar.
- Why do cancer cells upregulate glutamine transporters? Because glutamine becomes a second fermentable fuel for survival.
- Why do targeted genetic therapies so often fail? Because they treat downstream symptoms of energy collapse, not its origin.

- Why does mitochondrial dysfunction appear across nearly every cancer type? Because cancer begins in the mitochondria, not the nucleus.

Energy explains behavior.
Energy explains structure.
Energy explains survival.
Energy explains mutation.
Energy explains malignancy.

Cancer is not the result of random genetic chaos. It is the predictable expression of a cell trying to survive after its most fundamental energy pathway has failed.

Why the Metabolic View Unifies the Puzzle

The metabolic origin of cancer offers something the genetic model never fully achieved: **coherence**. It ties together decades of observations that once appeared unrelated or paradoxical. It explains why different cancers behave similarly despite entirely different genetic profiles. It accounts for why cancer cells exhibit metabolic rigidity while healthy cells retain flexibility. It also clarifies why treatments that damage mitochondria—such as chemotherapy and radiation—can be followed by more aggressive recurrences. Most importantly, this perspective reframes cancer in a way that opens new therapeutic possibilities—not by attacking the genome, but by restoring the energetic environment in which healthy biology can function.

Metabolic Rigidity vs. Metabolic Flexibility

Healthy cells possess a remarkable property that often goes unnoticed until it is lost: metabolic flexibility. This flexibility is the hallmark of a resilient, well-functioning bioenergetic system. A healthy cell can shift seamlessly between fuel sources based on its environment, workload, and available substrates. When glucose is abundant, it can use glucose. When glucose is scarce, it can burn fatty acids. During fasting or increased physical demand, it can switch to ketones—an efficient, clean-burning fuel that supports mitochondrial stability. When needed, certain tissues can also draw on amino acids or lactate as backup substrates.

In a metabolically flexible organism, cells are not locked into a single mode. This adaptability allows the body to withstand fasting, exercise, injury, stress, and fluctuations in nutrient supply. Flexibility is resilience. Flexibility is survival. Flexibility is health.

Cancer cells are the opposite. They are metabolically rigid. Their mitochondria—damaged, unstable, and unable to sustain oxidative phosphorylation—cannot burn fats effectively. They cannot use ketones efficiently. They cannot shift between fuel sources. Instead, they are trapped in a metabolic corner, forced to rely on a small set of primitive, fermentable substrates that require little or no mitochondrial participation:

- glucose
- glutamine

This rigidity is not a quirk—it is the defining vulnerability of cancer biology. Fermentation is an ancient fallback system, preserved from a time before oxygen was abundant on Earth. It produces just enough ATP to keep a cell alive, but not enough to sustain a differentiated, well-regulated identity. Cancer cells are stuck in this metabolic basement. They cannot climb back up without repairing the very respiratory machinery that has failed.

This dependency is visible in real time. Every PET scan on Earth is a metabolic portrait. It does not illuminate mutated genes or abnormal chromosomes; it highlights glucose uptake. Tumors glow because they burn sugar at ferocious rates to feed their fermentative machinery. The brightness seen on a PET scan is not detecting a mutation—it is detecting a metabolic addiction.

When oncology acknowledges this—even implicitly through imaging—it unintentionally validates the metabolic interpretation of cancer. We are already measuring cancer's energy signature. The question is why we are not treating it accordingly.

Reinterpreting the Hallmarks of Cancer Through a Metabolic Lens

Conventional oncology describes cancer through a familiar set of hallmarks: uncontrolled growth, angiogenesis, genomic instability, resistance to apoptosis, and others. These features are typically presented as downstream consequences of damaged DNA. When viewed through an energetic framework, however, a far more coherent pattern emerges.

Uncontrolled Growth: A Symptom of Energy Desperation

Fermentation is profoundly inefficient. A cell relying on glycolysis must consume large quantities of glucose to generate even a fraction of the ATP produced by oxidative phosphorylation. To survive this deficit, cancer cells increase substrate uptake, upregulate glucose transporters, and accelerate division. Rapid proliferation is not an act of genetic aggression—it is a compensatory response to energetic crisis.

Angiogenesis: A Metabolic Lifeline

Fermenting cells cannot survive without a continuous supply of fuel. They must induce new blood vessels to deliver the glucose and glutamine they require. Conventional oncology often frames angiogenesis as a genetically programmed act of conquest—as if cancer cells deliberately build supply lines. Through a metabolic lens, however, angiogenesis appears as a predictable, almost mechanical adaptation. The cell is starving; new vasculature becomes its only chance to survive.

Resistance to Apoptosis: A Mitochondrial Failure, Not a Genetic Strategy

Apoptosis—the programmed self-destruction of damaged cells—is governed largely by mitochondrial signaling. When mitochondrial function fails, apoptosis can no longer proceed. Cells that should die instead persist. This apparent "immortality" is not the result of a mutation that disables the cell's self-destruct mechanism; it is the consequence of mitochondrial collapse.

Genomic Instability: The Debris of Energy Failure

Fermentation disrupts redox balance and depletes ATP and NAD+. DNA repair enzymes require energy to maintain genomic integrity. When mitochondrial function declines, repair capacity declines with it. Mutations accumulate not because they are driving cancer, but because the cell no longer has the energy to fix them. Genetic chaos is the downstream wreckage of metabolic collapse. Viewed through this lens, cancer behaves less like a rogue genetic entity and more like a cell caught in a desperate, energy-driven survival program—a last-ditch metabolic adaptation to mitochondrial injury.

The Shift in Perspective

Seeing cancer as a metabolic disease is not an act of rebellion against scientific consensus. It is a return to biological first principles—a recognition that life is built on energy, that metabolism underlies every cellular function, and that mitochondria are not passive spectators but architects of cellular behavior. When respiration fails, the cell's internal order collapses. When energy falters, integrity falters—genomic, structural, epigenetic, and functional.

This perspective does not deny that genes matter; it simply rearranges the hierarchy. Genes respond to metabolic signals. They adapt to redox shifts—the balance between reduction (gaining electrons) and oxidation (losing electrons) inside the cell that shapes how efficiently mitochondria generate energy and how stable cellular structures remain. Genes react to changes in ATP availability, oxygen levels, and mitochondrial metabolites. DNA is dynamic—not a tyrant, but a participant. It reflects the cellular environment; it does not dictate it.

Under this metabolic framework:

- cancer is not destiny
- genes are not fate
- mutations are not the root cause
- mitochondria are central, not peripheral
- metabolism shapes genomics, not the reverse
- energy determines cellular behavior

This reframing does not trivialize cancer; it clarifies it. It organizes decades of confusing and contradictory data into a coherent picture. It offers testable hypotheses, measurable biomarkers, and therapeutic strategies that address cancer where it begins—not where it ends.

The metabolic interpretation reconnects cancer to the broader landscape of chronic disease. It reveals that cancer does not emerge in isolation, but arises from the same metabolic stresses that drive diabetes, Alzheimer's disease, obesity, fatty liver disease, neurodegeneration, and accelerated aging. When the metabolic groundwork deteriorates, cancer finds fertile soil. In this sense, modern metabolic biology echoes a much older insight—that disease reflects adaptation to a damaged internal environment, not the sudden appearance of something foreign or random.

This chapter has laid the emotional and conceptual foundation—bringing cancer down from the abstract realm of genes and placing it back into the living, dynamic world of cellular biology. But this is only the beginning. What follows goes deeper. Much deeper.

It explores the modern revival of the Warburg hypothesis and the emergence of the mitochondrial metabolic theory of cancer—sharpened, expanded, and refined through the 2025 synthesis. It examines:

- how lipid droplets signal chronic mitochondrial stress
- how cells transition from acute to chronic energetic failure
- how fermentative rigidity traps cancer in an ancient metabolic state
- how glutamine becomes a second indispensable fuel
- how mitochondrial dysfunction generates the genomic chaos seen in tumors
- how metabolism—not mutation—predicts malignancy
- and what this means for therapy, prevention, and survival

What comes next does not simply add more detail. It provides resolution. It completes the picture. It bridges a century of science, from Warburg to the present, and reveals a cohesive, unifying metabolic framework that makes sense of cancer in a way genetics alone never has. The story now moves from the surface to the core. From observation to mechanism. From isolated findings to an integrated theory of disease.

That journey begins next.

CHAPTER TWENTY-EIGHT

THE WARBURG REVIVAL

*"The prime cause of cancer is the replacement
of the respiration of oxygen in normal body
cells by a fermentation of sugar."*
OTTO WARBURG

For nearly a century, cancer biology has been shaped by two competing narratives, two fundamentally different explanations for why a normal cell loses its identity and becomes malignant. The first, the genetic narrative, came to dominate the late twentieth century. It framed cancer as a disease of damaged DNA, drew billions in research funding, and guided the development of nearly every modern therapy. The second narrative, the metabolic one, predates the genetic revolution by decades. Rooted in the early laboratories of biochemistry and physiology, it was articulated most clearly by **Otto Warburg**, who argued that the defining feature of cancer is a fundamental breakdown in cellular energy production.

For a long time, this metabolic view receded into the background. The excitement surrounding genes, mutations, and sequencing technologies swept the field into a new era, and metabolism was increasingly treated as a relic of an earlier scientific age. But scientific ideas do not disappear simply because priorities shift or funding follows new tools. They resurface when evidence demands it. And over the last two decades, that evidence has accumulated—quietly at first, and now unmistakably.

As genetic research delivered diminishing explanatory returns, mitochondrial biology advanced, and powerful new technologies emerged: metabolomics, live-cell imaging, isotope tracing, redox mapping, a more complete picture began to take shape. What once appeared to be a collection of genetic accidents now looks deeply tied to a failure of cellular energy systems. Hallmarks of cancer that seemed unrelated now converge into a coherent metabolic pattern. Across disciplines as diverse as biochemistry, physiology, oncology, and systems biology, the same conclusion has emerged: the metabolic explanation was never wrong; it was incomplete. Today, armed with tools Warburg could never have imagined, that theory is reassembling into something far more expansive, and far more precise.

This chapter builds that modern picture. It is not a nostalgic return to old ideas, nor an attempt to resurrect a discarded theory. It is the recognition that Warburg's original insight pointed toward a deeper truth—one contemporary science is finally equipped to illuminate. What follows is not merely a retelling of history, but the unveiling of a modern, mechanistic understanding of cancer as a disorder rooted in energy, metabolism, and cellular resilience.

A Return to First Principles—Warburg Rediscovered

Modern cancer biology often speaks as if the genetic narrative has always occupied the center of the field. But long before DNA sequencing or oncogene catalogues, cancer was approached through a different doorway—one grounded not in genes, but in energy. That doorway was opened by Otto Warburg. Working in the early twentieth century with scientific instruments that now appear primitive, Warburg made observations of such clarity and precision that they continue to shape the foundations of cancer research, even when the field is reluctant to acknowledge them. He asked a simple question: *How does a cancer cell make energy?* In answering it, he uncovered a truth that remains relevant to this day.

Warburg demonstrated, through direct biochemical measurements, that cancer cells suffer from a defect in cellular respiration. Even in the presence of abundant oxygen, a condition under which healthy cells rely on mitochondrial oxidative phosphorylation, cancer cells failed to use that pathway efficiently. Their oxygen consumption was abnormally low. Respiration remained suppressed regardless of how much oxygen was supplied. Instead, these cells generated energy through fermentation, producing large quantities of lactic acid despite full access to oxygen. This was not a minor quirk or an occasional anomaly. It was consistent across nearly every tumor Warburg examined. Aerobic fermentation became the defining metabolic signature of cancer.

Otto Heinrich Warburg (1883–1970). Public domain.

THE PARADOX

This core observation has been replicated for nearly a century. One of the most widely used diagnostic tools in oncology—the PET scan—is built directly on Warburg's discovery. PET imaging highlights tissues that consume glucose at unusually high rates. Tumors "light up" because they are consuming and fermenting glucose at extraordinary speed. When a PET scan reveals a bright hotspot, it is not detecting a genetic mutation. It is detecting the same metabolic phenomenon Warburg described in the 1920s. Few scientific discoveries have translated so directly into routine clinical practice.

What is especially remarkable is how much of Warburg's framework has endured. He argued that defective respiration was not merely associated with cancer but central to its origin—that once a cell could no longer generate energy efficiently through oxygen-dependent pathways, it reverted to a more primitive metabolic program. Modern mitochondrial biology has repeatedly confirmed this view. Malignant cells exhibit compromised respiratory function across tumor types. Electron transport chain activity is diminished. Mitochondrial membrane potential is altered. Structural abnormalities appear consistently under electron microscopy. These features are observed across tissues, species, and stages of disease. Their persistence over decades of research stands as one of the strongest validations of Warburg's early work.

Yet Warburg worked at the limits of his era. He could measure oxygen consumption and lactic acid production, but he could not map the intricate metabolic networks operating inside the cell. He did not know about glutamine metabolism, amino-acid flux, mitochondrial DNA damage, or lipid-driven signaling. He could not quantify reactive oxygen species or trace carbon atoms through biochemical pathways. As a result, parts of his interpretation were incomplete—not incorrect, but constrained by the tools available to him.

For example, Warburg correctly identified glucose fermentation as a defining metabolic hallmark of cancer, but he underestimated the role of glutamine—an amino acid now known to be essential for tumor survival. Modern metabolic tracing has revealed that glutamine sustains the TCA cycle, fuels nucleotide synthesis, maintains redox balance, and supports biomass production. If glucose is the spark that illuminates the PET scan, glutamine is the steady fuel that keeps the malignant engine running. Warburg could not have detected this dependency; the tools of stable isotope tracing and metabolomics did not yet exist.

Warburg also believed mitochondrial damage to be irreversible and absolute—a permanent collapse of respiration. Today, the picture is more nuanced. Some defects are structural and profound. Others are functional, involving enzymes, membrane lipids, or redox imbalances. Certain aspects of mitochondrial function can partially recover under low-glucose or ketotic conditions, while others deteriorate under chronic oxidative stress. Modern research shows that mitochondrial failure is not a binary on–off event, but a progressive decline across multiple interacting systems. This more nuanced view does not contradict Warburg; it extends him.

Another area beyond Warburg's reach was lipid metabolism. He did not know how heavily cancer cells depend on lipid remodeling, how lipid droplets accumulate as early indicators of metabolic distress, or how cardiolipin, an essential mitochondrial membrane lipid, deteriorates under oxidative pressure. These lipid-centric insights help explain why respiration collapses, why redox balance becomes unstable, and why fermentation becomes a fixed metabolic identity rather than a temporary adaptation.

Warburg also lacked a modern understanding of how metabolic dysfunction drives genetic instability. He observed the chaotic mutational landscape within tumors but could not mechanistically link it to energy failure. Today, mitochondrial research shows that impaired respiration increases reactive oxygen species, disrupts redox regulation, destabilizes nucleotides, impairs DNA-repair enzymes, and corrupts the epigenetic landscape. The genomic chaos of cancer is not the cause of malignant transformation, but its downstream consequence. Warburg recognized the pattern; modern science now explains the mechanism.

What makes Warburg's work so relevant today is not that he had all the answers, but that he asked the right question. He understood cancer as a disease of disordered energy metabolism long before molecular biology shifted the field toward genes. Now, with decades of research in mitochondrial biology, redox chemistry, metabolomics, and tumor microenvironment dynamics, the scientific community is returning to the conceptual ground he first mapped. The metabolic theory of cancer is no longer a historical curiosity—it is a modern framework supported across multiple branches of contemporary biology.

Warburg gave the field its first clear view of cancer's energetic core. Today, with far more powerful tools, we can see deeper into that core than he ever could. The refinements that have emerged—concerning glutamine, redox balance, lipid metabolism, mitochondrial structure, and genetic instability—do not replace his insight; they complete it.

These refinements form the launching point for the rest of this chapter. What follows is not a repetition of Warburg's ideas but an expansion—a modern synthesis that connects his early work to the latest discoveries in mitochondrial biology, metabolic flux, microenvironment dynamics, and therapeutic strategy. The return to first principles is not nostalgia. It is necessary. Only by understanding what Warburg truly saw—and what we can now see more clearly—can we move beyond genetic fragmentation toward metabolic coherence.

Chronic Mitochondrial Stress—The Long Descent into Substrate-Level Survival

One of the most important distinctions in modern cancer biology is the difference between stress a cell can recover from and stress it cannot. Acute mitochondrial stress—the kind produced by brief hypoxia, short-lived oxidative bursts, or temporary nutrient scarcity— is something healthy cells are designed to endure. These events activate repair: mitochondrial fusion and fission, autophagy, antioxidant deployment, and temporary shifts in fuel use. In a healthy organism, such stresses often increase resilience rather than cause lasting harm.

Cancer does not arise from these acute shocks. Its roots lie in chronic mitochondrial stress—slow, cumulative injury that erodes respiratory capacity over years or decades. This stress emerges from persistent hyperglycemia, chronically elevated insulin, low-grade inflammation, toxic exposures, disrupted circadian rhythms, fatty liver, chronic hypoxia, and the continuous production of reactive oxygen species that outpaces repair. None of these insults is catastrophic in isolation. Together, they steadily undermine the mitochondria's ability to sustain oxidative phosphorylation.

As damage accumulates, the electron transport chain becomes unstable, membrane potential declines, redox systems falter, and the cell can no longer meet its energetic demands through respiration alone. Initially, the cell compensates by drawing on backup pathways. But as respiratory capacity continues to erode, those compensations become the only option—and an irreversible shift occurs. The cell retreats into substrate-level phosphorylation: ATP production pathways that bypass the electron transport chain and extract energy directly from metabolic intermediates. This shift involves far more than classical "fermentation." It encompasses a suite of ancient survival pathways that together sustain malignant growth:

- **Cytosolic glycolysis**, generating ATP through phosphoglycerate kinase and pyruvate kinase when respiration is impaired.
- **Mitochondrial substrate-level phosphorylation (mSLP)** within the TCA cycle—particularly via succinyl-CoA synthetase—which becomes a critical ATP source when oxidative phosphorylation collapses.
- **Glutaminolysis**, supplying the TCA cycle and supporting ATP production independently of the electron transport chain.
- **Reductive carboxylation**, a reverse-flow pathway used to generate biomass and maintain redox balance when oxidative metabolism fails.

- **Amino-acid anaplerosis**, replenishing carbon skeletons to keep a truncated TCA cycle turning despite respiratory dysfunction.

None of these pathways is efficient. They are ancient, improvised mechanisms of survival. But together they generate just enough ATP—and just enough metabolic intermediates—to keep the cell alive and dividing. As substrate-level pathways dominate, the cell loses the structure and identity that depend on respiration. Differentiation erodes. Signaling becomes distorted. Redox balance collapses. DNA repair falters. The surrounding microenvironment grows acidic and inflamed. The cell no longer behaves as a modern, oxygen-dependent organism; it retreats into the metabolic logic of early life.

This descent is not triggered by a sudden burst of nuclear mutations. The chaotic genome that appears later reflects the consequences of respiratory failure, not its cause. The true beginning lies in the slow, chronic erosion of mitochondrial integrity—a prolonged collapse of energy metabolism that ultimately forces the cell into substrate-level survival.

THE PARADOX

Cancer begins not with a genetic decision, but with a metabolic surrender. It is not the result of a sudden rebellious act, but the endpoint of gradual energetic erosion that leaves the cell no alternative but to rely on primitive ATP-generating pathways that predate complex multicellular life.

Lipid Droplets—Early Signatures of Energetic Breakdown

One of the most underappreciated insights in modern cancer metabolism is the role of lipid droplets. For decades, these structures were viewed as little more than inert fat deposits—simple storage bins for excess fuel. Today, they are understood very differently. In cancer biology, lipid droplets are early, visible markers of mitochondrial distress, often appearing long before a cell becomes malignant. They accumulate not because a cell is "overfed," but because its energy machinery can no longer function properly.

In a healthy cell, fatty acids enter the mitochondria and are efficiently oxidized through β-oxidation and the electron transport chain. This process generates large amounts of ATP and feeds the respiratory network that sustains cellular identity. But when mitochondrial enzymes falter—due to chronic inflammation, oxidative damage, impaired cardiolipin structure, disrupted redox balance, or persistent nutrient overload—the cell loses its ability to burn fats cleanly. Fatty acids that would normally be oxidized begin to accumulate in the cytoplasm. Unable to process this incoming fuel, the cell sequesters it into lipid droplets as a protective response. This is not indulgence; it is triage.

Lipid droplet formation therefore signals that mitochondrial capacity has fallen below a critical threshold. The cell is receiving more substrate than it can oxidize. β-oxidation slows. The electron transport chain destabilizes. Redox systems can no longer keep pace—particularly when cardiolipin, the mitochondrial membrane lipid essential for respiratory enzyme function, has been damaged by reactive oxygen species. Overwhelmed by unburned fuel, the cell must divert it somewhere safe. Lipid droplets become that temporary refuge.

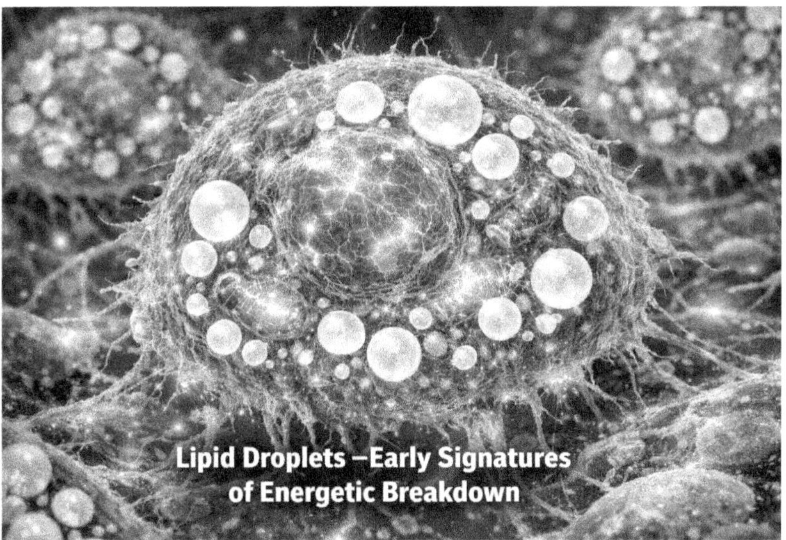

But lipid droplets are not inert. Once present, they begin to influence the cell's internal chemistry. They alter membrane composition, affect gene expression, and release signaling molecules that reshape the metabolic environment. They interfere with insulin signaling, modify the availability of lipid-derived intermediates, and participate in remodeling the tumor microenvironment. Over time, their presence reinforces the very metabolic rigidity that caused them to form, trapping the cell more deeply in a low-oxygen, low-respiration state. They become active participants in the descent into malignancy.

What makes lipid droplets especially important is their timing. They often appear before a cell becomes cancerous, serving as early metabolic fingerprints of mitochondrial decline. Their presence correlates with the severity of respiratory failure, predicts aggressiveness, and signals a shift toward substrate-level ATP production—precisely the metabolic compromise described in modern refinements of the Warburg model. They are not markers of overeating or caloric surplus. They are markers of energetic overload without oxidative capacity—cells drowning in fuel they cannot burn.

In this light, lipid droplets bridge historical observation and modern mechanism. Warburg recognized that cancer cells abandon oxidative phosphorylation early in transformation. Contemporary research now reveals one of the earliest structural consequences of that abandonment: the accumulation of lipids that should have been oxidized but could not be. Lipid droplets are not merely manifestations of metabolic failure—they are among its earliest visible warnings. They mark the point at which the cell's capacity to burn fuel begins to collapse, setting the stage for the fermentative, substrate-level survival strategies that follow.

THE PARADOX

Cancer cells are often described as "high-energy" because they consume large amounts of glucose. In reality, the opposite is true. Healthy cells generate energy efficiently through mitochondrial respiration, producing large amounts of ATP while maintaining redox balance, metabolic flexibility, and cellular order. Cancer cells, by contrast, consume fuel voraciously because their energy systems are inefficient. Fermentation and other substrate-level pathways generate ATP quickly but in very small amounts, forcing the cell to burn more fuel simply to survive. Metabolic rigidity is not a sign of excess energy–it is a sign of energetic failure.

Metabolic Rigidity—Trapping the Cell in Substrate-Level Survival

One of the defining features of healthy cellular metabolism is flexibility. A normal cell moves fluidly between fuel sources—glucose when it is abundant, fatty acids during fasting, ketones in low-insulin states, and amino acids during periods of repair or increased demand. This adaptability is not a luxury; it is foundational biology. Flexibility allows cells to withstand fluctuations in nutrient availability, environmental stress, and energetic demand. It is also a reliable signal that the mitochondria, the central hub of energy metabolism, are functioning properly.

Cancer cells, by contrast, lose this adaptability. As mitochondrial respiration declines under chronic stress, the cell becomes increasingly incapable of oxidizing fats or ketones. It can no longer transition smoothly between metabolic states. It becomes metabolically rigid. This loss of flexibility is not a secondary feature or a late-stage development. It is the core metabolic fingerprint of malignancy. A cell trapped in this rigid state can no longer rely on oxygen-dependent energy production. Instead, it must survive almost entirely through substrate-level pathways that bypass the electron transport chain.

This rigidity forces dependence on two primary substrates: glucose and glutamine. Glucose drives cytosolic glycolysis, producing ATP rapidly but inefficiently. Glutamine feeds the truncated TCA cycle, sustaining mitochondrial substrate-level phosphorylation and supplying the biosynthetic intermediates required for rapid proliferation. Together, these pathways form a metabolic lifeline. They generate just enough ATP to keep the cell alive and just enough carbon and nitrogen to support growth—but they never restore the efficient, integrated respiratory system the cell once possessed.

This dual dependency is not theoretical. It is observed consistently across metabolic analyses of tumors. PET imaging reveals relentless glucose uptake—one of the most reproducible clinical signatures of cancer. Tumor biopsies show elevated glutamine transporters and increased glutaminase activity. Isotope-tracing studies demonstrate accelerated flux through glycolysis, glutaminolysis, and mitochondrial substrate-level phosphorylation. Across tissues, species, and stages of disease, the same pattern emerges: respiratory capacity declines, while dependence on substrate-level energy production rises.

Crucially, this rigidity is not a minor inconvenience for the malignant cell—it is its greatest vulnerability. Healthy cells retain the ability to shift fuels as conditions change. They can increase fat oxidation, utilize ketones, slow glycolysis, or rely more heavily on amino acids when needed. Cancer cells cannot. Their damaged mitochondria prevent a return to oxidative phosphorylation even when oxygen is abundant. The fermentative and substrate-level pathways that sustain them are not adaptive choices; they are obligations. When glucose availability falls, when glutamine is restricted, or when redox stress exceeds the cell's buffering capacity, the malignant cell struggles to survive.

This inability to regain metabolic flexibility explains why metabolic therapies can exert selective pressure. Strategies that lower glucose, reduce insulin signaling, elevate ketones, or restrict glutamine create conditions that healthy cells tolerate but cancer cells cannot. Normal tissues shift fuels and maintain energy balance. Malignant cells, locked into rigid survival modes, lack that capacity. They remain trapped in an ancient metabolic corner—capable of persistence, but unable to return to the respiratory competence required for normal cellular identity.

THE PARADOX

In this sense, cancer does not simply grow; it retreats. It regresses toward a metabolic state reminiscent of early unicellular life, relying on primitive ATP-generating pathways that predate complex mitochondrial respiration. Metabolic rigidity is the signature of this regression. It explains why malignant cells proliferate rapidly yet inefficiently, resist apoptosis, acidify their surroundings, and reshape the microenvironment to secure fermentable fuels.

Metabolic rigidity is therefore not just a feature of cancer; it is the mechanism that binds the disease to its energetic origin. When respiration fails, flexibility disappears. When flexibility disappears, the cell becomes trapped in the very pathways that define malignancy. And once trapped, escape would require restoration of mitochondrial function—something a deeply damaged cancer cell is no longer capable of achieving.

Glutamine—The Missing Fuel Warburg Could Not See

Warburg's focus on glucose fermentation was revolutionary, but it reflected the scientific tools of his era. He could measure oxygen consumption and lactate production, yet he had no way to follow carbon or nitrogen atoms through the deeper architecture of cellular metabolism. Because of that limitation, he missed a second fuel source that modern science now recognizes as indispensable for cancer survival: glutamine.

Only with the advent of stable isotope tracing, metabolomics, and single-cell technologies did the centrality of glutamine become unmistakable. These tools revealed that glutamine is not a supplemental nutrient or a secondary contributor—it is a core pillar of malignant metabolism. As mitochondrial respiration falters, cancer cells become increasingly dependent on glutamine to keep the truncated TCA cycle turning. Glutamine replenishes carbon intermediates through anaplerosis, allowing mitochondrial substrate-level phosphorylation to continue even after oxidative phosphorylation has collapsed.

But glutamine's role extends far beyond ATP generation. It supplies the nitrogen required for rapid nucleotide synthesis, making it essential for DNA replication during uncontrolled proliferation. It supports lipid and membrane synthesis, enabling continuous expansion of cellular structure. It helps maintain redox balance by fueling glutathione production, buffering the oxidative stress generated by dysfunctional mitochondria. It also assists with ammonia handling, preventing toxic accumulation during accelerated amino-acid turnover. In parallel, glutamine-derived signals activate pathways that promote growth, survival, and adaptation within the evolving tumor microenvironment.

This deep entanglement between glutamine metabolism and cancer survival explains a long-standing paradox: restricting glucose alone often slows tumor growth but rarely collapses it. When glycolysis is constrained, malignant cells shift metabolic weight toward glutaminolysis, using glutamine to sustain both ATP production and biomass expansion. This narrow form of adaptability—flexibility within rigidity—is what gives cancer much of its resilience. Glucose and glutamine together form the dual lifeline of malignant cells: the metabolic scaffolding that sustains life once respiration fails.

Healthy cells do not share this dependence. They move fluidly across metabolic states, increasing fat oxidation, utilizing ketone bodies when glucose is scarce, and relying on intact respiration to maintain ATP production and redox balance. Cancer cells cannot. Their damaged mitochondria trap them in a state where glutamine becomes indispensable rather than optional. This asymmetry— metabolic flexibility in healthy tissue versus metabolic dependence in malignant cells—is the fundamental reason metabolic therapies can act selectively. By targeting fuels cancer cells cannot live without, while supporting the respiratory pathways healthy cells readily use, it becomes possible to exert pressure on tumors without harming metabolically adaptable tissues.

In this sense, glutamine is not merely a nutrient cancer happens to exploit. It is the molecule that reveals the deeper metabolic architecture Warburg could not see. It completes the picture he began, showing that malignant cells persist through a partnership of fermentable fuels—glucose driving glycolysis, glutamine sustaining the crippled TCA cycle, and both together enabling survival in the absence of functional respiration.

The Tumor Microenvironment—A Metabolic Ecosystem Built for Survival

Once a cell loses respiratory competence and shifts into substrate-level survival, the world around it begins to change. Cancer does not arise in isolation; it reshapes its surroundings so thoroughly that the environment becomes part of the disease itself. This evolving microenvironment reflects the same metabolic pressures unfolding within the malignant cell, forming a self-sustaining biochemical ecosystem that reinforces metabolic rigidity and survival without respiration.

One of the earliest consequences is acidification. Cells dependent on glycolysis and glutaminolysis generate large quantities of lactate and ammonia, which diffuse into the extracellular space. This acidic environment degrades collagen, weakens tissue architecture, and alters the behavior of neighboring cells. Acidification is not a side effect, it is a metabolic signature of substrate-level ATP production. It also shifts the balance of local control, because many normal cellular functions including immune surveillance perform poorly as pH falls.

As acidity increases, oxygen delivery becomes increasingly disordered. The abnormal vascular networks that form around tumors— thin, leaky, and poorly organized—reflect demand for nutrient influx rather than efficient perfusion. These vessels preferentially deliver glucose and glutamine while failing to restore oxygenation, deepening local hypoxia. Importantly, this hypoxia is not the cause of the fermentative phenotype but a consequence of it, even as it further entrenches reliance on oxygen-independent pathways.

The surrounding stromal cells undergo parallel reprogramming. Fibroblasts, normally responsible for maintaining tissue structure, are metabolically altered by signals released from the tumor. These cancer-associated fibroblasts export lactate, amino acids, and other intermediates that malignant cells can import and use. This is not a passive interaction but a coordinated metabolic coupling that supplies substrates the tumor cannot efficiently generate on its own. The tumor does not merely occupy tissue; it reorganizes it around its energetic needs.

Redox balance deteriorates across the microenvironment as well. Even as cancer cells rely on glutamine to buffer their own oxidative stress, they release reactive oxygen species and other chemically reactive intermediates into surrounding tissue. These molecules damage nearby cells, propagate mitochondrial stress, and expand the zone of energetic compromise. The microenvironment becomes a region of chronic injury rather than resolution.

Lipid metabolism shifts in parallel. Tumors recruit or stimulate nearby adipocytes, drawing in fatty acids they cannot efficiently oxidize but can repurpose for membrane synthesis, signaling platforms, and anabolic growth. Lipid-rich stromal environments are consistently associated with faster tumor progression and increased metastatic potential. The failure of β-oxidation does not render lipids useless; it redirects them from energy production toward structural and signaling roles that support proliferation.

Together, these changes give rise to a metabolic ecosystem—acidic, hypoxic, inflamed, redox-imbalanced, and structurally disordered. This environment is not orchestrated by genetic intent but emerges from the energetic constraints of cells that can no longer respire. Every feature supports substrate-level survival while suppressing the respiratory metabolism of healthy cells and immune defenses. The tumor microenvironment is therefore not a passive consequence of cancer. It is the terrain required for life without respiration. And once established, it becomes extraordinarily difficult to reverse, because each metabolic distortion reinforces the others.

In this way, the microenvironment is both a mirror and a multiplier of mitochondrial failure. It reflects the metabolic desperation of malignant cells and then amplifies it outward, spreading the same energetic logic into surrounding tissue. Cancer is not merely a cell gone wrong—it is a metabolic territory shaped by chronic energetic collapse.

At this point, an essential distinction must be made.
In healthy tissue, the immune system functions as both guardian and caretaker. Cytotoxic T cells identify and eliminate abnormal cells before they can establish dominance. Macrophages patrol tissues, clear debris, coordinate repair, and help restore order after injury. These roles are protective by design, guided by local signals that normally indicate infection, damage, or healing.

Within the tumor microenvironment, however, those same signals are distorted. The metabolic conditions created by respiratory failure—acidification, nutrient depletion, redox imbalance, and chronic mitochondrial stress—reshape immune behavior at its energetic core. Cells meant to destroy threats lose effectiveness, while cells meant to repair tissue are redirected. The immune system does not simply fail to stop cancer; it is coerced into responding as if cancer were a wound to be healed rather than a threat to be eliminated.

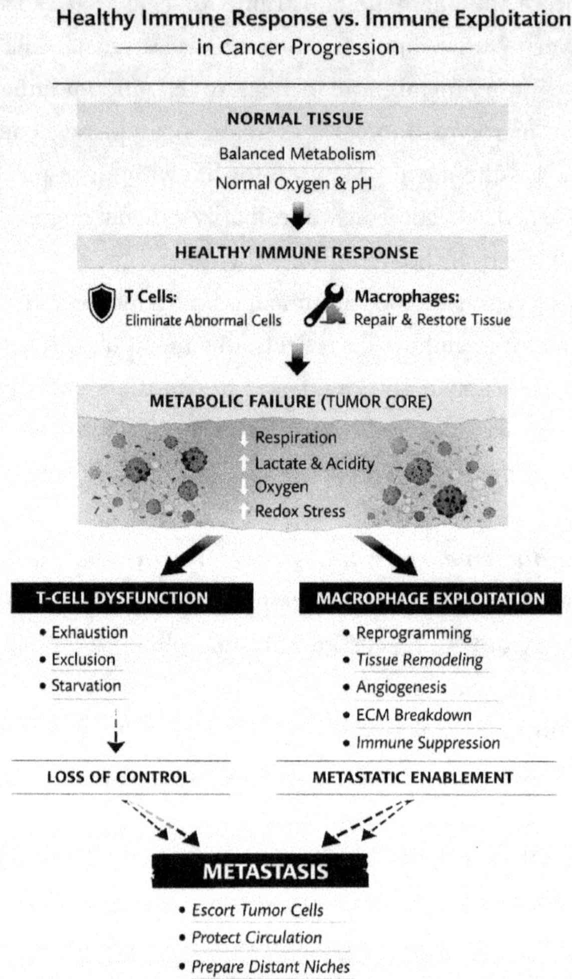

Exploiting Immune Cells for Metastasis— How Tumors Corrupt Defense into Infrastructure

Macrophages are uniquely susceptible to exploitation because they belong to the innate immune system and are designed to respond to environmental context rather than fixed instructions. In healthy physiology, this plasticity allows them to adapt to injury, coordinate repair, clear debris, and restore tissue integrity. Cytotoxic T cells, by contrast, operate under far stricter metabolic and signaling constraints. They require sustained energy availability, intact mitochondrial function, and precise antigen recognition to carry out targeted killing. Within the tumor microenvironment, this difference becomes decisive. What is normally a strength for macrophages becomes a liability.

As metabolic stress intensifies, macrophages are progressively reprogrammed by signals that reflect energetic collapse rather than true tissue damage. Lactate accumulation, low oxygen tension, altered lipid availability, and chronic redox stress shift macrophages away from defensive or cytotoxic roles and toward phenotypes that suppress immunity and remodel tissue. These signals do not announce malignancy; they mimic the metabolic cues of chronic injury. Tumor-associated macrophages (TAMs) do not merely tolerate cancer under these conditions—they actively enable its spread. **This distinction helps explain a long-standing paradox in cancer treatment.**

Tumors suppress immune function locally by reshaping their metabolic environment, but that suppression is often incomplete and potentially reversible. Cytotoxic T cells may become exhausted, excluded, or metabolically constrained, yet they frequently remain present within or near tumors and can regain function if energetic conditions improve. By contrast, systemic therapies that damage DNA through chemotherapy or radiation can profoundly deplete T cells throughout the body, impair mitochondrial function, and limit immune replenishment for extended periods. In this setting, immune pressure may collapse more completely than the tumor itself, subtly altering the balance between containment and escape.

In practical terms, a tumor does not need to be completely eliminated to gain an advantage; it only needs to persist while the forces holding it in check are weakened. These forces include immune surveillance, metabolic resilience, and the structural integrity of surrounding tissue. When treatment disrupts this balance—reducing tumor burden while also weakening the body's normal restraining systems, including immune surveillance—even a smaller amount of remaining disease may encounter less resistance than before. Cells that survive, particularly those already adapted to metabolic stress, can exploit this instability, shifting from a state of partial containment to renewed spread. In this context, outcome is shaped not by tumor size alone, but by the balance between forces that restrain disease and those that allow it to move.

One of the earliest barriers a tumor must overcome is structural. Healthy tissues are constrained by an organized extracellular matrix that limits cell movement and preserves tissue boundaries. TAMs dismantle this barrier by secreting enzymes that degrade collagen and matrix proteins, loosening the architecture that normally confines cells. This degradation does not produce random destruction. It creates defined migration paths—channels through which malignant cells can move with direction rather than chance.

TAMs simultaneously promote angiogenesis. By releasing vascular growth factors, they stimulate the formation of new, fragile blood vessels that penetrate the tumor mass. These vessels are poorly organized and leaky, but they serve a critical purpose: they bring circulation to the tumor rather than forcing tumor cells to breach intact vessel walls. Access replaces invasion.

Macrophages also act as guides. Malignant cells have been observed moving in close physical association with TAMs, following chemokine gradients toward blood vessels and lymphatic channels. Rather than escaping blindly, tumor cells are escorted. Macrophages sense tissue terrain, navigate structural boundaries, and effectively lead cancer cells to exits they could not locate independently.

Once tumor cells enter the circulation, macrophage-derived signals continue to provide protection. Circulating tumor cells are vulnerable to immune attack and mechanical stress. TAM-associated factors suppress immune recognition, dampen cytotoxic responses, and enhance survival during transit, increasing the likelihood that tumor cells reach distant tissues intact.

Even before arrival, macrophage-like cells help prepare meta-static niches. They alter distant tissue environments—adjusting immune tone, extracellular structure, and nutrient availability—so that arriving tumor cells encounter permissive terrain rather than resistance. Colonization becomes continuation rather than invasion.

Throughout this process, macrophages also suppress cytotoxic immune responses more broadly. T-cell activity is dampened, antigen presentation is altered, and immune tolerance is reinforced. Physical barriers are removed while immune barriers are silenced. Escape, travel, and settlement become coordinated steps within a single system.

THE PARADOX

In this context, metastasis is not a solitary act of malignant cells. It is a cooperative process driven by immune cells responding to a distorted metabolic environment. Tumor-associated macrophages function not as defenders, but as infrastructure—breaking walls, opening roads, escorting travelers, and preparing destinations. Cancer does not spread despite the immune system. It spreads, in part, through it.

From Fragmentation to Coherence

Understanding this changes how we think about cancer. It reveals the disease not as a collection of genetic accidents, but as a predictable response to sustained mitochondrial injury. Respiration falters. The cell retreats into substrate-level survival. Metabolic rigidity sets in. The surrounding microenvironment reorganizes itself around these energetic failures, reinforcing the malignant state. Beneath the genomic noise, the same metabolic architecture appears again and again. Seen through this lens, the genome is downstream. Energy is upstream. Disease follows the gradients of metabolism, not the blueprints of DNA.

What this model ultimately demands is not merely a new interpretation of cancer, but a reconsideration of how intervention should work when energy—not mutation—is the organizing principle. The next chapter moves from mechanism to application, translating this metabolic framework into the practical questions patients and clinicians must confront.

The metabolic story is not a rejection of the last century of science. It is its integration. It reconnects physiology with genetics, energy with signaling, environment with disease. And it returns us to a truth the earliest physiologists understood intuitively: when energy falters, biology follows.

CHAPTER TWENTY-NINE

PRESS–PULSE METABOLIC THERAPY FOR CANCER

"Biology adapts to rhythm, not to constant force."
HORMESIS RESEARCH

Therapeutic Implications— From Mechanism to Strategy

If cancer is organized by energy failure rather than genetic accident, then treatment cannot be judged solely by how effectively it destroys cells. It must be judged by how it reshapes the energetic environment in which those cells survive.

For decades, cancer therapy has been guided by a simple assumption: remove enough tumor, damage enough malignant cells, and the disease will retreat. But this logic presumes that cancer exists independently of its surroundings—that it can be attacked in isolation without consequence to the system that contains it. The metabolic model reveals a different reality. Cancer persists within a fragile equilibrium, stabilized by altered respiration, distorted fuel availability, and a microenvironment that adapts around energetic failure.

Once this is understood, a more precise question replaces the old one. Not how much tumor can be eliminated, but what happens to the system when intervention disrupts the balance that has been holding disease in check?

> ***To make this shift in thinking more intuitive, it helps to step briefly outside medicine.*** *Ecologists learned long ago that extinguishing every visible fire does not protect a forest. Suppressing flames without managing fuel load, underbrush, and soil health allows pressure to accumulate beneath the surface. When fire inevitably returns, it does so with greater intensity and less restraint. Modern fire management therefore focuses less on eliminating every blaze and more on reshaping the landscape—reducing excess fuel, restoring balance, and applying controlled burns under conditions that prevent catastrophic spread.*

Some therapies apply constant force—chronic toxicity, continuous pathway inhibition, sustained cellular stress. Others act in pulses—brief disruptions applied against a background that has already been reshaped. These two modes of intervention are not equivalent. When energy is the organizing principle, timing, sequence, and metabolic context matter as much as intensity.

The **Press–Pulse framework** arises from this recognition. It describes cancer therapy not as a single decisive strike, but as a coordinated strategy operating across time. One component establishes a sustained metabolic background that favors healthy cellular respiration and constrains fermentative survival. The other introduces intermittent, targeted stressors that exploit the rigid dependencies created by respiratory failure.

This approach does not seek to overpower cancer through escalating damage. It seeks to outmaneuver it—by strengthening the systems malignant cells cannot use and destabilizing the pathways they cannot escape. Healthy cells, metabolically flexible and respiratorily intact, adapt and recover. Cancer cells, locked into fragile survival modes, do not.

What follows is not a protocol and not a promise. It is a framework for thinking clearly about intervention when cancer is understood as a disease of constrained energy rather than uncontrolled growth. Before examining specific therapies, agents, or tools, the logic itself must be made explicit—because when energy is upstream, strategy must change.

The Metabolic Logic Behind Treatment

Cancer cells survive within narrow energetic constraints. Once oxidative respiration is impaired, they lose access to the flexible fuel economy that sustains healthy tissue. They cannot freely oxidize fats or ketones, cannot efficiently process glucose through the electron transport chain, and cannot tolerate wide fluctuations in energy availability. Survival depends instead on a limited set of conditions that must remain continuously favorable. In practice, this means malignant cells require:

- reliable glucose availability
- steady glutamine flux
- continuous substrate-level phosphorylation
- restricted oxygen utilization
- a hypoxic or pseudo-hypoxic microenvironment
- strong redox buffering to contain excess oxidative stress

These conditions are not optional. They are the scaffolding that allows fermentative survival to persist in the absence of functional respiration. Two fermentative systems provide the bulk of ATP under these constraints. One operates in the cytoplasm through glycolysis. The other operates within dysfunctional mitochondria through glutamine-driven substrate-level phosphorylation at succinyl-CoA. Together, these pathways allow short-term survival, but at the cost of rigidity. Cancer cells become metabolically trapped within the very systems that keep them alive. Healthy cells operate under a different set of rules. With intact mitochondria and functional respiration, they retain metabolic flexibility. They can switch between glucose, fatty acids, and ketones; tolerate fasting; adapt to fluctuations in oxygen availability; and restore redox balance without loss of identity or viability. This divergence—flexibility versus rigidity—is the central vulnerability Press–Pulse therapy exploits.

The Press: Building a Landscape Where Cancer Cannot Thrive

The Press establishes the metabolic terrain in which fermentative survival becomes increasingly difficult and respiratory metabolism is favored. It is neither extreme nor aggressive. It is physiological— designed to restore the conditions under which healthy cells function efficiently and malignant cells operate at a disadvantage. At its core, the Press reshapes fuel availability, signaling, and energetic flexibility. Rather than targeting cancer directly, it alters the environment that cancer depends on.

Fuel Restriction and Low-Insulin Physiology

Reducing dietary carbohydrates lowers circulating glucose and insulin, diminishing the primary signals that drive fermentative metabolism and anabolic growth. As insulin falls, healthy cells regain access to alternative fuels and restore metabolic flexibility. Cancer cells, by contrast, remain dependent on a continuous supply of fermentable substrates.

Therapeutic Ketosis

Ketones are efficient respiratory fuels for healthy cells and largely unusable by cells constrained to fermentation. As ketones rise, redox balance improves, mitochondrial efficiency increases, and the energetic gap between healthy and malignant cells widens. What nourishes one system destabilizes the other.

Fasting and Time-Restricted Feeding

Intermittent fasting further deepens this divide. Periods without nutrient intake lower insulin, reduce inflammatory signaling, restrict glutamine availability, and activate autophagy. Even modest fasting windows shift the metabolic landscape away from constant substrate supply and toward oxidative efficiency. Together, these elements do not starve the body. They restore access to stored energy and reintroduce variability into a system that cancer requires to remain fixed.

Metabolic Switching: The Functional Outcome of the Press

Metabolic switching—the ability to move seamlessly between glucose, fatty acids, and ketones—is a defining feature of metabolic health. Healthy cells retain this capacity. Cancer cells do not. The Press is designed to widen this divide. As metabolic flexibility returns, healthy tissues grow more resilient, adaptive, and energetically efficient. Cancer cells, locked into rigid fermentative pathways, are pushed closer to their limits. Under Press conditions, strength accumulates on one side of the equation. Constraint accumulates on the other.

The Pulse: Intermittent Strikes Against Fermentation

Once the Press has reshaped the metabolic landscape—lowering glucose, elevating ketones, restoring flexibility, and strengthening healthy cells—the Pulse introduces intermittent, targeted stresses that exploit the rigid dependencies of malignant metabolism. The Pulse is not a single intervention, but a category of timed metabolic stressors applied against specific points of fermentative dependence. These pulses are not designed to poison cells. They are applied briefly and deliberately, at moments when healthy tissues are metabolically prepared to adapt and recover. Cancer cells are not. Locked into fermentative survival and unable to shift fuels or restore respiratory balance, they experience these stresses as destabilizing rather than hormetic. The Pulse therefore functions as strategic intensification, not escalation. Its effectiveness depends on timing, metabolic context, and selectivity—not constant force.

Oxygen-Based Pulses: Revealing the Respiratory Defect

Healthy cells respond to increased oxygen availability by elevating oxidative phosphorylation and improving ATP efficiency. Cancer cells cannot. When oxygen rises, defective mitochondria generate oxidative stress instead of usable energy, exposing the respiratory weakness that fermentative metabolism normally conceals.

Hyperbaric oxygen therapy (HBOT) transiently increases tissue oxygenation well above atmospheric levels. When applied within a Pressed metabolic environment—characterized by low glucose and elevated ketones—it creates a selective energetic challenge. Healthy mitochondria increase ATP production and strengthen membrane potential, while cancer cells experience oxidative stress they cannot buffer as pseudo-hypoxic signaling is disrupted and respiratory defects are unmasked rather than compensated.

In this context, HBOT acts as a Pulse amplifier, increasing metabolic pressure precisely when malignant cells are least capable of adaptation. Unlike cytotoxic therapies that weaken both host and tumor, oxygen-based pulses strengthen one while destabilizing the other.

Glutamine Modulation: Targeting Mitochondrial Fermentation

Glutamine supports mitochondrial substrate-level phosphorylation, redox buffering, and biomass production in cells with impaired respiration. Interrupting glutamine availability places direct pressure on the fermentative machinery that malignant cells depend on for survival.

Agents explored in the literature illustrate how this pressure can be applied without functioning as traditional cytotoxins. One such example is DON (6-diazo-5-oxo-L-norleucine), a glutamine antagonist investigated for its ability to disrupt glutamine-dependent cancer metabolism. When glutamine flux is restricted in a Pressed environment, cancer cells lose a critical compensatory pathway. Healthy cells, supported by functional respiration and alternative fuels, tolerate this stress.

Glycolysis Modulation: Destabilizing Cytosolic ATP

Cytoplasmic glycolysis provides rapid ATP and lactate production, supporting both energy needs and the acidic microenvironment that protects tumors from immune detection. Disrupting glycolysis removes this buffer and destabilizes cytosolic ATP supply. Agents studied in this context include dichloroacetate (DCA), which activates pyruvate dehydrogenase and shifts metabolism away from lactate fermentation toward mitochondrial oxidation, and benzimidazole compounds such as **fenbendazole and mebendazole**, which in research models impair glucose handling and increase oxidative stress in malignant cells. When glycolysis falters, cancer cells lose cytosolic ATP support. Healthy cells compensate through oxidative metabolism.

Redox-Active Pulses: Pressuring a Fragile Balance

Cancer cells operate near the limits of redox tolerance, maintaining survival through heightened buffering systems in the face of chronic oxidative stress. Pulses that alter the NAD+/NADH ratio or glutathione cycling selectively destabilize this balance.

Healthy cells, supported by ketones and intact mitochondria, adapt. Cancer cells do not. Agents explored in this context include **ivermectin**, shown to disrupt mitochondrial membrane potential and redox signaling in malignant cells, and **metformin**, which lowers insulin, reduces hepatic glucose output, and mildly inhibits complex I, adding metabolic strain without direct toxicity.

Repurposed Metabolic Agents: Exploiting Mismatch

A recurring theme in metabolic oncology is the effectiveness of low-toxicity agents that do not act as poisons, but as stressors applied to a rigid system. Their value lies not in lethality, but in mismatch. Healthy cells shift, adapt, and recover. Cancer cells fracture under pressure.

> **The objective is not indiscriminate cell killing. It is the progressive cornering of fermentation.**

The Press sets the foundation. The Pulse applies pressure at exactly the right moment. Together, they exploit the rigidity of malignant metabolism—and the adaptability of healthy cells.

The GKI: A Metabolic Compass

The glucose–ketone index (GKI) functions as a directional marker, not a treatment. It translates two simple measurements—blood glucose and blood ketones—into a single value that reflects the body's current metabolic terrain.

By condensing fuel availability and alternative energy use into one number, the GKI provides a practical way to observe how the internal environment is shifting in response to the Press–Pulse framework. It does not diagnose disease, predict outcomes, or prescribe action.

> **It simply reveals where the system is operating along the spectrum from fermentative dependence to respiratory flexibility.**

The calculation itself is straightforward—glucose divided by ketones, both measured in mmol/L—but modern meters automate the process, displaying the GKI directly and logging trends over time. This removes arithmetic from the equation and allows attention to remain on pattern rather than precision.

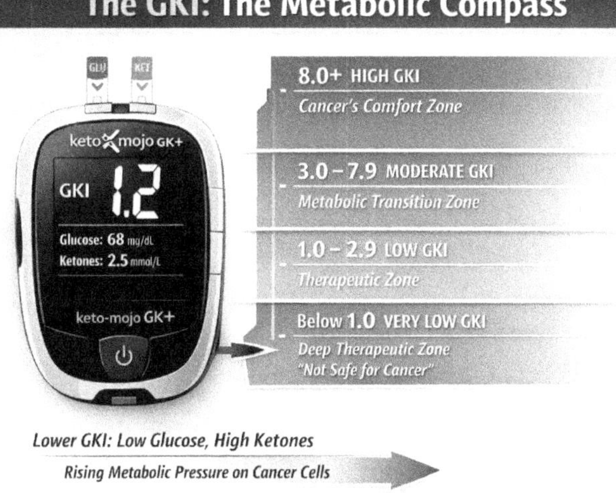

What makes the GKI useful is not the number itself, but what it represents. A **lower GKI** reflects a metabolic state in which glucose is constrained, insulin signaling is quiet, and ketones are available—conditions that support healthy cellular respiration while placing pressure on fermentation-dependent survival. A **higher GKI** reflects the opposite: an environment in which cancer cells operate with relative comfort.

Because metabolism is dynamic, the GKI is most informative when observed over time. Many patients measure it in the morning, when overnight fasting reveals baseline physiology. During fasting cycles or on days when metabolic pulses such as hyperbaric oxygen are applied, additional readings may illustrate how context and timing influence pressure on malignant metabolism.

As patterns emerge, individuals begin to see how daily rhythms—food choices, fasting duration, sleep quality, stress, movement, and recovery—shape the metabolic landscape. These influences are no longer abstract. They leave measurable fingerprints. In this way, the GKI becomes a compass rather than a target. A high GKI signals fermentative ease. A downward trend signals rising metabolic constraint. Sustained low values mark terrain in which malignant cells struggle to adapt. Used in this way, the GKI anchors Press–Pulse therapy in feedback rather than force. It allows both patients and clinicians to orient decisions within the biology of energy, without reducing treatment to protocols or numbers alone.

The Immune System: Recovering What Cancer Suppresses

A healthy metabolic environment restores immune function as a downstream consequence of energetic repair. Cytotoxic T lymphocytes (CD8+ T-cells), natural killer (NK) cells, and oxidative macrophages depend on intact mitochondrial respiration to activate, proliferate, and carry out cytotoxic work. Cancer suppresses these functions not by evasion alone, but by reshaping the metabolic environment in which immune cells operate.

Tumors create conditions that are acidic, hypoxic, glucose-hungry, and redox-stressed—an environment hostile to immune surveillance. Under these constraints, immune cells are deprived of oxygen, fuel flexibility, and respiratory capacity. Activation falters. Recognition weakens. Elimination fails.

When the metabolic landscape shifts, these barriers begin to dissolve. As fermentative pressure is reduced and respiration is supported, lactate levels fall, oxygenation improves, pH normalizes, and redox balance stabilizes. Mitochondrial respiration in immune cells increases, antigen presentation improves, and cytotoxic signaling reactivates. Immune cells regain the energetic capacity required to recognize and remove abnormal cells.

What emerges is not an artificially stimulated immune response, but the recovery of one that had been metabolically suppressed. Surveillance returns not through force, but through restoration.

From Metabolic Strategy to Systemic Coherence

Taken together, the Press, the Pulse, and the metabolic feedback provided by the GKI reshape more than isolated pathways. They alter the internal environment that governs cellular behavior across systems. As energy availability stabilizes and fermentative dependence is constrained, immune restraint reappears as a natural consequence of restored physiology.

This shift marks a transition—from fighting cancer as an isolated target to restoring coherence across the biological system that contains it. Strength returns not by escalation, but by alignment. The terrain changes. The rules change. And with them, the body's capacity to participate in its own defense. It is at this point that the framework moves from strategy into lived reality.

Living the Therapy: Rebuilding the
Daily Metabolic Landscape

Metabolic therapy is not confined to isolated interventions. It unfolds through daily rhythms—meals, fasting windows, movement, oxygenation, light exposure, stress cycles, and sleep. These patterns quietly shape the internal environment hour by hour, determining whether physiology supports respiratory flexibility or drifts back toward fermentative dependence. What follows is not a regimen, but a way of restoring alignment between biology and daily life.

Entering Therapeutic Ketosis

The most durable metabolic shifts occur through steady change rather than abrupt restriction. As sugars and refined carbohydrates are gradually reduced, insulin signaling quiets and access to stored energy returns. Nutrient-dense fats, adequate protein, and low-glycemic vegetables replace constant glucose intake, allowing metabolic switching to re-emerge naturally rather than by force. Ketosis, in this context, is not a goal to be chased but a state that arises when fuel availability and hormonal signaling are restored to physiologic ranges.

Fasting as Metabolic Renewal

Fasting is best understood not as deprivation, but as periodic metabolic renewal. Time without nutrient intake lowers insulin, reduces inflammatory signaling, activates autophagy, and shifts the body toward oxidative metabolism. Many individuals begin with a modest overnight fast and extend duration gradually as hunger stabilizes and metabolic flexibility improves. These intervals of restraint interrupt constant substrate delivery and reintroduce variability—conditions that healthy cells tolerate and cancer cells do not.

Oxygenation Through Movement and Timing

Movement supports oxygen delivery, mitochondrial efficiency, and metabolic signaling. Gentle activity after meals, structured exercise, and nasal breathing improve tissue oxygenation and reinforce respiratory metabolism. When oxygen-based pulses such as hyperbaric oxygen are used, their effect is greatest when timed within low-GKI states, where fermentative compensation is least available. Here again, timing matters more than intensity.

Nutrition That Supports Respiration

A respiratory-supportive diet emphasizes whole, minimally processed foods that stabilize energy delivery rather than spike it. Low-glycemic vegetables, quality fats, adequate protein, and mineral-rich broths support mitochondrial function while reducing reliance on fermentable substrates. Highly refined sugars, seed oils, and ultra-processed foods distort signaling and reintroduce the very conditions metabolic therapy seeks to correct.

Circadian Rhythm and Stress

Mitochondrial repair is inseparable from circadian alignment. Deep sleep, consistent timing, morning light exposure, and nighttime darkness regulate hormonal rhythms that govern glucose control, oxidative capacity, and cellular repair. Stress reduction—through breathwork, prayer, meditation, time in nature, and quiet—further stabilizes the internal environment by lowering counter-regulatory glucose signaling. Together, these rhythms do not constitute a separate therapy. They are the terrain on which metabolic strategy becomes lived reality.

Metabolic Therapy Versus Standard of Care

Most conventional cancer therapies are built around a difficult trade-off: malignant cells must be damaged or destroyed, even if healthy tissue is harmed in the process. This logic has shaped decades of treatment—chemotherapy, radiation, and many targeted agents that interfere with cell division or signaling. These approaches can reduce tumor burden, but they often do so by creating a harsh internal environment that affects all metabolically active cells, not malignant cells alone.

Metabolic therapy begins from a different organizing principle. Rather than attempting to injure cells directly, it focuses on reshaping the internal environment so that healthy physiology is supported while malignant survival becomes increasingly difficult. This selectivity does not depend on targeting specific genes, mutations, or receptors. It depends on metabolic flexibility—a capacity healthy cells retain and cancer cells largely lack.

Within a stable metabolic landscape, healthy cells function efficiently. They switch fuels, tolerate fasting, recover from stress, and preserve normal structure and signaling. Cancer cells cannot. Locked into fermentative metabolism and rigid fuel dependence, they struggle as glucose availability falls, insulin signaling quiets, and alternative energy sources dominate. Pressure is applied not through toxins, but through conditions the malignant cell cannot adapt to.

This difference carries practical consequences. Because metabolic therapy supports rather than undermines healthy tissue, it is generally non-toxic to the host and often improves energy, sleep quality, cognitive clarity, and physical resilience rather than degrading them. At the same time, it frequently addresses metabolic conditions—such as insulin resistance, fatty liver disease, obesity, and hypertension—that are common in cancer patients and are often exacerbated by conventional treatment.

Stabilizing metabolism also reduces physiological stress. As metabolic signaling quiets, cortisol output falls and excess glucose availability declines—important shifts in an environment where chronic stress and elevated glucose favor fermentative survival. Immune function benefits as well. Immune cells require energy to move, recognize threats, and execute cytotoxic work. When treatments exhaust the host, immune surveillance falters. When physiology is restored, immune restraint returns.

Equally important, metabolic therapy changes the patient's relationship to treatment. Daily rhythms—food timing, fasting intervals, sleep, stress regulation, and movement—become measurable influences on the internal environment rather than abstract recommendations. Tools such as the glucose–ketone index (GKI) make these effects visible, allowing both patients and clinicians to orient decisions within the biology of energy rather than guesswork.

Where standard-of-care treatments often weaken the host in order to damage the tumor, metabolic therapy aims to strengthen the system that contains the disease while selectively constraining malignant survival.

The shift is not from treatment to non-treatment, but from escalation to alignment, from force applied against cells to strategy applied to terrain.

Living the Press–Pulse Framework

In daily life, the Press–Pulse framework unfolds as a structured, sustainable rhythm rather than a constant state of crisis. It does not demand perpetual struggle or heroic endurance. Instead, it restores the conditions under which the body can participate in its own repair.

The Press forms the metabolic background—a steady physiology that reshapes the internal environment day after day. It is built from low-insulin living, therapeutic ketosis, daily fasting rhythms, nutrient-dense whole foods, consistent movement, improved oxygen-ation, and regular metabolic switching between fed and fasted states. Sleep and circadian alignment reinforce this foundation: morning light exposure, consistent sleep timing, and nighttime darkness stabilize hormonal signaling, reduce glucose variability, and support mitochon-drial repair.

These are not ancillary lifestyle suggestions or acts of self-denial. They are the biological terrain upon which all metabolic pressure is applied. Over time, the Press restores flexibility, resilience, and energetic efficiency in healthy tissues while quietly narrowing the survival options available to malignant cells.

The Pulse introduces intermittent intensification. These are time-limited, strategically timed stressors—hyperbaric oxygen sessions, fasting amplification, targeted modulation of glycolysis or glutamine metabolism, and carefully supervised redox challenges. Pulses are applied only when the body is metabolically prepared for them, not chronically or indiscriminately. Their purpose is not to exhaust the patient, but to exploit rigidity at moments when healthy systems can adapt and recover.

The logic is deliberate: The Press builds capacity. The Pulse exploits rigidity. Healthy cells adapt and recover. Cancer cells cannot.

Monitoring anchors the entire process. Modern meters automati-cally measure glucose, ketones, and the glucose–ketone index (GKI), translating daily choices into visible metabolic data without math or guesswork. Alongside these objective markers, subjective signals—energy, sleep quality, mental clarity, stress tolerance, and recovery—help fine-tune the pattern to the individual.

Over time, the GKI becomes more than a number. It becomes feedback, orientation, and agency—a metabolic compass guiding both patient and clinician.

Why This Matters

Cancer is frightening not only because of the disease itself, but because of the system that surrounds it. From the moment of diagnosis, many patients are drawn into a narrow pathway defined by urgency, authority, and fear. Options are often presented as fixed. Time is framed as scarce. The implicit message is that survival requires immediate surrender of agency and acceptance of harm as the unavoidable price of care.

THE FINAL PARADOX

Within this culture, patients are not simply treated; they are positioned. Strength becomes synonymous with compliance. Endurance is framed as virtue. Decline in energy, immune function, and overall vitality is normalized as evidence that treatment is "working." The patient's role quietly shifts from active participant to passive recipient—expected to tolerate toxicity, accept systemic injury, and hope that destruction of disease outweighs destruction of self.

Metabolic therapy begins from a fundamentally different premise. It does not require the patient to weaken in order for treatment to succeed. It does not assume that immune suppression, exhaustion, or metabolic collapse are necessary costs of care. Instead, it seeks to restore the internal conditions under which healthy physiology, and immune restraint, can function.

Crucially, this framework does not require rejection of conventional treatment. Metabolic therapy is not an ideological alternative and not an all-or-nothing proposition. It stands as a coherent strategy in its own right, and it may also be used alongside standard-of-care treatments when patients choose to do so. Preclinical models and emerging clinical experience suggest that improving metabolic stability can enhance treatment tolerance, preserve immune function, and reduce collateral injury during chemotherapy, radiation, or other interventions—without diminishing their intended effect.

The point is not that standard care must be accepted or rejected. The point is that patients should not be forced into a false binary. A healthcare system worthy of trust should support metabolic therapy whether it is used alone or in combination, and should educate clinicians to understand, study, and refine this paradigm rather than dismiss it as incompatible with existing practice.

By reshaping the internal environment rather than escalating injury, metabolic therapy reduces toxicity, preserves physiological capacity, and strengthens immune function instead of suppressing it. The patient is no longer treated as collateral damage in a war against disease, but as the terrain in which outcomes are decided. This shift changes the experience of illness as much as the biology. Fear gives way to orientation. Helplessness is replaced by feedback. The body, once experienced as hostile or failing, becomes intelligible again. Choice re-enters the picture—not as defiance, but as informed participation.

In this way, metabolic therapy reframes cancer care from enforced sacrifice to supported agency. It does not promise certainty or eliminate risk. But it restores something that too often disappears early in the cancer journey: the understanding that healing does not require the systematic weakening of the person it is meant to save.

A Call for Education and Discovery

If metabolic therapy is to mature responsibly, it must be studied, taught, and refined within mainstream medicine rather than remaining at its margins. Long before cancer was fragmented into pathways and mutations, early physiologists such as Otto Warburg recognized the central role of respiration and metabolic state in disease. Modern research has brought these questions back into focus, yet education has lagged behind discovery. Clinicians deserve training that treats nutrition, fasting, and therapeutic metabolic states not as peripheral lifestyle topics, but as core elements of physiology that shape treatment response, immune function, and resilience. A healthcare system committed to learning should support rigorous investigation of metabolic strategies—alone and in combination—so that future patients are guided not by fear or false binaries, but by evidence, understanding, and choice.

Returning to Design

CHAPTER THIRTY

A KETOGENIC START

*"All that man needs for health and healing
has been provided by God in nature; the
challenge of science is to find it."*
PARACELSUS

A New Beginning

There came a point when I realized I was searching for something deeper than another solution. Looking back now, I see that what I was really seeking was control over my own energy—clarity, steadiness, and health that didn't require layering one intervention on top of another. I wanted an approach that made sense biologically and felt right physically.

What I didn't know at the time was that the answer wasn't new at all. It wasn't invented in a lab or packaged in a supplement. It was already built into the human design—an ancient metabolic pathway that had simply faded from view. As I learned more, a question began to form—not as a challenge, but as an invitation to look again. What if fat and fasting weren't obstacles to health, but part of how the body was meant to function? What if the issue wasn't that the body was broken, but that it had been fueled in a way that no longer fit?

We decided to try something different. We didn't know if it would work long-term, and we didn't know how it would feel. But we were open to an approach grounded in human physiology rather than habit or assumption. We began a ketogenic lifestyle. And almost immediately, everything shifted.

The First Steps We Took: A Look Back

Although I touched on parts of our transition earlier in the book, the beginning matters. This is where uncertainty is highest—and where the most important shifts quietly take place. Looking back, our early steps were simple, almost deceptively so. Yet those small changes set the foundation for everything that followed.

The first change was removing processed foods from our home. We didn't ease into it or negotiate with ourselves. We cleared the shelves of the foods that had quietly shaped our hunger, cravings, and energy for years—boxes, bags, cereals, crackers, snacks engineered to be hyper-palatable and hard to stop eating. Once they were gone, something unexpected happened. The mental noise softened. The environment itself changed. Without constant visual triggers, better choices required far less effort.

Next came filling the kitchen with real food—food with texture, color, and nourishment. Avocados, leafy greens, eggs, meat, olives, berries, cheese, butter, olive oil. It wasn't a rigid list or a prescribed plan; it was simply a return to foods our bodies recognized. Meals shifted away from calorie anxiety and toward nutrient density. Learning to cook with fat—to enjoy it rather than avoid it—felt unfamiliar at first, but also freeing. It challenged habits that had been reinforced for decades.

Another shift followed naturally: learning to trust satiety again. For years, hunger had been something to manage, override, or anticipate. But once processed carbohydrates were removed, eating began to settle into a calmer rhythm. Three real meals a day—without snacking—became sufficient. Not through discipline, but because the meals themselves were satisfying. Hunger softened. Cravings loosened. Even emotional eating began to fade into the background.

There was also a short period of adaptation as the body shifted from glucose dependence to fat metabolism. We felt it—several days of fatigue and sugar cravings, but it passed quickly. What stood out most was how brief the discomfort was compared to the years of instability that had come before. Within weeks, clarity returned. Exercise capacity followed. Over time, it improved beyond what we had known.

Throughout this period, learning played a supportive role. We read, watched, and talked through what we were experiencing. Early feedback helped as well—not obsessively, but enough to confirm what we were feeling. Energy was stabilizing. Hunger was changing. The system was shifting.

None of these steps were dramatic, and none required perfection. They were simply intentional choices, made one day at a time, that aligned with how human metabolism is designed to work. And as ordinary as they may have seemed, they formed the foundation for everything that came next—understanding ketosis, embracing metabolic flexibility, and discovering a way of fueling the body that created stability rather than chaos. This was the beginning of our return to energy.

What Ketosis Really Is

Before moving further, it helps to clarify what ketosis actually is—and what it is not. The word itself often carries unnecessary weight, sometimes confused with extreme or dangerous states. Much of that confusion comes from conflating nutritional ketosis with diabetic ketoacidosis, a rare and serious condition that occurs only in uncontrolled type 1 diabetes. Despite sharing a name, the two are fundamentally different.

Nutritional ketosis is simply the body's ability to shift from burning glucose to burning fat. It is a form of metabolic flexibility built into human physiology—one that sustained our species for most of its history. When carbohydrates are limited, the liver converts fatty acids into ketone bodies—acetoacetate, beta-hydroxybutyrate, and acetone—which circulate through the bloodstream and provide a steady, efficient fuel for the brain, heart, and muscles.

Most people today rarely experience ketosis because modern eating patterns deliver a constant stream of carbohydrates. Frequent meals, snacks, sugar, and refined grains keep insulin elevated and the body locked into glucose-burning mode. As long as insulin remains high, fat stays stored and largely inaccessible.

When carbohydrates are reduced—through diet or fasting— insulin falls and fat metabolism reopens. Ketone production resumes. In healthy individuals, this shift is tightly regulated and stable, producing ketone levels well within a normal physiological range. The body responds quickly once given the opportunity. Within days, ketones appear. Within weeks, tissues become more efficient at using them. Hunger steadies. Energy smooths out. Cravings fade. Many people notice improvements in mood, focus, and sleep.

Ketosis is often described as a metabolic exhale—as if the body stops struggling against its environment and begins working with it again. This isn't because something new has been introduced, but because something familiar has been restored. Ketosis reflects the body remembering how to fuel itself during times when food is inconsistent, seasonal, or scarce.

It is also important to distinguish ketosis from starvation. In starvation, energy intake collapses and muscle tissue is broken down to survive. In nutritional ketosis, dietary fat and protein remain available, and ketones help preserve lean mass by reducing the body's need to convert muscle into glucose. This is why many people experience improved body composition, reduced inflammation, and greater physical resilience.

Physiologically, ketosis reveals how adaptable and resilient human metabolism truly is. It supports the brain under stress, sustains endurance when glycogen runs low, and delivers energy in a clean, efficient form that reduces oxidative strain.

In short, ketosis is not new. It is not experimental. It is ancient. It is normal. It is part of our biology. And when we step into it, we are not forcing the body into an unnatural state—we are allowing it to return to one.

The History of Keto and Fasting: A Return to Ancestral Health

To understand why this path felt so steady and familiar, it helps to look back. Long before the word *ketosis* existed, humans moved in and out of it as a normal part of survival. The expectation that carbohydrates must appear at every meal is a recent development—shaped by agriculture, abundance, and modern food systems. For most of human history, food availability rose and fell with seasons, migrations, and circumstance. Periods of abundance were naturally followed by periods of scarcity, and human biology adapted accordingly.

In those environments, ketosis wasn't a strategy or a protocol. It was simply part of being human. When food was limited or unpredictable, the body shifted to burning fat without hesitation. Ketones became a primary fuel for the brain and body, sustaining energy through lean times and ensuring survival.

Even in recorded history, fasting was widely recognized not only as a necessity, but as a source of healing and clarity. Ancient Greek physicians observed its benefits in conditions like seizures. Across spiritual traditions—Christian, Jewish, Islamic, and Buddhist—fasting was practiced as a means of renewal, focus, and resilience. What was happening beneath those traditions was physiological: the natural rise of ketones, supporting stable energy and mental clarity.

By the time modern medicine began to take shape, this wisdom had already been practiced for thousands of years. In the early 20th century, it was formally studied when Dr. Russell Wilder at the Mayo

Clinic sought to replicate the therapeutic effects of fasting in children with epilepsy. By recognizing that fasting worked through ketosis, he developed a dietary approach that produced the same metabolic state without complete food abstinence. The ketogenic diet was born—and remains a recognized treatment for drug-resistant epilepsy today.

This rediscovery was not isolated. Decades earlier, William Banting had described dramatic improvements in weight and health by eliminating sugar and starch. Later researchers and clinicians—studying sugar, insulin resistance, endurance performance, ketone metabolism, fasting, and chronic disease—arrived at the same conclusion from different angles: when the human body burns fat and produces ketones, it becomes more stable, resilient, and metabolically flexible. These insights did not introduce something new. They helped us remember something old.

Traditional cultures around the world reflect this pattern. The Inuit thrived on diets dominated by fat and protein. The Maasai consumed milk, meat, and blood with minimal carbohydrates. Other hunter-gatherer societies ate seasonally, relying heavily on fat to endure harsh climates and long periods without food. These cultures were not following a "diet." They were living within the boundaries of nature—and their metabolism reflected it.

The real departure came with the rapid rise of refined carbohydrates, sugar, and processed foods in the late 19th and early 20th centuries. These foods were inexpensive, shelf-stable, and highly rewarding. They spread quickly through industrialized societies, and the metabolic consequences followed just as steadily.

The ketogenic lifestyle, as it exists today, is not a novelty. It is a reclamation of a metabolic rhythm humans relied on for most of their existence. It honors the body's adaptability. It reflects a design that allows us to move through feast and famine, abundance and scarcity, strength and recovery. And for many people now, it represents something even simpler: a return to energy, clarity, and metabolic honesty in a world that has drifted far from its ancestral patterns.

Why Ketones Matter

Up to this point, we've explored what ketones are and how the body enters ketosis. But what ultimately matters is *why* they change the experience of energy so profoundly. Ketones are not simply an alternative fuel. They act as stabilizers, signals, and protectors—quietly restoring functions most people never access in a lifetime dominated by glucose.

When fat becomes the primary fuel, something deeper shifts. Energy steadies. Recovery improves. Inflammation quiets. Cells begin repairing rather than reacting. This is where ketosis stops being a way of eating and becomes a different way of powering the body—one that supports resilience, clarity, and long-term stability from the inside out.

Ketones as a Stable, High-Octane Energy Source

When the body produces ketones—primarily beta-hydroxybutyrate and acetoacetate—it gains access to a fuel source that is steady, efficient, and remarkably consistent. Unlike glucose, which rises and falls with meals, stress, and insulin swings, ketones provide a smooth stream of energy drawn from stored fat. Once fat adaptation takes hold, the brain, heart, and muscles are no longer dependent on frequent eating to function well.

This is why many people describe a new kind of calm energy after several weeks in ketosis. For the first time in years, energy is no longer chased or rationed. It is simply available. Mitochondria are supplied continuously, drawing from a deep internal reserve rather than waiting on the next meal. The result is sustained energy—quiet, reliable, and durable, rather than the spikes and crashes most people have come to accept as normal.

Metabolic Flexibility:
The Engine of Human Resilience

One of the subtler losses of modern life is metabolic flexibility—the ability to move smoothly between burning glucose and burning fat. When eating patterns keep insulin elevated, the body becomes increasingly dependent on glucose, and energy begins to rise and fall with its availability. Over time, that dependence narrows the range of what the metabolism can do well.

Ketosis restores this lost range. As fat adaptation develops, the body regains its ability to shift between fuels, drawing primarily on fat and ketones while reserving glucose for brief, high-intensity demands. This flexibility underpins human resilience. It steadies energy, quiets unnecessary inflammation, supports mitochondrial health, stabilizes mood, and extends endurance. What feels like a breakthrough is simply the body returning to a broader, more capable operating range it once knew well.

Ketones as Signaling Molecules
That Protect and Repair

One of the most easily missed aspects of ketosis is that ketones do more than supply energy. They act as signals—quiet messages that tell the body it is safe to repair. In the presence of ketones, inflammatory pathways soften, antioxidant defenses strengthen, and damaged cellular components are cleared away. Mitochondria become more efficient, not through force, but through relief.

At a deeper level, ketones communicate directly with the systems that regulate cellular stress and renewal. They activate pathways involved in detoxification, repair, and resilience, and they encourage the kind of internal housekeeping that modern life rarely allows. This is why ketosis is often accompanied by improvements in inflammation, immune balance, and neurological stability.

Ketosis is not simply a change in fuel. It is a signal that shifts the body out of constant reaction and back toward restoration. Oxidative stress quiets. Repair accelerates. Balance returns. The body is no longer scrambling to keep up—it is given space to heal.

The Hormonal Shift: Lower Insulin, Higher Sensitivity

At the center of metabolic strain is a breakdown in how the body responds to insulin. Over time, frequent spikes narrow the body's ability to regulate energy smoothly. Ketosis shifts this pattern by allowing insulin levels to settle, giving cells the opportunity to respond more clearly and consistently again. As that pressure eases, stored energy becomes accessible rather than locked away.

With fewer swings, hormonal rhythms begin to recalibrate. Hunger signals soften. Satiety becomes more reliable. Stress hormones steady. Reward pathways quiet. What many people notice first is not a laboratory number, but a change in how the body feels—fewer cravings, more emotional steadiness, and a clearer, calmer mental state. Seen this way, ketosis is more than a change in food choices. It is a hormonal rebalancing that allows signals to speak honestly again, restoring coherence between appetite, energy, mood, and metabolism.

Supporting Brain Energy and Cognitive Performance

The brain is one of the body's most energy-demanding organs, requiring a constant and reliable fuel supply. Yet over time, many people experience a subtle decline in how well the brain uses glucose. The result is familiar—foggy thinking, uneven focus, mood swings, and mental fatigue that seem difficult to explain or correct.

Ketones offer the brain another option. They cross easily into the brain and provide a steady source of energy even when glucose delivery becomes less reliable. For many, this shift brings a noticeable sense of clarity and calm. Thoughts feel sharper. Focus lasts longer.

Emotional responses feel more regulated and less reactive. For some, the change is profound—not dramatic or stimulating, but steady. The mind feels supported rather than strained. In that stability, creativity often returns, concentration deepens, and mental effort begins to feel sustainable again. It can feel like the brain is finally being fueled in a way it recognizes.

Endurance, Physical Performance, and Recovery

Ketosis changes the way the body approaches endurance. As fat adaptation develops, performance becomes less dependent on a small, fragile reserve of stored carbohydrates and more supported by a deep, reliable internal fuel supply. Energy steadies. Effort becomes more predictable. The constant need to refuel fades into the background.

What researchers such as Jeff Volek have shown is that fat-adapted athletes draw far more energy from fat at the same intensity, preserving glycogen for moments when it truly matters. The result is not just endurance, but composure—the ability to sustain effort without chasing the next surge of sugar.

Recovery shifts as well. With fewer fuel swings and less oxidative strain, muscles experience less inflammation and rebound more quickly. Many athletes notice they can train again sooner, with less soreness and less cumulative fatigue. Performance no longer feels like a cycle of depletion and repair, but a steadier rhythm the body can maintain over time.

In this way, ketosis doesn't just extend endurance—it reshapes it. Movement becomes something the body supports rather than resists, allowing effort and recovery to work together instead of against each other.

Why Ketosis Feels Different:
A Biological Shift, Not a Diet

When people describe "something different" happening in ketosis—less hunger, steadier energy, emotional calm, clearer focus—they are noticing the combined effect of several systems finally working in harmony. Energy delivery stabilizes. Cellular function becomes more efficient. Hormonal signals quiet. Inflammation eases. What had felt scattered begins to feel coherent.

Ketosis realigns metabolism with a deeper, older design—one built for adaptability rather than constant stimulation. This is why so many people describe the experience as feeling "turned back on." Not energized in a forced way, but supported. Grounded. Reliable.

Ketosis is not a trick or a shortcut. It is a return to metabolic rhythms humans lived within for most of their history. When those rhythms are restored, the body doesn't need to be pushed. It begins to function as it was meant to—steady, resilient, and awake.

Why Ketones Matter: The Big Picture

To understand why ketones matter, it helps to zoom out. A ketogenic state restores capabilities the body was designed to have but rarely uses in modern life. Metabolic flexibility returns. Mitochondria function more efficiently. Oxidative stress quiets. Energy and mood stabilize. Cognitive function sharpens. Endurance becomes more sustainable, and recovery more complete. Hormonal and immune signaling fall back into rhythm rather than constant correction.

Seen this way, ketosis is not simply a tool for weight loss. Its influence extends far beyond the scale, touching neurological stability, metabolic health, and cellular resilience. It is not about restriction or control. It is about allowing the body to return to a fuel system it still remembers how to use, even if it has gone unused for years. Ketones matter because they restore order—steady energy instead of volatility, repair instead of reaction, resilience instead of depletion. In that order, the body finds calm. And in that calm, health has space to return.

Beyond the Waves: Finding Calm in Ketosis

What makes the start of a ketogenic lifestyle challenging isn't the food itself—it's the fear of what might happen without constant carbohydrates. For years, people have been taught that hunger is something to avoid, that a dip in energy signals danger, and that eating all day keeps metabolism alive. Letting go of those beliefs can feel like stepping away from solid ground, even when that ground has quietly kept them stuck. That fear exists largely because most people have never stayed in ketosis long enough to feel what comes next.

I often describe the early days of keto as stepping into the ocean. At first, the water is shallow and gentle. You wade in—curious, cautious. Then the waves begin to break. One after another. Stronger than expected. They knock you back, sting your eyes, and fill your ears with noise. This is the period many call the "keto flu"—the moment the body begins releasing its reliance on sugar and searching for a steadier fuel it hasn't used in years.

Those early waves are uncomfortable. And for many people, this is where they turn back—not because they can't continue, but because they don't trust what lies beyond the turbulence. Anyone who has swum past a breaking surf knows a quiet truth: If you push just a little farther, the entire ocean changes.

Beyond the waves, the water becomes calm and expansive. The noise fades. You can float, breathe deeply, and feel supported rather than battered. What once looked threatening from shore reveals itself as a narrow threshold into a different world—one that was always there. Ketosis is like that.

Most people never reach the calm because they retreat at the first surge of fatigue or craving. They assume the body is rejecting the process, when in truth it is simply reorganizing its priorities. Sugar burns fast and loud. Fat burns steady and quiet. The body is not failing—it is learning.

If you stay with it, something subtle but profound begins to happen. Hunger softens. Energy steadies. Mental fog lifts. Cravings lose their urgency. The body stops shouting for glucose and begins speaking in a calmer, older language—one it has always known. This is the calm beyond the waves. And once you experience it, your understanding of energy, hunger, and health changes for good. The first week of keto isn't about doing everything perfectly. It's about trusting that the calm exists—trusting that your biology has a second gear, a deeper rhythm, an older design waiting to come alive. Your only task is to swim past the waves.

Finding Your Way Into Calm

Beginning doesn't require perfection. It requires patience. Most people struggle at the start not because they are doing something wrong, but because they expect certainty too soon. Ketosis is not a process that responds to force or precision. It unfolds as the body is given room to adapt, recalibrate, and remember how to regulate energy on its own.

The simplest place to begin is by shaping the environment rather than relying on willpower. When the foods that keep energy volatile are no longer in front of you, the process becomes quieter. From there, hunger begins to change in its own time. What once felt urgent often softens as the body regains access to a steadier fuel.

The calm does not arrive instantly. It comes after a brief period of uncertainty, just beyond the breaking waves. If you stay with the process long enough to reach it, much of what once required effort begins to resolve naturally. You don't need to get it right. You only need to stay with it long enough to let the body do what it already knows how to do.

Why Keto Works When Nothing Else Has

For many people, keto isn't the first path they've explored in search of better health. It often comes after years of trying—diets, exercise plans, calorie tracking, and well-intended advice that never quite addressed what felt wrong beneath the surface. The struggle wasn't a lack of effort or discipline. It was a mismatch between biology and fuel.

Ketosis changes the context. As insulin levels settle, stored energy becomes accessible again. Hunger softens. Appetite signals regain clarity. Many people notice, often with surprise, that food no longer occupies their thoughts throughout the day. The body begins regulating intake in a way that feels natural rather than enforced.

Alongside this shift, inflammation quiets and cellular energy production becomes more efficient. As the background strain eases, mental clarity improves and mood steadies. The constant noise—fatigue, irritability, fog—begins to fade, not because it's being managed, but because the underlying signal has changed.

Perhaps most noticeable is what happens to the drive for constant stimulation. With steadier energy and more reliable fuel, the urgency to snack, chase sugar, or override satiety weakens. The body stops asking for more because it is finally receiving what it needs.

This is why keto rarely feels like restriction once the transition is complete. It feels like relief. Like exhaling after years of effort. Like the body settling into a rhythm it recognizes. For many, this is the first approach that makes sense not just in theory, but in lived experience. It doesn't push against biology. It works with it.

The Return to Energy

Ketosis is a return—to the metabolism our ancestors relied on for survival, to the steady energy the body was designed to produce, to the clarity that emerges when fuel and hormones fall back into rhythm. It is a return to a way of being that modern life has obscured, but never erased.

For many, entering ketosis is the first moment in years when the body begins to feel trustworthy again. Hunger quiets. Cravings loosen their grip. Thoughts sharpen. Moods settle. The daily highs and lows of glucose dependence fade, replaced by a stability that feels unfamiliar at first—only because it has been absent for so long. What people often attribute to "willpower" is simply biology finally working with them instead of against them.

More than anything, ketosis offers a glimpse of what healing can feel like. Not the dramatic, overnight kind, but the grounded, cellular kind—the healing that begins with stable energy, efficient mitochondria, balanced hormones, and reduced inflammation. The kind that starts at the root rather than on the surface.

Ketosis is not a trend or a trick. It is a way home—back to metabolic stability, back to trust in your own physiology, back to the energy that was always yours.

ACKNOWLEDGMENTS

This book exists because of the work of scientists, physicians, clinicians, and researchers—past and present—who have dedicated their lives to understanding human health, metabolism, and energy. Their careful observations and willingness to question assumptions laid the foundation for everything explored here.

My wife and I are especially grateful to those who not only advanced this knowledge, but shared it beyond traditional academic and clinical settings, making complex ideas accessible to people seeking better health.

We remain deeply thankful to the professionals whose insights reshaped how we understand health and energy, and whose work changed our lives in lasting ways.

GLOSSARY

A

Adaptive Stress Response Protective cellular changes initiated in response to environmental or metabolic stress.

Adipose Dysfunction Impaired fat tissue function characterized by energy trapping and inflammatory signaling.

Aerobic Capacity The ability to generate energy through oxygen-dependent pathways.

Aerobic Glycolysis Glucose metabolism that produces lactate despite adequate oxygen availability.

Amyloid Plaque Extracellular aggregates of misfolded amyloid-β protein that accumulate between neurons when normal production–clearance balance is disrupted; in healthy physiology, the brain continuously clears amyloid via enzymatic degradation, glymphatic flow, and energy-dependent cellular processes, with plaque formation reflecting impaired clearance rather than overproduction alone.

Apoptosis A regulated, energy-dependent form of programmed cell death largely controlled by mitochondrial signaling pathways, enabling the orderly removal of damaged or dysfunctional cells without inflammation while preserving tissue integrity.

Astrocytes Supportive brain cells involved in energy distribution, neurotransmitter recycling, and metabolic regulation.

ATP (Adenosine Triphosphate) The primary energy currency of the cell required for biological work.

Autophagy A fundamental cellular recycling process that degrades and repurposes damaged proteins and organelles to maintain energy balance and cellular health; it is suppressed by high insulin and nutrient excess and activated during fasting and low-insulin states—a mechanism recognized by the **2016 Nobel Prize in Physiology or Medicine**.

Axonal Transport Energy-dependent movement of molecules along neuronal axons.

B

Basal Insulin The background insulin level present between meals.

Body Mass Index (BMI): A population-level screening measure calculated as weight (kg) ÷ height² (m²), classifying body size as **underweight** <18.5, **normal** 18.5–24.9, **overweight** 25.0–29.9, and **obesity** ≥30.0; it estimates relative body mass but does not assess metabolic health or body composition.

Blood–Brain Barrier A selective barrier regulating which substances enter the brain from circulation.

Brain Glucose Hypometabolism Reduced cerebral glucose utilization despite adequate glucose availability.

Brain Insulin Resistance Reduced responsiveness of brain cells to insulin signaling, impairing glucose uptake.

C

Calcium Signaling Intracellular calcium regulation influencing mitochondrial energy production and cell signaling.

Caloric Restriction Reduction of energy intake below usual consumption.

Cardiometabolic Disease A cluster of metabolic and cardiovascular disorders influenced by metabolic dysfunction.

Carbohydrate Intolerance An impaired ability to metabolize dietary carbohydrates without excessive insulin or glucose response.

Cell Danger Response (CDR) A conserved cellular metabolic state prioritizing defense over growth and repair.

Cholesterol Hypothesis The theory linking dietary fat and cholesterol to cardiovascular disease risk.

Chronic Inflammation Persistent inflammatory signaling that disrupts metabolic and vascular function.

Cortisol A stress hormone that increases glucose availability and influences energy distribution.

D

Dyslipidemia An abnormal lipid profile characterized by elevated triglycerides, altered LDL particles, low HDL, or adverse lipoprotein ratios, reflecting disrupted lipid transport and fuel handling commonly associated with insulin resistance and metabolic dysfunction.

E

Ectopic Fat Fat accumulation in non-adipose tissues such as liver, muscle, pancreas, or heart.

Electron Transport Chain (ETC) A series of mitochondrial protein complexes that drive ATP synthesis.

Endoplasmic Reticulum (ER) An organelle involved in protein folding, lipid synthesis, and calcium regulation.

Endurance Performance The capacity to sustain prolonged physical activity.

Energy Balance Model The theory that weight change results solely from caloric intake exceeding expenditure.

Epinephrine A hormone and neurotransmitter that mobilizes energy during acute stress.

Exercise Economy The energy cost of performing a given workload.

Extracellular ATP ATP released outside the cell that functions as a danger signal.

F

Fat Adaptation A metabolic state characterized by efficient use of fatty acids and ketones as primary fuels.

Fat Oxidation The mitochondrial process of converting fatty acids into usable energy.

Fat Oxidation Rate The rate at which fatty acids are used for energy during rest or exercise.

Fasting A metabolic state characterized by low insulin and reliance on stored energy.

Fermentation ATP production through glycolysis and related pathways independent of oxygen.

Fuel Partitioning Hormonal regulation determining whether energy is stored or oxidized.

Fuel Switching The ability to alternate between glucose and fat as primary fuels.

G

Ghrelin A hormone that stimulates hunger and food intake.

GLP-1 (Glucagon-Like Peptide-1) An incretin hormone that enhances insulin secretion and suppresses appetite.

Glucagon A hormone that promotes glucose release from the liver and supports fat mobilization.

Glucose–Fatty Acid Cycle Competition between glucose and fat for oxidation within cells.

Glucose Dependence Reliance on glucose as a primary energy substrate.

Glycogen A stored form of glucose in liver and muscle.

Glycogen Sparing Reduced reliance on glycogen due to increased fat and ketone utilization.

Glycolysis The metabolic pathway that converts glucose into energy.

Glycosuria Presence of glucose in the urine.

H

Healing Cycle The progression from cellular defense back to growth, repair, and differentiation.

Homeostasis The maintenance of stable internal biological conditions.

Hyperglycemia Elevated blood glucose levels.

Hyperinsulinemia Persistently elevated insulin levels.

Hypoglycemia Abnormally low blood glucose levels.

Hypometabolic State A reduction in cellular energy production associated with prolonged stress.

I

Industrial Food Processing Manufacturing practices that increase exposure to oxidized fats.

Inflammasome A multi-protein complex that activates inflammatory responses.

Innate Immune Activation Rapid, non-specific immune responses triggered by cellular stress.

Insulin A hormone that regulates energy storage and release.

Insulin Discovery (1921) The isolation of insulin as a therapeutic hormone.

Insulin Resistance Reduced cellular responsiveness to insulin.

Islets of Langerhans Clusters of endocrine cells in the pancreas responsible for hormone secretion.

K

Ketoacidosis A dangerous metabolic state marked by uncontrolled ketone production and metabolic acidosis, typically occurring in absolute or severe insulin deficiency (most commonly type 1 diabetes), and distinct from the normal, regulated ketosis seen during fasting or ketogenic diets.

Ketogenic Diet A high-fat, low-carbohydrate diet developed to induce ketosis.

Ketone Bodies Fat-derived energy substrates usable by many tissues, including the brain.

Ketone Metabolism Cellular utilization of ketone bodies as an energy source.

Ketone Transporters (MCTs) Proteins facilitating ketone transport across membranes.

Ketosis A metabolic state characterized by elevated circulating ketone bodies.

Ketosis (Historical Context) Early recognition of ketosis during fasting or carbohydrate restriction.

L

Leptin A hormone produced by adipose tissue that signals energy sufficiency to the brain.

Linoleic Acid (LA) An omega-6 polyunsaturated fatty acid abundant in seed oils.

Lipid Peroxidation Oxidative degradation of lipids causing cellular damage.

Lipolysis The breakdown and release of stored fat for energy use.

Low-Fat Diet Paradigm Dietary guidance emphasizing fat reduction.

M

Macular Degeneration Degenerative retinal disease associated with oxidative lipid injury.

Medium-Chain Triglycerides (MCTs) Triglycerides composed of medium-chain fatty acids (typically 6–12 carbons) that are rapidly absorbed and metabolized for energy, often converted to ketone bodies by the liver.

Metabolic Efficiency The ability to perform work with lower energy cost.

Metabolic Flexibility The capacity to adapt fuel utilization to changing conditions.

Metabolic Inflexibility Reduced ability to adapt fuel utilization.

Metabolic Neuroprotection Preservation of neuronal function through optimized energy metabolism.

Metabolic Reset Restoration of normal cellular metabolism following danger resolution.

Metabolic Syndrome A constellation of insulin resistance, dyslipidemia, hypertension, and central obesity.

Metabolic Theory of Cancer The hypothesis that cancer originates from disordered energy metabolism.

Metabolic Theory of Disease A framework proposing chronic disease originates from disrupted cellular energy.

Metabolic Therapy Interventions aimed at altering fuel availability to target metabolism.

Metabolic Trap Failure to exit a persistent cell danger response.

Mitohormesis Adaptive benefits from mild mitochondrial stress.

Mitochondria Organelles that generate ATP and coordinate energy signaling.

Mitochondrial Biogenesis The formation of new mitochondria.

Mitochondrial Resilience The capacity of mitochondria to tolerate metabolic stress.

Mitochondrial Signaling Communication from mitochondria to other cellular systems.

Mitochondrial Uncoupling Partial dissipation of the proton gradient reducing ATP efficiency.

Muscle Sparing The preservation of lean muscle mass during fasting or carbohydrate restriction, achieved through adequate protein intake, low insulin, and increased reliance on fat and ketones for energy, reducing the need for amino acids to be used as fuel.

N

Neuroenergetics The study of energy production and use in the nervous system.

Neuroinflammation Inflammatory signaling within the nervous system.

Neuronal Energy Failure Insufficient ATP to support neuronal function.

Norepinephrine A hormone and neurotransmitter involved in alertness and energy mobilization.

Nuclear–Mitochondrial Communication Bidirectional signaling between nucleus and mitochondria.

O

Obesity A chronic disorder of energy regulation marked by excess adipose storage alongside impaired access to stored fuel, driven by insulin resistance, hormonal dysregulation, and disrupted fuel partitioning, rather than excess caloric intake alone.

Omega-3 Fatty Acids Polyunsaturated fats involved in anti-inflammatory signaling.

Omega-6 Fatty Acids Polyunsaturated fats that become harmful in excess.

Oncogenic Signaling Cell signaling pathways promoting growth and proliferation.

Oxidative Phosphorylation ATP generation through oxygen-dependent mitochondrial processes.

Oxidative Stress Cellular damage from excess reactive oxygen species.

Oxidized Lipids Damaged fats formed through oxidation.

P

Pancreatectomy Surgical removal of the pancreas.

PUFAs (Polyunsaturated Fatty Acids) Fats containing multiple double bonds prone to oxidation.

Purinergic Signaling Cell communication mediated by extracellular nucleotides.

R

Radioimmunoassay A technique for measuring hormone concentrations.

Reactive Oxygen Species (ROS) Highly reactive metabolic byproducts.

Redox Balance Maintenance of appropriate oxidative–reductive states.

Redox Signaling Cellular communication via redox state changes.

Respiratory Exchange Ratio (RER) A measure estimating fat versus carbohydrate oxidation.

Retrograde Signaling Mitochondria-to-nucleus communication.

S

Satiety The physiological state of fullness after eating, regulated by hormonal and neural signals—primarily leptin, insulin, peptide YY (PYY), cholecystokinin (CCK), and GLP-1—that signal sufficient energy availability and suppress further food intake.

Saturated Fats: Fatty acids with no double bonds, typically solid at room temperature, that provide a stable, energy-dense fuel; their metabolic effects depend on insulin status and fuel partitioning, with oxidation favored under low-carbohydrate, low-insulin conditions rather than storage.

Somatic Mutation Theory The theory that cancer originates from genetic mutations.

Starvation Therapy Pre-insulin diabetes treatment involving severe caloric restriction.

Subcutaneous Fat Fat stored beneath the skin that serves as a relatively safe, expandable energy reservoir; when metabolically healthy, it buffers excess energy and is less strongly associated with insulin resistance and cardiometabolic risk than visceral or ectopic fat.

Substrate Utilization The proportion of energy derived from fat versus carbohydrate.

Syndrome X An early term for metabolic syndrome.

T

Tau A microtubule-associated protein that stabilizes neuronal structure and intracellular transport; when abnormally phosphorylated, tau detaches, misfolds, and aggregates into neurofibrillary tangles, reflecting disrupted cellular energy balance and impaired protein regulation rather than normal physiological function. **Thermogenesis** Heat production from metabolic processes.

Total Cholesterol The sum of cholesterol carried in all major lipoproteins (LDL, HDL, and VLDL) measured in blood, used as a broad screening marker but limited in assessing cardiometabolic risk without context such as triglycerides, HDL, particle number, and metabolic health.

Triglycerides The primary circulating form of stored fat in the blood, composed of three fatty acids and glycerol; elevated levels reflect excess energy availability, impaired fat oxidation, and insulin resistance, and are a key marker of metabolic dysfunction.

Tumor Microenvironment The local metabolic and cellular environment surrounding tumors.

Type 1 Diabetes An autoimmune condition characterized by insulin deficiency.

Type 2 Diabetes A metabolic disorder characterized by insulin resistance.

U

Uncoupling Proteins (UCPs) Proteins regulating mitochondrial proton leak.

V

Visceral Fat Fat stored within the abdominal cavity surrounding internal organs, strongly associated with insulin resistance, inflammation, and cardiometabolic risk due to its high metabolic activity and direct drainage into the portal circulation.

VLDL (Very Low-Density Lipoprotein) A triglyceride-rich lipoprotein produced by the liver to transport endogenous fat to peripheral tissues; elevated VLDL reflects increased hepatic fat and insulin resistance and contributes to atherogenic dyslipidemia through downstream LDL formation.

VO$_2$ Max The maximum rate of oxygen consumption during exercise.

W

Warburg Effect Preference of cancer cells for glycolysis despite oxygen availability.

Western Diet A dietary pattern high in refined carbohydrates and industrial fats.

READING LIST

Apple, Sam—*Ravenous: Otto Warburg, the Nazis, and the Search for the Cancer-Diet Connection* (2013)

Berger, Amy—*The Alzheimer's Antidote* (2017)

Bikman, Benjamin—*Why We Get Sick* (2020)

Bernstein, Richard K.—*Dr. Bernstein's Diabetes Solution* (latest edition)

Bredesen, Dale—*The End of Alzheimer's* (2017); *The End of Alzheimer's Program* (2020)

Christofferson, Travis—*Tripping Over the Truth* (2014)

Cunnane, Stephen—*Ketones: The Fourth Fuel* (2013)

Eaton, S. Boyd; Konner, Melvin—*The Paleolithic Prescription* (1985)

Fung, Jason—*The Obesity Code* (2016)

Fung, Jason; Ramos, Jimmy—*The Complete Guide to Fasting* (2016)

Know, Lee—*Mitochondria and the Future of Medicine* (2018)

Knobbe, Chris A.—*The Ancestral Diet Revolution* (2020)

Lieberman, Daniel E.—*The Story of the Human Body* (2013)

Ludwig, David—*Always Hungry?* (2016)

Lustig, Robert H.—*Metabolical* (2021)

Lysiak, Matthew—*Fiat Food* (2024)

Margulis, Lynn—*Symbiotic Planet* (1998)

Montgomery, David R.—*Dirt: The Erosion of Civilizations* (2007); *Growing a Revolution* (2017)

Moss, Michael—*Salt Sugar Fat* (2013); *Hooked* (2021)

Nestle, Marion—*Food Politics* (2002)

Newport, Mary T.—*What If There Was a Cure?* (2011/2013); *Clearly Keto* (2022)

Noakes, Tim (et al.)—*Ketogenic: The Science of Therapeutic Carbohydrate Restriction in Human Health*

Ohlson, Kristin—*The Soil Will Save Us* (2014)

Palmer, Christopher M.—*Brain Energy* (2022)

Phillips, Matthew C. L.—*Metabolic Strategies in Healthcare: A New Era* (2024)

Poff, Angela M.; Koutnik, Andrea P.; D'Agostino, Dominic P.—*The Ketogenic Diet and Metabolic Therapies* (2020)

Pollan, Michael—*In Defense of Food*

Ramos, Amanda—*The Complete Ketogenic Diet for Beginners*

Seyfried, Thomas N.—*Cancer as a Metabolic Disease* (2012)

Taubes, Gary—*Good Calories, Bad Calories* (2008); *The Case Against Sugar* (2016); *Rethinking Diabetes* (2024)

Teicholz, Nina—*The Big Fat Surprise* (2014)

Volek, Jeff S.; Phinney, Stephen D.—*The Art and Science of Low Carbohydrate Living* (2011)

Volek, Jeff S.; Phinney, Stephen D.—*The Art and Science of Low Carbohydrate Performance* (2012)

Volek, Jeff S.; Phinney, Stephen D.; Westman, Eric C.—*The New Atkins for a New You* (2010)

Westman, Eric C.—*End Your Carb Confusion*

Yudkin, John—*Pure, White, and Deadly* (1972)

REFERENCES

Introduction

McKeown, T. (1976). *The Role of Medicine: Dream, Mirage, or Nemesis?*

Cutler, D., & Miller, G. (2005). The role of public health improvements in health advances: the 20th-century United States. *Demography.*

Chapter 4

Warburg, O. (1956). On the origin of cancer cells. *Science,* 123(3191), 309–314.

Krebs, H. A., & Johnson, W. A. (1937). The role of citric acid in intermediate metabolism in animal tissues. *Enzymologia,* 4, 148–156.

Cahill, G. F. (1970). Starvation in man. *New England Journal of Medicine,* 282, 668–675.

Cahill, G. F., & Owen, O. E. (1968). Starvation and survival. *Transactions of the American Clinical and Climatological Association.*

Veech, R. L. (2004). The therapeutic implications of ketone bodies. *Prostaglandins, Leukotrienes and Essential Fatty Acids,* 70(3), 309–319.

Veech, R. L., et al. (2001). Ketone bodies, potential therapeutic uses. *IUBMB Life,* 51(4), 241–247.

Cox, P. J., et al. (2016). Ketone bodies as a fuel for the brain. *Journal of Cerebral Blood Flow & Metabolism,* 36(6), 1021–1031.

Mizushima, N., & Komatsu, M. (2011). Autophagy: Renovation of cells and tissues. *Cell,* 147(4), 728–741.

Palikaras, K., Lionaki, E., & Tavernarakis, N. (2018). Mechanisms of mitophagy in cellular homeostasis. *Nature Cell Biology,* 20, 1013–1022.

López-Otín, C., et al. (2013). The hallmarks of aging. *Cell*, 153(6), 1194–1217.

Hallberg, S. J., et al. (2018). Effectiveness and safety of a novel care model for type 2 diabetes. *Diabetes Therapy*, 9(2), 583–612.

Reaven, G. M. (1988). Role of insulin resistance in human disease. *Diabetes*, 37(12), 1595–1607.

Chapter 6

Jenkins, D. J. A., et al. (1981). Glycemic index of foods: A physiological basis for carbohydrate exchange. *American Journal of Clinical Nutrition*.

Ludwig, D. S. (2002). The glycemic index: Physiological mechanisms and clinical implications. *JAMA*.

Holt, S. H., Brand-Miller, J., et al. (1999). The relationship between blood glucose response and mood. *Appetite*.

Schwartz, M. W., et al. (2000). Central nervous system control of food intake and body weight. *Nature*.

Batterham, R. L., et al. (2002). Gut hormone peptide YY regulates appetite and food intake. *Nature*.

Flint, A., et al. (2003). Ghrelin and the regulation of appetite. *Journal of Clinical Endocrinology & Metabolism*.

Neary, N. M., & Batterham, R. L. (2005). The actions of GLP-1 on appetite and weight regulation. *Endocrine Reviews*.

Lustig, R. H. (2017). Processed food addiction: Foundations, environmental factors, and prevention. *Nature Reviews Endocrinology*.

Cunnane, S. C., et al. (2020). Fuel metabolism, brain energy, and stability. *Frontiers in Neuroscience*.

Bikman, B. T. (2018–2022). Insulin resistance / insulin–lipolysis / fatty-acid availability (lectures & selected reviews). (As cited.)

Chapter 7

Saklayen, M. G. (2018). The global epidemic of the metabolic syndrome. *Current Hypertension Reports*, 20(2), 12.

Alberti, K. G. M. M., et al. (2009). Harmonizing the metabolic syndrome. *Circulation*, 120(16), 1640–1645.

International Diabetes Federation. (2023). *IDF Consensus Worldwide Definition of the Metabolic Syndrome.*

Holloszy, J. O. (2006). Regulation of mitochondrial biogenesis in mammals. *Exercise and Sport Sciences Reviews*, 34(1).

San-Millán, I., & Brooks, G. A. (2020). Reexamining metabolic flexibility: An evolutionary perspective. *Metabolism*, 113, 154–167.

Phillips, M. C. L. (2024). *Metabolic Strategies in Healthcare: A New Era.* (As cited.)

Chapter 8

Alberti, K. G. M. M., et al. (2009). Harmonizing the metabolic syndrome. Circulation.

Grundy, S. M. (2016). Metabolic syndrome update. Trends in Cardiovascular Medicine.

O'Hearn, M., et al. (2022). Trends in cardiometabolic health among U.S. adults, 1999–2018. Journal of the American College of Cardiology.

Chalasani, N., et al. (2018). The diagnosis and management of nonalcoholic fatty liver disease: Practice guidance from the AASLD. Hepatology.

Younossi, Z. M., et al. (2016). Global epidemiology of nonalcoholic fatty liver disease. Hepatology.

Taylor, R. (2013). Type 2 diabetes and remission: Practical management and mechanisms. Diabetologia.

Lean, M. E. J., et al. (2018). Primary care-led weight management for remission of type 2 diabetes (DiRECT). The Lancet.

Hallberg, S. J., et al. (2018). Effectiveness and safety of a novel care model for type 2 diabetes at 1 year. Diabetes Therapy.

Reaven, G. M. (1988). Role of insulin resistance in human disease. Diabetes.

Hall, J. E., et al. (2015). Obesity-induced hypertension: Interaction of neurohumoral and renal mechanisms. Circulation Research.

Brownlee, M. (2005). The pathobiology of diabetic complications: A unifying mechanism. Diabetes.

Katz, P. O., et al. (2022). ACG clinical guideline for the diagnosis and management of GERD. American Journal of Gastroenterology.

Ness-Jensen, E., et al. (2016). Lifestyle intervention in gastroesophageal reflux disease. Clinical Gastroenterology and Hepatology.

Johnston, C. S., et al. (2004). Vinegar improves insulin sensitivity to a high-carbohydrate meal. Diabetes Care.

David, L. A., et al. (2014). Diet rapidly and reproducibly alters the human gut microbiome. Nature.

Sonnenburg, J. L., & Bäckhed, F. (2016). Diet–microbiota interactions as moderators of human metabolism. Nature.

Lacy, B. E., et al. (2021). ACG clinical guideline: Management of irritable bowel syndrome. American Journal of Gastroenterology.

Teede, H. J., et al. (2018). International evidence-based guideline for the assessment and management of polycystic ovary syndrome. Human Reproduction.

Dunaif, A. (1997). Insulin resistance and the polycystic ovary syndrome: Mechanism and implications. Endocrine Reviews.

Arnold, S. E., et al. (2018). Brain insulin resistance in Alzheimer's disease and related disorders. Nature Reviews Neuroscience.

Hotamisligil, G. S. (2006). Inflammation and metabolic disorders. Nature.

Berenbaum, F. (2013). Osteoarthritis as an inflammatory disease. Osteoarthritis and Cartilage.

Courties, A., & Sellam, J. (2016). Obesity and osteoarthritis: More than mechanical stress. Joint Bone Spine.

Punjabi, N. M. (2008). The epidemiology of adult obstructive sleep apnea. Proceedings of the American Thoracic Society.

Peppard, P. E., et al. (2013). Increased prevalence of sleep-disordered breathing in adults. American Journal of Epidemiology.

Thomas, M. C., et al. (2015). Diabetic kidney disease. Nature Reviews Disease Primers.

KDIGO. (2013). Clinical practice guideline for the evaluation and management of chronic kidney disease. Kidney International Supplements.

Hall, K. D., et al. (2019). Ultra-processed diets cause excess calorie intake and weight gain. Cell Metabolism.

Monteiro, C. A., et al. (2019). Ultra-processed foods: What they are and how to identify them. Public Health Nutrition.

Cahill, G. F. (2006). Fuel metabolism in starvation. *Annual Review of Nutrition.*

Volek, J. S., et al. (2016). Metabolic characteristics of keto-adapted ultra-endurance runners. *Metabolism.*

Picard, M., et al. (2018). Mitochondrial psychobiology. *Biological Psychiatry.*

Panda, S. (2016). Circadian physiology of metabolism. *Science.*

Naviaux, R. K. (2014). Metabolic features of the cell danger response. *Mitochondrion.*

Chapter 9

Biblical texts: Daniel 1; Isaiah 58; Matthew 4.

Hippocrates. *On Regimen in Acute Diseases.* (4th century BC).

Tanner, H. S. (1880). *The Fasting Cure: Forty Days Without Food.*

Dewey, E. H. (1895). *The No-Breakfast Plan and the Fasting Cure.*

Wheless, J. W. (2008). History of the ketogenic diet. *Epilepsia,* 49(Suppl 8), 3–5.

Wilder, R. M. (1921). The effects of ketonemia on the course of epilepsy. *Mayo Clinic Proceedings*, 2, 307–314.

Wilhelmi de Toledo, F., et al. (2013). Fasting therapy—an update. *Research in Complementary Medicine*, 20(6), 444–453.

Ohsumi, Y. (2001). Autophagy: from phenomenology to molecular understanding in less than a decade. *Nature Reviews Molecular Cell Biology*, 2(3), 211–216.

Ohsumi, Y. (2014). Historical landmarks of autophagy research. *Cell Research*, 24(1), 9–23.

Nobel Prize in Physiology or Medicine. (2016). The Nobel Prize in Physiology or Medicine 2016 – Yoshinori Ohsumi. *NobelPrize.org.*

Cahill, G. F. (1970). Ketosis. *Journal of Clinical Investigation*, 49(11), 2017–2025.

Longo, V. D., & Mattson, M. P. (2014). Fasting: molecular mechanisms and clinical applications. *Cell Metabolism*, 19(2), 181–192.

Brandhorst, S., et al. (2015). Fasting-mimicking diet and markers/risk factors for aging, diabetes, cancer, and cardiovascular disease. *Science Translational Medicine*, 7(296), 296ra111.

Mitchell, S. J., et al. (2018). Autophagy, fasting, and longevity: the molecular pathways. *Nature Reviews Genetics*, 19, 579–593.

Kossoff, E. H., et al. (2009). Optimal clinical management of children receiving the ketogenic diet: recommendations of the International Ketogenic Diet Study Group. *Epilepsia*.

Chapter 10

Miller, M., et al. (2011). Triglycerides and cardiovascular disease: A scientific statement from the AHA. *Circulation*.

Gaziano, J. M., et al. (1997). Triglycerides, HDL cholesterol, and coronary heart disease. *Circulation*.

Ravnskov, U., et al. (2016). LDL cholesterol and mortality in the elderly. *BMJ Open*.

Feinman, R. D., et al. (2015). Dietary carbohydrate restriction as the first approach in diabetes management. *Nutrition*.

Athinarayanan, S. J., et al. (2019). Long-term effects of continuous remote care intervention including nutritional ketosis for type 2 diabetes. *Frontiers in Endocrinology*.

Kravitz, R. L., et al. (2018). N-of-1 trials in the era of precision medicine. *New England Journal of Medicine*.

Chapter 15

Phinney, S. D. (2004). Ketogenic diets and physical performance. *Nutrition & Metabolism*.

Noakes, T. D. (2014). Fat adaptation in endurance exercise. *Sports Medicine*.

Chapter 16

FAO. *The State of Food Security and Nutrition in the World* (annual reports).

FAO. (2015). *Status of the World's Soil Resources*.

Lal, R. (2020). Regenerative agriculture for food and climate. *Journal of Soil and Water Conservation*.

Teague, W. R., et al. (2016). The role of ruminants in reducing agriculture's carbon footprint. *Journal of Soil and Water Conservation.*

Davis, D. R., Epp, M. D., & Riordan, H. D. (2004). Changes in USDA food composition data. *Journal of the American College of Nutrition.*

Benbrook, C. M. (2016). Trends in glyphosate herbicide use. *Environmental Sciences Europe.*

Krüger, M., et al. (2013). Glyphosate affects microbiota. *Current Microbiology.*

Monteiro, C. A., et al. (2019). Ultra-processed foods and health. *BMJ.*

Popkin, B. M., Adair, L. S., & Ng, S. W. (2012). Global nutrition transition. *Nutrition Reviews.*

WHO Africa Region. *Noncommunicable Diseases Reports.*

IPES-Food. (2016). *From Uniformity to Diversity.*

Chapter 17

Keys, A. (1970). Coronary heart disease in seven countries. *Circulation, 41*(4 Suppl), I-1–I-211.

American Heart Association. (1961). Dietary fat and its relation to heart attacks and strokes. *Circulation, 23*(1), 133–136.

U.S. Senate Select Committee on Nutrition and Human Needs. (1977). *Dietary Goals for the United States.*

U.S. Department of Agriculture & U.S. Department of Health and Human Services. (1980). *Dietary Guidelines for Americans (1st ed.).*

USDA. (1992). *The Food Guide Pyramid.*

Yudkin, J. (1972). *Pure, White and Deadly.* London: Davis-Poynter.

Kearns, C. E., Schmidt, L. A., & Glantz, S. A. (2016). Sugar industry and coronary heart disease research: A historical analysis of internal industry documents. *JAMA Internal Medicine, 176*(11), 1680–1685.

Siri-Tarino, P. W., et al. (2010). Meta-analysis of prospective cohort studies evaluating the association of saturated fat with cardiovascular disease. *American Journal of Clinical Nutrition, 91*(3), 535–546.

de Souza, R. J., et al. (2015). Intake of saturated and trans unsaturated fatty acids and risk of all-cause mortality, CVD, and type 2 diabetes. *BMJ, 351*, h3978.

Te Morenga, L., Mallard, S., & Mann, J. (2013). Dietary sugars and body weight: systematic review and meta-analyses of randomized controlled trials and cohort studies. *BMJ, 346,* e7492.

Stanhope, K. L., et al. (2009). Consuming fructose-sweetened, not glucose-sweetened, beverages increases visceral adiposity and lipids in overweight adults. *Journal of Clinical Investigation, 119*(5), 1322–1334.

Bray, G. A., Nielsen, S. J., & Popkin, B. M. (2004). Consumption of high-fructose corn syrup in beverages may play a role in the epidemic of obesity. *American Journal of Clinical Nutrition, 79*(4), 537–543.

Chapter 18

Ramsden, C. E., et al. (2013). Re-evaluation of the traditional diet-heart hypothesis: Analysis of the Sydney Diet Heart Study. *BMJ.*

Hibbeln, J. R., et al. (2023). High dietary linoleic acid intake and chronic disease risk. *PNAS.*

Pillon, N. J., et al. (2014). 4-Hydroxy-2-nonenal and metabolic disease. *Journal of Lipid Research.*

Yin, H., & Porter, N. A. (2011). Free radical lipid peroxidation: Mechanisms and analysis. *Chemical Reviews.*

DiNicolantonio, J. J., & O'Keefe, J. H. (2018). Oxidized LDL and atherosclerosis: A link between vegetable oils and cardiovascular disease. *Open Heart.*

Kostner, G. M., et al. (2015). The role of oxidized LDL in cardiovascular disease. *Current Opinion in Lipidology.*

Paradies, G., et al. (2009). Role of cardiolipin in mitochondrial function and dysfunction in heart disease. *BBA – Bioenergetics.*

Chong, E. W.-T., et al. (2009). Dietary fatty acids and the risk of age-related macular degeneration. *Progress in Retinal and Eye Research.*

Chapter 19

Bloch, K. (1965). Cholesterol: Evolution of structure and function. *Science.*

Maxfield, F. R., & van Meer, G. (2010). Cholesterol, the central lipid of mammalian cells. *Current Opinion in Cell Biology.*

Dietschy, J. M., & Turley, S. D. (2004). Cholesterol metabolism in the brain. *Journal of Lipid Research.*

Goldstein, J. L., & Brown, M. S. (2015). A century of cholesterol and coronaries. *Nature.*

Feingold, K. R., & Grunfeld, C. (1992). The role of lipoproteins in host defense. *Journal of Lipid Research.*

Chapter 20

Westman, E. C., et al. (2007). Low-carbohydrate nutrition and metabolism. *American Journal of Clinical Nutrition.*

Lean, M. E. J., et al. (2018). Primary care-led weight management for remission of type 2 diabetes (DiRECT): an open-label, cluster-randomised trial. *The Lancet.*

Taylor, R. (2008). Pathogenesis of type 2 diabetes: tracing the reverse route from cure to cause. *Diabetologia.*

Lim, E. L., et al. (2011). Reversal of type 2 diabetes following a very-low-calorie diet. *Diabetologia.*

Chapter 21

Randle, P. J., Garland, P. B., Hales, C. N., & Newsholme, E. A. (1963). The glucose–fatty acid cycle: its role in insulin sensitivity and the metabolic disturbances of diabetes mellitus. *The Lancet.*

McGarry, J. D., & Foster, D. W. (1980). Regulation of hepatic fatty acid oxidation and ketogenesis. *Annual Review of Biochemistry.*

McGarry, J. D. (1992). What if Minkowski had been ageusic? An alternative angle on diabetes. *Science.*

Newman, J. C., & Verdin, E. (2017). β-hydroxybutyrate: A signaling metabolite. *Annual Review of Nutrition.*

Müller, T. D., et al. (2019). Mechanisms of action of GLP-1 receptor agonists. *Nature Reviews Endocrinology.*

Holst, J. J. (2007). The physiology of glucagon-like peptide 1. *Physiological Reviews.*

Meier, J. J. (2012). GLP-1 receptor agonists for individualized treatment of type 2 diabetes. *Diabetes Research and Clinical Practice.*

Nauck, M. A., & Meier, J. J. (2018). Incretin hormones and their role in side effects of GLP-1 therapy. *Diabetologia.*

Marathe, C. S., Rayner, C. K., et al. (2020). Gastric emptying effects of GLP-1 receptor agonists: implications for clinical practice. *Diabetes Care.*

Liu, J., et al. GLP-1–based therapies and risk of gallbladder disease: systematic review and meta-analysis. *JAMA Internal Medicine.* (Year not provided.)

Chapter 22

Naviaux, R. K., et al. (2016). Metabolic features of chronic fatigue syndrome. *Proceedings of the National Academy of Sciences.*

Naviaux, R. K., et al. (2017). Antipurinergic therapy corrects autism-like features in a poly(I:C) mouse model and in children with autism. *Annals of Clinical and Translational Neurology.*

Naviaux, R. K. (2019). Metabolic features and regulation of the healing cycle: A new model for chronic disease pathogenesis and treatment. *Mitochondrion.*

Naviaux, R. K. (2019). Oxidative shielding and the healing cycle. *Metabolic Syndrome and Related Disorders.*

Chapter 23

Demine, S., Renard, P., & Arnould, T. (2019). Mitochondrial uncoupling: A key controller of biological processes in physiology and diseases. *Cells*, 8(8), 795.

Ma, D., et al. (2018). Ketones drive mitochondrial uncoupling in adipose tissue. *FASEB Journal* (Suppl.).

Maalouf, M., Rho, J. M., & Mattson, M. P. (2009). The neuroprotective properties of the ketogenic diet. *Brain Research Reviews*, 59(2), 293–315.

Sullivan, P. G., et al. (2004). The ketogenic diet increases mitochondrial uncoupling protein levels and activity. *Annals of Neurology*, 55(4), 576–580.

Pickles, S., Vigié, P., & Youle, R. J. (2018). Mitophagy and quality control mechanisms in mitochondrial maintenance. *Current Biology*, 28(4), R170–R185.

Brand, M. D. (2016). Mitochondrial uncoupling: Insights into mechanisms and physiology. *Nature Reviews Molecular Cell Biology*, 17(10), 605–620.

Cannon, B., & Nedergaard, J. (2004). Brown adipose tissue: Function and physiological significance. *Physiological Reviews*, 84(1), 277–359.

Ballabio, A., & Bonifacino, J. S. (2020). Lysosomes as dynamic regulators of cell and organismal homeostasis. *Nature Reviews Molecular Cell Biology*, 21(2), 101–118.

Kozak, L. P., & Harper, M. E. (2000). Mitochondrial uncoupling proteins in energy expenditure. *Annual Review of Nutrition*, 20, 339–363.

Green, D. R., & Kroemer, G. (2004). The pathophysiology of mitochondrial cell death. *Science*, 305(5684), 626–629.

Ristow, M., & Schmeisser, S. (2014). Mitohormesis: Promoting health and lifespan by increased levels of reactive oxygen species. *Nature Reviews Endocrinology*, 10(6), 336–342.

Mattson, M. P., & Liu, D. (2002). Mitochondrial regulation of neuronal plasticity. *Neurochemistry Research*, 27(10), 957–972.

Schrauwen, P., & Hesselink, M. K. (2004). Oxidative capacity, lipotoxicity, and mitochondrial uncoupling in skeletal muscle as factors in insulin resistance and type 2 diabetes. *Journal of Physiology*, 541(1), 301–308.

Dulloo, A. G., Samec, S., & Seydoux, J. (2001). Uncoupling protein 2 and obesity resistance: A current perspective. *International Journal of Obesity*, 25(Suppl 5), S45–S57.

López-Lluch, G., & Navas, P. (2016). Mitochondrial bioenergetics and longevity. *Aging*, 8(3), 260–261.

Weinberg, S. E., Sena, L. A., & Chandel, N. S. (2015). Mitochondria in the regulation of innate and adaptive immunity. *Immunity*, 42(3), 406–417.

Brand, M. D. (2000). Uncoupling to survive? The role of mitochondrial inefficiency in ageing. *Experimental Gerontology*, 35(6–7), 811–820.

Toime, L. J., & Brand, M. D. (2010). Uncoupling protein-3 lowers reactive oxygen species production in isolated mitochondria. *Free Radical Biology & Medicine*, 49(4), 606–611.

Nedergaard, J., Bengtsson, T., & Cannon, B. (2007). Unexpected evidence for active brown adipose tissue in adult humans. *American Journal of Physiology – Endocrinology and Metabolism*, 293(2), E444–E452.

Speakman, J. R., & Mitchell, S. E. (2011). Caloric restriction. *Molecular Aspects of Medicine*, 32(3), 159–221.

Cadenas, S. (2018). Mitochondrial uncoupling, ROS generation and cardioprotection. *BBA – Bioenergetics*, 1859(9), 940–950.

Tonkonogi, M., et al. (2000). Mitochondrial function and antioxidative defence in human muscle: Effects of endurance training and oxidative stress. *Journal of Physiology*, 528(2), 379–388.

Jones, T. E., et al. (2003). Exercise induces an increase in muscle UCP3 as a component of the adaptation to training. *American Journal of Physiology – Endocrinology and Metabolism*, 284(1), E96–E101.

Okamatsu-Ogura, Y., et al. (2020). Cold exposure induces recruitment of brown adipose tissue and enhances nonshivering thermogenesis in humans. *Frontiers in Endocrinology*, 11, 222.

Mattson, M. P., et al. (2018). Intermittent metabolic switching, neuroplasticity and brain health. *Nature Reviews Neuroscience*, 19, 63–80.

Baur, J. A., & Sinclair, D. A. (2006). Therapeutic potential of resveratrol: The in vivo evidence. *Nature Reviews Drug Discovery*, 5(6), 493–506.

Silva, J. E. (2006). Thermogenic mechanisms and their hormonal regulation. *Physiological Reviews*, 86(2), 435–464.

López-Lluch, G., et al. (2006). Calorie restriction induces mitochondrial biogenesis and bioenergetic efficiency. *Proceedings of the National Academy of Sciences*, 103(6), 1768–1773.

Chapter 24

Cunnane, S. C., et al. (2016). Can ketones compensate for deteriorating brain glucose uptake during aging? *Annals of the New York Academy of Sciences*, 1367(1), 12–20.

Castellano, C.-A., Fortier, M., St-Pierre, V., et al. (2015). Ketone body uptake into the brain in mild cognitive impairment: A PET study. *Neurobiology of Aging*, 36(9), 2319–2326.

Mosconi, L. (2008). Brain glucose metabolism in the early and specific diagnosis of Alzheimer's disease. *European Journal of Nuclear Medicine and Molecular Imaging*, 35(6), 1039–1055.

Reiman, E. M., et al. (2004). Functional brain abnormalities in young adults at genetic risk for late-onset Alzheimer's dementia. *Proceedings of the National Academy of Sciences*, 101(1), 284–289.

Cunnane, S. C., et al. (2011). Brain fuel metabolism, aging, and Alzheimer's disease. *Nutrition*, 27(1), 3–20.

Castellano, C.-A., et al. (2015). Lower brain glucose metabolism but normal ketone metabolism in mild Alzheimer's disease: A dual tracer PET study. *Brain*, 138(2), 390–402.

Castellano, C.-A., et al. (2015). Brain energy deficit in Alzheimer's disease is not related to reduced cerebral blood flow. *Journal of Cerebral Blood Flow & Metabolism*, 35(3), 409–416.

Fortier, M., et al. (2019). A ketogenic drink improves brain energy and some measures of cognition in mild cognitive impairment. *Alzheimer's & Dementia*, 15(5), 625–634.

Croteau, E., et al. (2018). Ketogenic medium-chain triglycerides increase brain energy metabolism in mild cognitive impairment. *Alzheimer's Research & Therapy*, 10(1).

Henderson, S. T., et al. (2009). Study of the ketogenic agent AC-1202 in mild to moderate Alzheimer's disease. *Nutrition & Metabolism*, 6, 31.

Owen, O. E., et al. (1967). Brain metabolism during fasting. *Journal of Clinical Investigation*, 46(10), 1589–1595.

Murray, A. J., et al. (2016). Ketogenic diets and the ketogenic agent ketone bodies as therapies for Alzheimer's disease. *Neurotherapeutics*, 13(4), 795–806.

Andrews, S. J., et al. (2020). Mitochondrial dysfunction in mental health disorders: A systematic review of depression, bipolar disorder, schizophrenia, and autism. *Translational Psychiatry*, 10(1).

Stork, C., & Renshaw, P. F. (2005). Mitochondrial dysfunction in bipolar disorder: Evidence from magnetic resonance spectroscopy research. *Current Psychiatry Reports, 7*, 37–41.

Jain, F. A., et al. (2020). Mitochondrial dysfunction in major depressive disorder: Evidence, theory and translational implications. *Neuropsychiatric Disease and Treatment, 16*, 2539–2552.

Scaini, G., et al. (2020). Mitochondrial dysfunction in bipolar disorder: A meta-analysis of proteomic and mitochondrial DNA studies. *Neuroscience & Biobehavioral Reviews, 112*, 134–147.

Kraeuter, A.-K., et al. (2020). The therapeutic potential of ketogenic diets in neuropsychiatric disorders. *Molecular Psychiatry, 25*(9), 2344–2346.

Bostock, E. C. S., et al. (2020). The current status of the ketogenic diet in psychiatry. *Frontiers in Psychiatry, 11*, 598373.

Heneka, M. T., et al. (2015). Neuroinflammation in Alzheimer's disease. *Lancet Neurology, 14*(4), 388–405.

Niraula, A., et al. (2022). Glial shutdown: The critical link between mitochondrial dysfunction and brain inflammation. *Nature Reviews Neuroscience, 23*, 43–57.

Chapter 25

Phillips, M. C. L., & Picard, M. (2024). Neurodegenerative disorders, metabolic icebergs, and mitohormesis. *Translational Neurodegeneration, 13*, 46.

Johri, A., & Beal, M. F. (2012). Mitochondrial dysfunction in neurodegenerative diseases. *Journal of Pharmacology and Experimental Therapeutics, 342*(3), 619–630.

Alqahtani, T., et al. (2023). Mitochondrial dysfunction and oxidative stress in Alzheimer's disease, Parkinson's disease, Huntington's disease and amyotrophic lateral sclerosis—An updated review. *Mitochondrion, 71*, 83–92.

Beal, M. F. (2004). Mitochondrial dysfunction and oxidative damage in neurodegenerative diseases. *Trends in Neurosciences, 27*(10), 595–602.

Wang, W., et al. (2020). Mitochondria dysfunction in the pathogenesis of Alzheimer's disease: Recent advances. *Molecular Neurodegeneration*, 15, 30.

Yan, M. H., et al. (2013). Mitochondrial defects and oxidative stress in Alzheimer disease and Parkinson disease. *Free Radical Biology & Medicine*, 62, 90–101.

Bose, A., & Beal, M. F. (2016). Mitochondrial dysfunction in Parkinson's disease. *Journal of Neurochemistry*, 139(S1), 216–231.

Hoekstra, J. G., et al. (2011). Mitochondrial therapeutics in Alzheimer's disease and Parkinson's disease. *Current Pharmaceutical Design*, 17(31), 3374–3380.

Cozzolino, M., & Carri, M. T. (2012). Mitochondrial dysfunction in amyotrophic lateral sclerosis (ALS). *Progress in Neurobiology*, 97(2), 54–66.

Shi, P., et al. (2010). Mitochondrial dysfunction in amyotrophic lateral sclerosis. *BBA – Molecular Basis of Disease*, 1802(1), 45–51.

Parvanovova, P., et al. (2024). Mitochondrial dysfunction in sporadic amyotrophic lateral sclerosis. *Biomedicines*, 12(6), 1294.

Johri, A., et al. (2013). PGC-1α, mitochondrial dysfunction, and Huntington's disease. *Free Radical Biology & Medicine*, 62, 37–46.

Costa, V., & Scorrano, L. (2012). Shaping the role of mitochondria in the pathogenesis of Huntington's disease. *EMBO Journal*, 31(8), 1853–1864.

Ristow, M., & Zarse, K. (2010). How increased oxidative stress promotes longevity and metabolic health: The concept of mitochondrial hormesis (mitohormesis). *Experimental Gerontology*, 45(6), 410–418.

Yun, J., & Finkel, T. (2014). Mitohormesis. *Cell Metabolism*, 19(5), 757–766.

Radak, Z., et al. (2016). Mitohormesis in exercise training. *Journal of Sports Science and Medicine*, 15(1), 1–8.

Picard, M., Trumpff, C., & Burelle, Y. (2019). Mitochondrial psychobiology: Foundations and applications. *Current Opinion in Behavioral Sciences*, 28, 142–151.

Picard, M., & McEwen, B. S. (2018). Psychological stress and mitochondria: A conceptual framework. *Psychosomatic Medicine*, 80(2), 126–140.

Picard, M., et al. (2018). An energetic view of stress: Focus on mitochondria. *Frontiers in Neuroendocrinology, 49*, 72–85.

Chapter 26

Volek, J. S., et al. (2024). The KIND Study: Ketogenic dietary intervention for mental health outcomes in college students. *Nutrients, 16*(3), 450–468.

Klement, R. J., & Brehm, N. (2020). Ketogenic diets and psychiatric disorders: A review of current evidence. *Frontiers in Psychiatry, 11*, 565.

Stafstrom, C. E., & Rho, J. M. (2012). The ketogenic diet as a treatment paradigm for diverse neurological disorders. *Frontiers in Pharmacology, 3*, 59.

Morris, G., et al. (2017). Why should neuroscientists worry about mitochondria? *Frontiers in Psychiatry, 8*, 40.

Bough, K. J., & Rho, J. M. (2007). Anticonvulsant mechanisms of the ketogenic diet. *Epilepsia, 48*(1), 43–58.

Stubbs, B. J., et al. (2017). On the metabolism of exogenous ketones in humans. *Frontiers in Physiology, 8*, 848.

Yuen, A. W., & Sander, J. W. (2014). Rationale for using intermittent calorie restriction as a dietary treatment for drug resistant epilepsy. *Epilepsy Research, 108*(6), 1002–1006.

Brietzke, E., et al. (2018). Ketogenic diet as a metabolic therapy for mood disorders: Evidence and developments. *Trends in Psychiatry and Psychotherapy, 40*(5), 446–454.

Paoli, A., et al. (2013). Beyond weight loss: A review of the therapeutic uses of very-low-carbohydrate (ketogenic) diets. *European Journal of Clinical Nutrition, 67*(8), 789–796.

Ruskin, D. N., & Masino, S. A. (2012). The nervous system and metabolic dysregulation: Emerging evidence converges on ketogenic diet therapy. *Frontiers in Neuroscience, 6*, 33.

Redlich, R., et al. (2018). The role of metabolic dysfunction in depression: From pathophysiology to novel treatments. *European Archives of Psychiatry and Clinical Neuroscience, 268*, 5–21.

Rho, J. M., & Stafstrom, C. E. (2014). The ketogenic diet: From molecular mechanisms to clinical effects. *Epilepsy Research*, 100(3), 205–210.

Brietzke, E., et al. (2020). Ketogenic diets and mood disorders: Translating promise into clinical practice. *Bipolar Disorders*, 22(5), 492–495.

Achanta, L. B., & Rae, C. D. (2017). β-hydroxybutyrate in the brain: One molecule, multiple mechanisms. *Neurochemistry International*, 100, 46–59.

Morris, G., et al. (2020). The interplay between oxidative stress and bioenergetics in the pathophysiology of psychiatric disorders. *Molecular Neurobiology*, 57, 2069–2087.

Westman, E. C., Volek, J. S., & Phinney, S. D. (2020). Low-carbohydrate diets: Evidence for their role in mood and mental health. *Current Opinion in Psychiatry*, 33(6), 531–537.

Chapter 28

Vander Heiden, M. G., Cantley, L. C., & Thompson, C. B. (2009). Understanding the Warburg effect: The metabolic requirements of cell proliferation. *Science*, 324(5930), 1029–1033.

Klement, R. J., & Kämmerer, U. (2011). Is there a role for carbohydrate restriction in cancer therapy? *Nutrition & Metabolism*, 8(1), 75.

Poff, A. M., et al. (2014). Ketone supplementation in cancer: Evidence for metabolic modulation. *International Journal of Cancer*, 135(9), 2013–2020.

Chapter 29

Seyfried, T. N., Shelton, L. M., & Mukherjee, P. (2017). Press–Pulse: A novel therapeutic strategy for managing cancer metabolism. *Nutrition & Metabolism*, 14(1), 19.

Klement, R. J., Brehm, N., & Sweeney, R. A. (2020). Ketogenic dietary therapy in oncology: Current evidence and controversies. *Clinical Nutrition*, 39(1), 7–14.

Weber, D. D., Aminzadeh-Gohari, S., & Feichtinger, R. G. (2020). Ketogenic diet in cancer therapy. *Aging (Albany NY)*, 12(12), 11936–11940.

Di Biase, S., et al. (2016). Fasting-mimicking diet reduces IGF-1 and protects from chemotherapy toxicity. *Science Translational Medicine*, 8(332), 332ra44.

Caffa, I., et al. (2023). Fasting-mimicking diet and cancer: Molecular mechanisms and clinical applications. *Nature Reviews Cancer*, 23, 130–150.

Longo, V. D., & Panda, S. (2016). Fasting, circadian rhythms, and time-restricted feeding for healthy lifespan. *Cell Metabolism*, 23(6), 1048–1059.

D'Agostino, D. P., & Dean, J. B. (2020). Hyperbaric oxygen therapy and ketone metabolism. *Journal of Applied Physiology*, 128(3), 684–692.

Moen, I., & Stuhr, L. E. B. (2012). Hyperbaric oxygen therapy and cancer—a review. *Targeted Oncology*, 7(4), 233–242.

Bonnet, S., et al. (2007). A mitochondria–K+ channel axis is suppressed in cancer and its normalization promotes apoptosis. *Cancer Cell*, 11(1), 37–51.

Michelakis, E. D., Webster, L., & Mackey, J. R. (2008). Dichloroacetate (DCA) as a potential metabolic-targeting therapy for cancer. *British Journal of Cancer*, 99(7), 989–994.

Yoo, H. C., et al. (2020). Glutamine reliance in cancer cells: Overview and therapeutic targeting. *Nature Reviews Molecular Cell Biology*, 21, 492–509.

Wang, J. B., et al. (2010). Targeting mitochondrial glutaminase activity with DON inhibits cancer cell growth. *Cancer Research*, 70, 7155–7163.

Pollak, M. (2012). Metformin and cancer: Mechanisms and clinical outcomes. *Nature Reviews Cancer*, 12, 93–96.

Dowling, R. J. O., et al. (2011). Metformin in cancer: Translational challenges. *Cancer Prevention Research*, 4(3), 352–364.

Juarez, M., et al. (2018). Ivermectin as an anti-cancer agent: Mechanisms and implications. *Frontiers in Pharmacology*, 9, 789.

Hou, Z., et al. (2018). Mebendazole induces apoptosis and inhibits cancer cell proliferation and tumor growth. *Journal of Experimental & Clinical Cancer Research*, 37, 266.

Dogra, N., et al. (2022). Fenbendazole's metabolic effects in malignant cells. *Scientific Reports*, 12, 1124.

Navarro-Yepes, J., et al. (2014). Oxidative stress, redox signaling, and autophagy in cancer: Therapeutic targets. *Antioxidants & Redox Signaling*, 21, 870–891.

Trachootham, D., Alexandre, J., & Huang, P. (2009). Targeting cancer cells by ROS-mediated mechanisms: A radical therapeutic approach. *Nature Reviews Drug Discovery*, 8(7), 579–591.

O'Sullivan, D., & Pearce, E. L. (2018). T cell metabolism in cancer immunity and immunotherapy. *Nature Reviews Immunology*, 18, 618–631.

Buck, M. D., O'Sullivan, D., & Pearce, E. L. (2016). Mitochondrial metabolism governs T cell fate and function. *Immunity*, 44, 406–417.

Chang, C.-H., et al. (2015). Metabolic competition in the tumor microenvironment suppresses T cell function. *Cell*, 162, 1229–1241.

Meidenbauer, J. J., Roberts, M. N., & Seyfried, T. N. (2015). The glucose–ketone index (GKI) as a biomarker for therapeutic ketosis. *Nutrition & Metabolism (London)*, 12, 12.

Koutnik, A. P., et al. (2020). Practical strategies for monitoring ketosis in metabolic oncology. *Metabolites*, 10(12), 510.

Manoogian, E. N. C., & Panda, S. (2017). Circadian rhythms, time-restricted feeding, and metabolic health. *Cell Metabolism*, 26, 801–819.

Ramsey, K. M., et al. (2007). The circadian clock and metabolism. *Science*, 316, 582–587.

Chapter 30

Perry, R. J., et al. (2019). Ketone bodies suppress appetite and improve metabolic health. *Nature Medicine*, 25, 1406–1415.

ABOUT THE AUTHOR

MICHAEL MACDONALD, CPA, is a public accountant based in Michigan with more than 30 years of experience. His early career in the Detroit metropolitan area included auditing public company financial statements, where he saw firsthand that numbers often tell a different story once examined closely. That perspective later led him to look more closely at health narratives surrounding nutrition and chronic disease.

He was raised in the small town of Goodrich during the 1980s. Growing up on ten acres along a dirt road with his two older brothers, he was shaped by parents who valued time outdoors. His father spent weekends with his sons running local road races, riding motocross, and later skydiving. His mother shared a love of animals and the outdoors, connecting the family to 4-H and raising pygmy goats.

A graduate of Western Michigan University, he met his wife, Carol, a proud Yooper from Michigan's Upper Peninsula. For the past fourteen years, they have worked side by side at a CPA firm in Mount Pleasant. They are the parents of two children, Megan and Stuart, and spent many years immersed in youth sports, figure skating, volleyball, and hockey, where Michael also coached his son's teams.

In 2019, Michael reached a personal turning point as his own energy declined and Carol faced many of the same underlying struggles common to modern life. Their search for answers led them deeper into metabolism and the body's capacity to heal when given the right conditions. That journey became the foundation for Return to Energy.

Today, with their children grown, Michael enjoys running, open-water swimming, and training for endurance events. Carol joins him on the trails, and together they participate in charitable events including the Mackinac Island Swim.